MW01000100

DIGITAL SIGNAL PROCESSING

An Overview of Basic Principles

DIGITAL SIGNAL PROCESSING
An Overview of Basic Principles

Jack Cartinhour
Oklahoma State University

Prentice Hall
Upper Saddle River, New Jersey   *Columbus, Ohio*

Library of Congress Cataloging-in-Publication Data

Cartinhour, Jack.
    Digital signal processing: An overview of basic principles / Jack
Cartinhour.
        p.   cm.
    Includes bibliographical references and index.
    ISBN 0-13-769266-8
    1. Signal processing—Digital techniques.   2. Discrete time
systems.   I. Title.
TK5102.9.C36   2000   98-53218
621.382'2—dc21   CIP

Cover photo: H. Armstrong Roberts
Publisher: Charles E. Stewart, Jr.
Production Editor: Alexandrina Benedicto Wolf
Production Coordination: Custom Editorial Productions, Inc.
Cover Design Coordinator: Karrie Converse-Jones
Cover Designer: Curt Besser
Production Manager: Deidra M. Schwartz
Marketing Manager: Ben Leonard

This book was set in Times by Custom Editorial Productions, Inc., and was printed and bound by R.R. Donnelley & Sons Company. The cover was printed by Phoenix Color Corp.

© 2000 by Prentice-Hall, Inc.
Pearson Education
Upper Saddle River, New Jersey 07458

Printed in the United States of America

10 9 8   7 6 5 4 3 2 1

ISBN: 0-13-769266-8

Prentice-Hall International (UK) Limited, *London*
Prentice-Hall of Australia Pty. Limited, *Sydney*
Prentice-Hall of Canada, Inc., *Toronto*
Prentice-Hall Hispanoamericana, S. A., *Mexico*
Prentice-Hall of India Private Limited, *New Delhi*
    ˜tice-Hall of Japan, Inc., *Tokyo*
        ˜ˉll (Singapore) Pte. Ltd., *Singapore*
            ˈˈˉll do Brasil, Ltda., *Rio de Janeiro*

PREFACE

This book was written to present the basic principles of digital signal processing (DSP) to seniors in electrical engineering technology programs. Seniors in the Department of Electrical Engineering Technology (Division of Engineering Technology) at Oklahoma State University formed the initial audience when a shorter and unpublished version of this book was first used in 1994.

Having taught DSP to electrical engineering students for seven years, I recognize that presenting this material to engineering technology students presents a different sort of challenge. Although the approach in this book is not as rigorous as the approach taken in DSP books for engineering students, it is written at a level that an engineering student would find useful.

This book emphasizes the theoretical aspects of digital signal processing. Although there is no detailed discussion of the specialized microprocessors that are popularly known as DSP chips, real time implementation issues are considered. Most of the emphasis is on the theory of *discrete time signal processing*, which is basically DSP theory minus the consideration of various finite precision effects (e.g., A/D conversion quantization noise and errors introduced by finite precision processor arithmetic). These finite precision effects are important and are dealt with at various sections in the book.

This book explains how the coefficients of the discrete time system difference equation are selected to implement the desired digital filter. It is worth noting that compared with a continuous time system, a discrete time system is much more flexible with respect to changing conditions and/or needs, since changing the system response is only a matter of changing some numbers in a program. This is one of the main reasons why DSP is such a powerful tool.

In studying circuit theory and control systems, electrical engineering technology students are exposed to much of the theory needed to understand *continuous time* (as opposed to *discrete time*) signal processing theory. I assume that the student has been exposed to linear constant coefficient differential equations, the Laplace transform, the Fourier series, and the steady-state response of a linear circuit to a sinusoidal input—and that more advanced topics such as convolution, system impulse response, and the Fourier transform have not been introduced to the student in previous courses. I believe that these advanced topics must be introduced before beginning a study of discrete time signal processing, as they are crucial to the understanding of DSP. The discrete time versions of these advanced topics are better understood if the analogous continuous time concepts are understood. Also, the important relationship between continuous time and discrete time systems is best understood in the Fourier transform domain. Therefore, the first three chapters of this book are devoted to an overview of continuous time system theory, which includes a review of some familiar concepts as well as the introduction of advanced ones.

Digital signal processing requires a way of looking at things that is perhaps more theoretical or mathematically intense than electrical engineering technology students are used to. For example, students can learn much about a continuous time filter circuit without ever considering the differential equation that describes it. The differential equation is one theoretical description of the filter and can be used to gain insight into the behavior of the filter, but this equation is not directly used to implement the filter. The bottom line is that the continuous time filter can be thought of first and foremost as a *circuit*, and its response can be tinkered with by changing things such as resistor and capacitor values. On the other hand, a discrete time system must be thought of first and foremost as a *mathematical algorithm*, since the *difference equation* that describes the system is directly used to implement the system by means of a computer program. To implement this algorithm in real time, a complicated circuit involving a microprocessor or specialized DSP chip, A/D and D/A converters, and so on, must be built; however, one does not tinker with the hardware as such to change the system response. The response of a discrete time system depends on the computer program that implements the system difference equation; the coefficients of this equation determine whether the system is, for example, a digital lowpass filter or a digital highpass filter, and what the cutoff frequency is.

The reader (or instructor) who is anxious to get to the heart of the book—DSP itself—can go directly from Chapter 3 to Chapter 5. However, the material in Chapter 4 on the frequency response and transfer functions of Butterworth and Chebyshev filters should be covered before Chapter 10.

The heart and soul of this book are contained in Chapters 5 through 10. Chapter 5 introduces the concept of a linear, shift-invariant (LSI) discrete time system and the difference equation that describes it. Important topics such as the discrete time system impulse response, discrete time convolution, the Z transform, and the discrete time system transfer function are also presented. Instructors may reduce the amount of time devoted to certain Z transform topics, such as finding the inverse Z transform using partial fraction expansion, which is covered in Section 5–6, but is not needed in the chapters that follow. Chapter 6 covers the discrete time Fourier transform (DTFT) and frequency domain analysis of discrete time signals and systems and discusses the relationship between continuous time and discrete time signal processing. Chapter 7 presents the details of the relationship between the Fourier transform and the DTFT, which is initially presented on an "it can be shown" basis in Chapter 6. The sampling theorem is also presented in Chapter 7, along with a discussion of ideal versus real-world digital-to-analog conversion. In Chapter 8, the student is introduced to the discrete Fourier transform (DFT) and the fast Fourier transform (FFT). Chapter 9 discusses the design of finite impulse response (FIR) filters using the Kaiser window method and the Parks-McClellan algorithm. Chapter 10 discusses the design of infinite impulse response (IIR) filters using the bilinear transformation.

Chapter 11 presents a topic that best illustrates the power of DSP—adaptive FIR filters based on the LMS algorithm. Chapter 12 begins by considering random signals and their associated power spectral density functions, and then shows how *oversampling* can be used to improve the signal-to-noise ratio at the output of an A/D converter. Also covered in Chapter 12 are downsampling (decimation) and upsampling (interpolation). The topics in Chapters 11 and 12 are relatively advanced; presenting this material on a level appropriate for this book requires more than the usual number of "it can be shown" arguments, nonrigorous analogies, and appeals to intuition.

I strongly believe that students should be given computer assignments in which they implement a DSP algorithm (for example, the difference equation for a lowpass filter). These programs can operate in non-real time in the sense that they process an input data file containing samples of a previously recorded signal and then write the processed output to another file. A commercially available sound card can be used to record the original signal and then play back the processed signal. If this is done, a student can listen to and compare before-processing and after-processing signals and actually hear the difference; it can be a convincing demonstration of the power of DSP, even though it isn't done in real time. The programs can be written in any high-level language (e.g., BASIC, C, or Fortran). Some of the computer assignments that I have used are included in the instructor's manual. Appendix B discusses programming FIR filter algorithms in a high-level language.

Problems are included at the end of all chapters. Many of the problems are designed to reinforce the material presented in the associated chapter. Some of the problems are more challenging because they are used to introduce new (but related) concepts. Solutions to selected problems are included in the instructor's manual.

I would like to thank the reviewers of the original manuscript, all of whom took the time to write an objective, interesting, and valuable review: Richard Baraniuk, Rice University; Alan Proffitt, Arkansas State University; Wayne Padgett, Rose Hulman Institute of Technology; and Bruce Fritchman, Lehigh University. Many of their suggestions have been incorporated, although perhaps not in the style that was recommended.

BRIEF CONTENTS

CONTENTS

10   DESIGN OF IIR FILTERS USING THE BILINEAR TRANSFORMATION   301

11   ADAPTIVE FIR FILTERS USING THE LMS ALGORITHM   327

12   RANDOM SIGNALS AND POWER SPECTRA, A/D CONVERSION NOISE, AND OVERSAMPLING   341

APPENDICES

BIBLIOGRAPHY   391

INDEX   393

INTRODUCTION

When engineers talk about digital signal processing (DSP), they are essentially discussing *discrete time* signal processing, as opposed to *continuous time* (analog) signal processing. (Discrete time signal processing theory is basically DSP theory minus the explicit consideration of various nonideal "finite precision" effects due to A/D conversion quantization error, finite precision computer arithmetic, etc.) The purpose of this brief introduction is to describe the difference between continuous time and discrete time signal processing and to explain some of the advantages of DSP. It is assumed that the reader is already familiar with continuous time signal processing, having previously studied a variety of circuits that carry out this kind of processing (for example, an active lowpass filter circuit).

Continuous time signal processing is represented by the system shown in Figure I–1. The input signal, $x_a(t)$, is a continuous function of time (t), usually a time-varying voltage at the output of a transducer of some sort (e.g., a microphone). The continuous time system is a circuit of some kind, such as an active lowpass filter circuit (built with opamps [operational amplifiers], resistors, and capacitors). The output of the system, $y_a(t)$, is also a continuous time signal.

The continuous time signal processing system of Figure I–1 should be compared with the digital signal processing system shown in Figure I–2. For the purposes of this discussion, let's assume that the two systems have been designed to achieve essentially the same purpose in that $x_a(t)$ is the same in both cases and the desired result with respect to $y_a(t)$ is also the same. In the DSP system, the continuous time input signal is *sampled* by an analog-to-digital (A/D) converter once every T seconds. At the output of the A/D converter, a new *number* appears once every T seconds. Therefore, a sequence of numbers $x(n)$ (where n is the sample number) appears at the output of the A/D converter. Assuming an *ideal* A/D converter, this sequence is related to the original continuous time signal as follows:

$$x(n) = x_a(nT) \qquad\qquad \textbf{(I–1)}$$

FIGURE I–1

Continuous time system.

Continuous Time System

$x_a(t)$ ⟶ | Circuit | ──── $y_a(t)$

   The sequence of numbers, $x(n)$, is called a *discrete time signal*. This discrete time signal is processed by a mathematical algorithm implemented by a computer program. The algorithm generates an output sequence of numbers, denoted as $y(n)$ in Figure I–2. The output sequence $y(n)$ is processed by a digital-to-analog (D/A) converter to create an output continuous time signal $y_a(t)$.

   Strictly speaking, the subject of "discrete time signal processing" pertains to the subsystem in Figure I–2 that is enclosed by the dashed line. A discrete time input signal $x(n)$ is processed by a discrete time system (which is the algorithm implemented by the computer) to create a discrete time output signal $y(n)$. The overall DSP system of Figure I–2 has a continuous time input and output and processes the continuous time signal using discrete time methodology.

   If the input sequence $x(n)$ is immediately processed by the computer, such that each time an input number is fetched from the A/D converter an output number is calculated and fed to the D/A converter, then the DSP system of Figure I–2 is operating in "real time." In some applications, it is permissible to operate in "non-real time," as depicted in Figure I–3. In the non-real time, or off-line processing scenario, the sequence of numbers $x(n)$ is stored in a computer file. This file is later accessed by the computer program, which generates the output sequence $y(n)$ without being forced to do so at the same speed that real-time processing would require. The output sequence is written to another computer file, which can later be accessed by a D/A converter or perhaps be subjected to additional processing. It should be emphasized here that the basic principles and theory underlying discrete time signal processing are the same regardless of whether the actual processing rate is real time or non-real time.

   There was a time not very many years ago when DSP was an esoteric subject studied at the graduate school level. There have been significant applications for DSP for many years, applications that span a number of different disciplines (for example, seismic exploration for oil and gas, speech and image processing, sonar and radar signal processing,

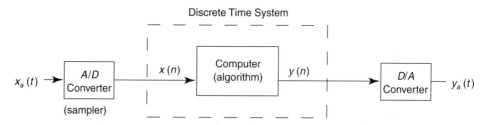

FIGURE I–2

Digital signal processing system. The subsystem within the dashed line is a discrete time system.

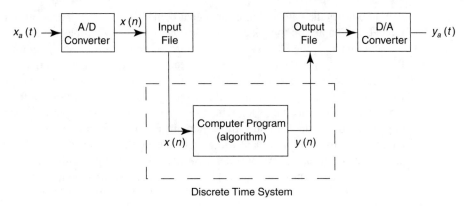

FIGURE I–3
Non-real time (off-line) DSP.

etc.). In earlier times (say 20 years ago), most DSP was carried out in non-real time. The theory was well developed, but the available digital computers and other supporting hardware (e.g., A/D converters) could not operate fast enough to make real-time DSP practical. In more recent years, rapid advances in integrated circuit technology have made real time DSP a realistic option for an increasing number of applications. DSP is now a buzzword even in the world of consumer electronics.

Let us once again compare the systems of Figure I–1 and Figure I–2 and assume that both systems have the same input signal and achieve essentially the same result with respect to the output signal. The DSP system certainly looks more complicated than the continuous time system; it is fair to ask what advantages are gained from such complexity. The main answer is that the DSP system has tremendous flexibility relative to the continuous time system. Suppose the purpose of these systems is to filter the input signal but that the exact filter requirements will vary according to changing conditions or needs. For example, suppose each of these systems represents a bandpass filter, but the bandwidth and center frequency may need to be changed by the user of the system. To change the continuous time system, various circuit components (resistors, capacitors, etc.) must be changed. To change the discrete time system, the algorithm must be changed; this is just a matter of changing some numbers in the program! In other words, the DSP system of Figure I–2 is programmable and can can be changed on the fly. In fact, there are *adaptive* filter algorithms that actually realize self-adjusting filters, such as a narrow bandstop (notch) filter that tracks an interfering sinusoid characterized by a changing frequency (this is discussed in Chapter 11). In addition to this flexibility, the DSP system can realize certain types of filters that are either impossible or impractical to implement with a continuous time system; an example is an extremely sharp lowpass or bandpass filter having an exact "linear phase" characteristic in the passband.

We don't mean to imply that DSP has made all continuous time filtering operations obsolete. As will be shown in Chapter 6, the usable bandwidth of a DSP system is equal to one-half of the sampling frequency. (If the A/D converter takes one sample every T seconds, the sampling frequency is $f_s = 1/T$ Hertz.) The sampling frequency must be greater than two times the bandwidth of the signal being sampled (again, for reasons that are explained in Chapter 6); this places a *lower bound* on the allowable sampling frequency. On the other hand, if the DSP system is to function in real time, the system must be able to fetch a sample from the A/D converter, execute the algorithm, and output a number to the D/A converter in T seconds or less; this puts an *upper bound* on the allowable sampling frequency. (There are also limits on how fast a given A/D or D/A converter can function.) Obviously, there can be cases in which the system requirements with respect to lower and upper bounds on the sampling frequency are mutually exclusive due to the input signal having an extremely large bandwidth or to the limitations on the speed of the proposed processor or A/D and D/A converters. Besides, there will probably always be some cases (at least in the lifetime of the author) in which an active filter circuit is a simpler, faster, and cheaper solution even if DSP could easily be used instead. Therefore, continuous time signal processing is still extremely important.

Figures I–2 and I–3 suggest that the desired output from a DSP system is always another continuous time signal, but this is not always the case. Sometimes the purpose of the DSP system is to *analyze* the input signal in some sense; in this case, the desired output is some form of information about the input signal. Figure I–4 shows the basic idea. A classic example is spectral analysis, for which the purpose of the system is to extract and display information about the frequency content, or *spectrum*, of the signal. In this case, the signal samples, $x(n)$, are processed by some kind of spectral analysis program, which is quite often based on the *Fast Fourier Transform* (FFT). (The FFT is covered in Chapter 8.)

Before going on to Chapter 1, read the Preface to this book if you have not already done so. Among other things, the Preface explains why Chapters 1, 2, 3, and 4 are devoted to continuous time signal processing instead of digital signal processing. There is good reason that the background information on continuous time signal and system theory must be covered before considering the DSP theory itself (which begins with Chapter 5).

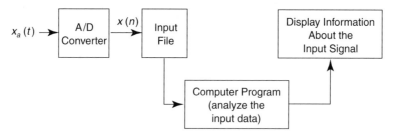

FIGURE I–4
Using DSP for signal analysis. A classic example is spectral analysis, in which the purpose is to determine the frequency content of the input signal.

1

LINEAR, SHIFT-INVARIANT CONTINUOUS TIME SYSTEMS

1–1   TIME DOMAIN DESCRIPTION

We begin by considering linear, shift-invariant (LSI) systems in a very general sense. Let $x(t)$ be the input signal to the system. (For example, $x(t)$ could be a time-varying voltage.) There is some operation, denoted $T[\ ]$, that creates an output signal $y(t)$:

$$y(t) = T[x(t)] \qquad (1\text{--}1)$$

Figure 1–1 illustrates the basic idea.

In this book it is assumed that the system under consideration is "at rest" before the input signal is applied. In other words, it is assumed that all of the initial conditions are zero. For example, if the system is a circuit having inductors and capacitors, the initial conditions are that all of the capacitors are discharged and all of the inductor currents are zero.

A system is said to be *linear* if superposition holds. That is, $T[\ ]$ represents a linear system if the following relationship holds:

$$T[a_1 x_1(t) + a_2 x_2(t)] = a_1 T[x_1(t)] + a_2 T[x_2(t)] \qquad (1\text{--}2)$$

A circuit having only passive components (resistors, capacitors, and inductors) is an example of a linear system. A circuit having active devices (such as transistors) could be either linear or nonlinear. One distinguishing characteristic of a linear system is that if the input signal is an undistorted sinusoid, the output signal will also be an undistorted sinusoid at the

FIGURE 1–1
Linear, shift-invariant system.

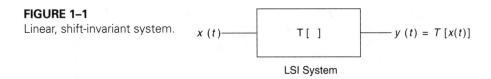

LSI System

5

same frequency, with possible changes in both amplitude and phase shift. If the input to a system is an undistorted sinusoid but the output is a distorted sinusoid, the system is non-linear. However, if the distortion is small, the system could be approximated as a linear system. In many cases, when a system is said to be linear, what is meant is that the system can be thought of as being linear for all practical purposes.

An example of a nonlinear system is a halfwave rectifier circuit. Another example of a nonlinear system is one that squares the input signal; that is, $T[x(t)] = [x(t)]^2$.

A system is *shift-invariant* if the operation $T[\ ]$ does not depend on time. (A shift-invariant system is also called a *time-invariant* system.) In other words, if the input is time shifted, the output will be time shifted by the same amount but is otherwise unchanged. Another way of saying the same thing is that for a shift-invariant system, $T[x(t)] = y(t)$ implies $T[x(t - \tau)] = y(t - \tau)$.

Many electrical circuits are essentially shift-invariant. For example, a lowpass filter that you just constructed will behave the same way an hour from now (it will have the same cutoff frequency, passband gain, etc.). Of course, one can think of exceptions. For example, certain circuit parameters might be temperature dependent, so that if the temperature changes with time, so will the behavior of the circuit. However, shift invariance is often assumed to be true for all practical purposes. Even in cases where a system is definitely not shift-invariant, such as a system powered by an array of solar cells, it can often be considered as essentially shift-invariant during a short time interval of interest.

The next concept that must be considered is *causality*. Consider a system initially at rest. Suppose a signal is first applied to the system input terminal at time $t = t_o$. If the system is *causal*, there will be no output signal until $t = t_o$ at the earliest. In other words, a causal system cannot see into the future and anticipate future inputs. Needless to say, in the real world, all systems are causal. However, it is possible to discuss noncausal systems, at least in theory. For example, in Chapter 3 it will be shown that a theoretical *ideal* lowpass filter is a noncausal system, and therefore cannot be realized.

The LSI systems we are generally interested in can be represented by a linear differential equation with constant coefficients, as given in Equation (1–3), where $x(t)$ and $y(t)$ are the system input and output signals, respectively:

$$y(t) = \sum_{k=0}^{M} b_k \frac{d^k}{dt^k}[x(t)] - \sum_{k=1}^{N} a_k \frac{d^k}{dt^k}[y(t)] \tag{1-3}$$

A simple example of such a system is the circuit shown in Figure 1–2, where $x(t)$ and $y(t)$ are time-varying voltages. Using standard circuit analysis techniques, we can find the differential equation that relates $x(t)$ and $y(t)$ and show that this equation is an example of Equation (1–3). To begin with, note that current $i(t)$ is related to output voltage $y(t)$ by

$$i(t) = C \frac{d}{dt} y(t) \tag{1-4}$$

The voltages across the resistor and inductor, respectively, are

$$v_R(t) = Ri(t) = RC \frac{d}{dt} y(t) \tag{1-5}$$

$$v_L(t) = L \frac{d}{dt} i(t) = LC \frac{d^2}{dt^2} y(t) \tag{1-6}$$

FIGURE 1–2
Example of an LSI system.

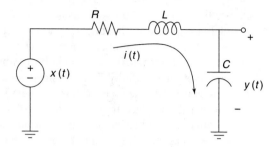

Using Kirchoff's voltage law, we obtain

$$y(t) = x(t) - RC\frac{d}{dt}y(t) - LC\frac{d^2}{dt^2}y(t) \tag{1-7}$$

Equation (1–7) is an example of Equation (1–3), where

$$M = 0$$
$$N = 2$$
$$b_0 = 1 \tag{1-8}$$
$$a_1 = RC$$
$$a_2 = LC$$

The actual circuit (Figure 1–2) is the real-world realization of the continuous time LSI system described by the differential equation (Equation [1–7]). It is worth noting here in passing that for discrete time systems, the *difference* equation (as opposed to *differential* equation) that describes the system is directly used to implement the system.

1–2 THE LAPLACE TRANSFORM AND SYSTEM TRANSFER FUNCTIONS

The Laplace transform $X(s)$ of a time domain function $x(t)$ is given by

$$X(s) = \int_{-\infty}^{\infty} x(t)e^{-st}dt \tag{1-9}$$

$x(t)$ and $X(s)$ form a Laplace transform "pair," which is often signified with the notation

$$x(t) \leftrightarrow X(s) \tag{1-10}$$

This notation also implies a reversible relationship: The *inverse* Laplace transform of $X(s)$ is $x(t)$.

In order for the Laplace transform to exist, it must be possible to define a region in the "s plane" (i.e., a range of values for s, where s is in general complex) where the Laplace transform integral of Equation (1–9) converges. It is possible to find such a region of convergence for the Laplace transform of any practical function $x(t)$ that interests us as engineers. Furthermore, the derivatives of any such practical signal will also be Laplace transformable. The region of convergence issue rarely impacts the utility of the Laplace transform, and so it is often neglected. However, this issue does make a brief appearance in Chapter 3, when the relationship between the Laplace and Fourier transforms is considered.

The reader is probably used to seeing the one-sided Laplace transform; that is, the Laplace transform defined with the lower limit of integration in Equation (1–9) set to zero instead of minus infinity. The one-sided version is well suited for solving differential equations starting at time $t = 0$, with some specified initial conditions; it is probably in this context that the Laplace transform was first presented to the reader. The two-sided version shown here is more general; it allows us to consider the Laplace transform of two-sided signals, such as the $x(t)$ shown in Figure P1–6 on page 18. Note that if $x(t)$ starts at time $t = 0$ or greater, the two-sided transform will result in the same $X(s)$ as the one-sided transform.

The other transforms considered in this book (Fourier transform, Z transform, and discrete time Fourier transform) are all two-sided (although a one-sided version of the Z transform exists). To be consistent and to help show how the various transforms are related, we have elected to use the two-sided Laplace transform here.

A short list of Laplace transforms is shown in Table 1–1. Several properties of the Laplace transform are also listed in Table 1–1. Note that $u(t)$ is the *unit step function*, defined as follows:

$$u(t) = \begin{cases} 1, t \geq 0 \\ 0, t < 0 \end{cases} \tag{1–11}$$

If a signal is multiplied by the unit step function, the effect is to set the signal equal to zero when $t < 0$. A signal having this property is sometimes called a *causal signal*. For example,

$$x(t) = e^{-at}u(t) = \begin{cases} e^{-at}, t \geq 0 \\ 0, t < 0 \end{cases} \tag{1–12}$$

The time-reversed unit step function $u(-t)$ is simply

$$u(-t) = \begin{cases} 0, t \geq 0 \\ 1, t < 0 \end{cases} \tag{1–13}$$

If a signal is multiplied by $u(-t)$, the effect is to set the signal equal to zero when $t \geq 0$. A signal having this property is sometimes called an *anticausal signal*. For example,

$$x(t) = e^{-at}u(-t) = \begin{cases} 0, t \geq 0 \\ e^{-at}, t < 0 \end{cases} \tag{1–14}$$

Table 1–1 shows Laplace transforms for various causal signals. For completeness, it should be mentioned that there is a theoretical "uniqueness problem" with the two-sided Laplace transform that does not exist with the one-sided transform. Any causal signal has an "anticausal twin" that has the same Laplace transform; the two transforms differ only with respect to the region of convergence. For example, it can be shown that

$$e^{-at}u(t) \leftrightarrow \frac{1}{s+a}, \quad \text{Re}\{s\} > -a$$

$$-e^{-at}u(-t) \leftrightarrow \frac{1}{s+a}, \quad \text{Re}\{s\} < -a \tag{1–15}$$

TABLE 1–1
Laplace Transforms

$x(t) \leftrightarrow X(s)$	Laplace transform properties:
$\delta(t) \leftrightarrow 1$	$Ax(t) \leftrightarrow AX(s)$
$u(t) \leftrightarrow \dfrac{1}{s}$	$x(t) + g(t) \leftrightarrow X(s) + G(s)$
$e^{-at}u(t) \leftrightarrow \dfrac{1}{s+a}$	$\int_{-\infty}^{t} x(\tau)d(\tau) \leftrightarrow \dfrac{X(s)}{s}$
$\left(1 - e^{-at}\right)u(t) \leftrightarrow \dfrac{a}{s(s+a)}$	$\dfrac{d^k}{dt^k} x(t) \leftrightarrow s^k X(s)$
$e^{-at}\cos(2\pi f_o t)u(t) \leftrightarrow \dfrac{s+a}{(s+a)^2 + (2\pi f_o)^2}$	$x(t - t_o) \leftrightarrow e^{-st_o} X(s)$
$e^{-at}\sin(2\pi f_o t)u(t) \leftrightarrow \dfrac{2\pi f_o}{(s+a)^2 + (2\pi f_o)^2}$	$e^{-at}x(t) \leftrightarrow X(s+a)$
$tu(t) \leftrightarrow \dfrac{1}{s^2}$	$y(t) = x(t) * g(t) \leftrightarrow Y(s) = X$

Re$\{s\}$ denotes "the real part of s." From a mathematician's perspective, one cannot find the inverse two-sided Laplace transform unless the region of convergence is specified. However, in an actual engineering application there are known constraints that make it unnecessary to formally consider the region of convergence. For example, if $1/(s + a)$ is the transfer function of a causal system, the system impulse response *must* be causal. (The impulse response is considered in Section 1–3.) Another example is the common case in which a signal $x(t)$ is first applied to a causal system at time $t = 0$; in other words, $x(t)$ is a causal signal. In this case, the output signal does not start until $t = 0$ at the earliest; therefore, the output signal *must* be a causal signal. In advanced electrical engineering courses, applications are considered in which anticausal time functions must be accounted for, but these cases will not be considered here.

Consider a continuous time LSI system with input signal $x(t)$ and output signal $y(t)$. Let $X(s)$ and $Y(s)$ denote the Laplace transforms of $x(t)$ and $y(t)$, respectively. The *transfer function $H(s)$* of the system is defined as

$$H(s) = \frac{Y(s)}{X(s)} \tag{1–16}$$

Rearranging Equation (1–16) shows that the Laplace transform of the output signal can be found by multiplying the Laplace transform of the input signal by the system transfer function as follows:

$$Y(s) = X(s)H(s) \tag{1–17}$$

It is important to note that when a system transfer function is defined, all of the initial conditions within the system (charges on capacitors, etc.) must be zero. In other words, when a system transfer function is defined, the system is assumed to be "at rest" before the input signal is applied.

There are three ways that the transfer function of a continuous time LSI system can be found:

1. Take the Laplace transform of both sides of the system differential equation (Equation [1–3]), and then solve for $Y(s)/X(s)$. To accomplish this, we must use the derivative property of the two-sided Laplace transform:

$$\frac{d^k}{dt^k} x(t) \leftrightarrow s^k X(s) \tag{1–18}$$

where $x(t) \leftrightarrow X(s)$.
2. Use circuit analysis to find $Y(s)/X(s)$ in terms of the s-domain circuit impedances.
3. $H(s)$ is the Laplace transform of the impulse response of the system.

The first two methods can be considered immediately, and method 2 is best shown by an example. Method 3 requires us to consider the concept of a system impulse response, which is the subject of the next section.

To use method 1, we start by taking the Laplace transform of the system differential equation (Equation [1–3]):

$$Y(s) = \sum_{k=0}^{M} b_k s^k X(s) - \sum_{k=1}^{N} a_k s^k Y(s) \tag{1–19}$$

Therefore

$$Y(s)\left[1 + \sum_{k=1}^{N} a_k s^k\right] = X(s) \sum_{k=0}^{M} b_k s^k \tag{1–20}$$

This leads to an expression for $H(s)$:

$$\frac{Y(s)}{X(s)} = H(s) = \frac{\displaystyle\sum_{k=0}^{M} b_k s^k}{1 + \displaystyle\sum_{k=1}^{N} a_k s^k} \tag{1–21}$$

The transfer function can also be expressed in factored form as follows:

$$H(s) = \frac{G \displaystyle\prod_{i=1}^{M} (s - z_i)}{\displaystyle\prod_{i=1}^{N} (s - p_i)} \tag{1–22}$$

where z_i, $i = 1, 2, \ldots, M$ are the "zeros" of the transfer function, and p_i, $i = 1, 2, \ldots, N$ are the "poles" of the transfer function. G is a gain constant.

In general, a transfer function may have poles and zeros that are complex numbers. (Complex numbers are reviewed in Chapter 2.) For example, a complex pole p_i can be expressed as

$$p_i = a_i + jb_i \tag{1–23}$$

where a_i and b_i are real numbers. The real part of p_i is a_i. The imaginary part of p_i is b_i. This is usually expressed as follows:

$$a_i = \text{Re}\{p_i\}$$
$$b_i = \text{Im}\{p_i\}$$

(1–24)

Poles and zeros can be plotted as points on the complex s plane. For example, suppose we have the following transfer function:

$$H(s) = \frac{(s-2)}{(s+1)(s^2 + 6s + 10)}$$

(1–25)

This system has a real zero at $s = 2$, a real pole at $s = -1$, and a pair of complex conjugate poles at $s = -3 \pm j$. Using the notation suggested by Equation (1–22), the poles and zeros can be listed as follows:

$$z_1 = 2$$
$$p_1 = -1$$
$$p_2 = -3 + j$$
$$p_3 = -3 - j$$

(1–26)

These poles and zeros are shown plotted on the complex s plane in Figure 1–3. The axis labeled Re{s} in Figure 1–3 is called the real axis. The axis labeled Im{s} in Figure 1–3 is called the imaginary axis. The region to the left of the imaginary axis (not including the imaginary axis) is called the left half of the s plane, or just the left-half plane. The left half of the s plane is defined by Re{s} < 0. The region defined by Re{s} > 0 is the right half of the s plane.

FIGURE 1–3
Poles and zeros on the
s plane.

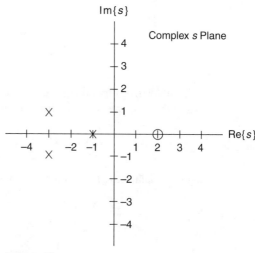

Poles: ✕
Zeros: ○

Assuming the coefficients a_k and b_k of Equation (1–21) are real (which is the case for any real-world system), complex poles must exist in complex conjugate pairs. The same thing holds for the zeros. Note that except for the gain constant G, an LSI system is completely represented by its poles and zeros.

The reader should note that if $H(s)$ is known (and has the form of Equation [1–21]), the system differential equation (in the form of Equation [1–3]) can be obtained. The method is as follows: Starting with a transfer function in the form of Equation (1–21), cross multiply and obtain an equation in the form of Equation (1–20). Rearrange the result to obtain an equation in the form of Equation (1–19). Finally, take the inverse Laplace transform term by term (applying the derivative property of the Laplace transform) to obtain a system differential equation.

Finding the transfer function using circuit analysis (method 2 above) can be illustrated using the R-L-C circuit of Figure 1–2 (for which we also have the system differential equation). The Laplace transforms $Y(s)$ and $X(s)$ are related by the voltage divider formula:

$$Y(s) = X(s)\left[\frac{1/sC}{R + sL + 1/sC}\right] \qquad (1\text{–}27)$$

where R, sL, and $1/sC$ are the s-domain impedances of the resistor, inductor, and capacitor, respectively. To show the relationship to the general form (Equation [1–21]), multiply the numerator and denominator of Equation (1–27) by sC and solve for $H(s)$:

$$H(s) = \frac{1}{1 + (RC)s + (LC)s^2} \qquad (1\text{–}28)$$

Equation (1–28) is the same as Equation (1–21) if we identify

$$
\begin{aligned}
M &= 0 \\
N &= 2 \\
b_0 &= 1 \\
a_1 &= RC \\
a_2 &= LC
\end{aligned}
\qquad (1\text{–}29)
$$

Let us verify that taking the Laplace transform of the system differential equation will yield the same results. The differential equation is repeated here for convenience:

$$y(t) = x(t) - RC\frac{d}{dt}y(t) - LC\frac{d^2}{dt^2}y(t) \qquad (1\text{–}30)$$

Taking the Laplace transform on both sides results in

$$Y(s) = X(s) - RCsY(s) - LCs^2Y(s) \qquad (1\text{–}31)$$

Therefore

$$Y(s)\left[1 + RCs + LCs^2\right] = X(s) \qquad (1\text{–}32)$$

Solving Equation (1–32) for $H(s) = Y(s)/X(s)$ results in the same transfer function found directly from circuit analysis (Equation [1–28]).

1–3   IMPULSE RESPONSE

We now turn our attention to the concept of a *system impulse response*. Consider an input signal $x(t)$, where $x(t)$ is an *impulse:*

$$x(t) = \delta(t) \tag{1–33}$$

The impulse $\delta(t)$ is sometimes called the *Dirac delta function*. Theoretically, an impulse could be formed as follows: We start with a rectangular pulse $p(t)$, centered on $t = 0$, with width τ and height $1/\tau$, as shown in Figure 1–4 . Note that the area of the rectangle is 1. The impulse is formed by letting τ approach zero "in the limit," that is,

$$\delta(t) = \lim_{\tau \to 0} p(t) \tag{1–34}$$

Thus an impulse is a rectangular pulse with an amplitude approaching infinity and a width approaching zero; however, the area of this impulse is still 1. An impulse $\delta(t)$ is usually represented graphically as shown in Figure 1–5(a). A time-shifted impulse $\delta(t - t_o)$ is shown in Figure 1–5(b).

Of course, no such "signal" actually exists. However, an impulse can be thought of as an approximation to a real-world pulse that is very short in duration relative to the time constant of the system in question.

It can be shown that the impulse function obeys the so-called "sifting property," that is,

$$\int_{-\infty}^{\infty} \delta(t - t_o)f(t)dt = f(t_o) \tag{1–35}$$

FIGURE 1–4
This pulse becomes a Dirac delta function (impulse) as $\tau \to 0$.

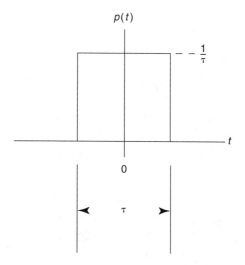

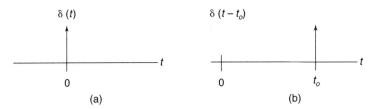

FIGURE 1–5
Graphical representation of impulses.

In fact, Equation (1–35) can be used as the defining property of an impulse. Figure 1–6 suggests an informal argument in favor of the sifting property. If the pulse $p(t)$ is very narrow, then the product of signal $f(t)$ with the shifted pulse $p(t - t_o)$ is approximately another narrow rectangular pulse with amplitude $f(t_o)/\tau$. The area of this pulse is approximately $\tau \times f(t_o)/\tau$. In the limit as τ approaches zero, the area is exactly $f(t_o)$.

   Figure 1–6 illustrates the fact that the limits of integration in Equation (1–35) need not be from minus infinity to plus infinity. Actually, all that is necessary is for the impulse

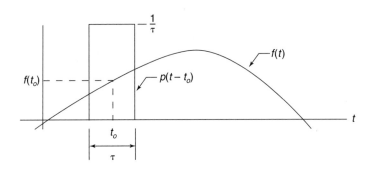

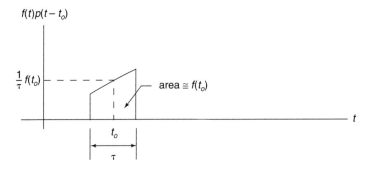

FIGURE 1–6
Explanation of the sifting property. The area of the pulse $f(t)p(t - t_o)$ becomes exactly $f(t_o)$ as $\tau \to 0$.

to be located between the upper and lower limits of integration. In other words, the sifting property can be expressed as follows:

$$\int_a^b \delta(t - t_o)f(t)dt = \begin{cases} f(t_o), & \text{provided } a < t_o < b \\ 0, & \text{if } t_o < a \text{ or } t_o > b \end{cases} \tag{1-36}$$

The Laplace transform of an impulse can be found by using the definition of the Laplace transform along with the sifting property of the impulse. Let $x(t) = \delta(t - t_o)$. Then:

$$X(s) = \int_{-\infty}^{\infty} \delta(t - t_o)e^{-st}dt = e^{-st_o} \tag{1-37}$$

For the special case where $t_o = 0$ (that is, for an impulse occuring at $t = 0$), Equation (1–37) reduces to

$$X(s) = \int_{-\infty}^{\infty} \delta(t)e^{-st}dt = 1 \tag{1-38}$$

Consider a linear, shift-invariant system with input $x(t)$ and output $y(t)$. If $x(t) = \delta(t)$, the output is said to be the *impulse response* of the system. In the Laplace transform domain, the output $Y(s)$ is given by $Y(s) = H(s)X(s)$; however, if $x(t) = \delta(t)$, then $X(s) = 1$, and we obtain

$$Y(s) = H(s)[1] = H(s) \tag{1-39}$$

Taking the inverse Laplace transform returns us to the time domain, where we have the impulse response, denoted $h(t)$:

$$h(t) \leftrightarrow H(s) \tag{1-40}$$

Thus, if the system impulse response $h(t)$ is known, the transfer function can be found by taking its Laplace transform.

If a system is *causal*, its impulse response doesn't start until the impulse is applied to the input. Therefore, for a causal system, $h(t) = 0$ for $t < 0$. In other words, if a system is causal, the impulse response of that system will be a causal time function. Any system having an impulse response that starts before $t = 0$ is noncausal and therefore not realizable.

1–4   CONVOLUTION

The system impulse response function is the key to another technique for finding the output $y(t)$, given the input $x(t)$. In the Laplace transform domain, the output is given as $Y(s) = H(s)X(s)$. Thus, if we take the inverse Laplace transform of the product $H(s)X(s)$, we should obtain $y(t)$. The following Laplace transform pair can be derived:

$$\int_{-\infty}^{\infty} h(\tau)x(t - \tau)d\tau \leftrightarrow H(s)X(s)$$

where

$$x(t) \leftrightarrow X(s) \tag{1-41}$$

$$h(t) \leftrightarrow H(s)$$

Therefore, $y(t)$ can be obtained by the following operation:

$$y(t) = \int_{-\infty}^{\infty} h(\tau)x(t - \tau)d\tau \tag{1-42}$$

Equation (1–42) is a *convolution* integral; it says that the system output $y(t)$ can be found by *convolving* the system impulse response $h(t)$ with the input $x(t)$. The convolution operation is often denoted by the symbol $*$, that is,

$$y(t) = h(t) * x(t) \tag{1-43}$$

It should be noted that convolution commutes, that is,

$$h(t) * x(t) = x(t) * h(t) - \int_{-\infty}^{\infty} x(\tau)h(t - \tau)d\tau \tag{1-44}$$

It should also be noted that the limits of integration shown in Equations (1–42) and (1–44) are intended to cover the general case. If the system is causal and $x(t)$ is first applied at time $t = 0$, the general case limits of integration can be changed from $(-\infty, \infty)$ to $(0, t)$:

$$y(t) = \begin{cases} \int_0^t h(\tau)x(t - \tau)d\tau = \int_0^t x(\tau)h(t - \tau)d\tau, t \geq 0 \\ 0, \hspace{5cm} t < 0 \end{cases} \tag{1-45}$$

For particular problems, the actual limits of integration must be carefully determined, as shown by example in Appendix A.

A detailed look at continuous time convolution is beyond the scope of this book, but we will need to refer to the convolution integral later on. (Appendix A has more information on convolution, including the proof for Equation [1–41].)

Convolution is a fundamental concept in system theory, communications theory, and signal analysis. The concept is not restricted to the particular context just described. As we will see in Chapter 3, convolution can be used to describe what happens to the spectrum (i.e., frequency content) when one time domain signal is multiplied by another.

1–5   BIBO STABILITY OF LSI SYSTEMS

A continuous time linear, shift-invariant (LSI) system is said to be *bounded input, bounded output* stable (BIBO stable) if for any amplitude bounded input $x(t)$, the output $y(t)$ is also amplitude bounded. Signals $x(t)$ and $y(t)$ are amplitude bounded if there exist finite real positive numbers α and β such that

$$|x(t)| < \alpha \hspace{0.5cm} \text{for all } t$$

$$|y(t)| < \beta \hspace{0.5cm} \text{for all } t \tag{1-46}$$

Essentially, BIBO stable means that the output of the system does not blow up when the input is excited by a signal that is bounded in amplitude. In this book, *stable* means BIBO stable.

It can be shown that for causal continuous time LSI systems represented by Equation (1–3), a system is BIBO stable if and only if its impulse response dies out as time goes on. That is, a system of this type is BIBO stable if and only if the following condition is satisfied:

$$\lim_{t \to \infty} h(t) = 0 \qquad (1\text{–}47)$$

It can also be shown that this condition will be satisfied only if all of the N poles of the system transfer function (Equation [1–22]) satisfy the following condition:

$$\text{Re}\{p_i\} < 0, i = 1, 2, \ldots, N \qquad (1\text{–}48)$$

In other words, BIBO stability requires that all of the poles be in the left half of the s plane.

PROBLEMS

1–1. Show that the system defined by $y(t) = T[x(t)] = x^2(t)$ is *not* linear.

1–2. Find the system differential equation for the circuit shown in Figure P1–2, without using Laplace transforms.

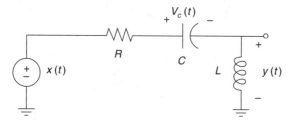

FIGURE P1–2

1–3. Let t_o be a positive constant. $u(t)$ is the unit step function. Sketch the following:
   a. $u(t - t_o)$
   b. $u(t + t_o)$
   c. $u(-t + t_o)$
   d. $u(-t - t_o)$

1–4. Let $x(t) = e^{-at}u(t)$, where $a > 0$. Let t_o be a positive constant. Sketch the following:
   a. $x(t - t_o)$
   b. $x(t + t_o)$
   c. $x(-t)$
   d. $x(-t + t_o)$
   e. $x(-t - t_o)$

1–5. The Laplace transform of $x(t - t_o)$ is $\int_{-\infty}^{\infty} x(t - t_o)e^{-st}dt$. Derive the time shifting property $x(t - t_o) \leftrightarrow e^{-st_o}X(s)$ by changing variables in this integral (let $t - t_o = \tau$).

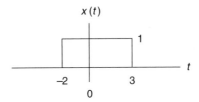

FIGURE P1–6

1–6. Let $x(t)$ be the pulse shown in Figure P1–6. Find $X(s)$. (Use Equation [1–9] directly.)

1–7. Let $x(t) = e^{-a(t-t_o)}u(t - t_o)$. Find $X(s)$. (Use the time shifting property.)

1–8. Given the system differential equation

$$y(t) = x(t) + \frac{1}{4}\frac{d}{dt}x(t) + \frac{1}{4}\frac{d^2}{dt^2}y(t)$$

    a. Find $H(s)$.

    b. Find the poles and zeros and plot them on the s plane.

    c. Is this system BIBO stable? Why or why not?

1–9. Given the system differential equation

$$y(t) = 4x(t) + 4\frac{d^2}{dt^2}x(t) - \left(2\sqrt{2}\right)\frac{d}{dt}y(t) - 4\frac{d^2}{dt^2}y(t)$$

    a. Find $H(s)$.

    b. Find the poles and zeros and plot them on the s plane.

    c. Is this system BIBO stable? Why or why not?

1–10. Consider a causal system having the transfer function

$$H(s) = \frac{3 + 2s + s^2}{1 + 10s + 3s^2 + 4s^3}$$

Find the system differential equation.

1–11. Consider a causal system having the transfer function

$$H(s) = \frac{10 + s^2}{1 + 0.5s + s^2}$$

    a. Find the system differential equation.

    b. Find the poles and zeros and plot them on the s plane.

    c. Is this system BIBO stable? Why or why not?

1–12. Consider a causal system having the transfer function

$$H(s) = \frac{a}{s + a}$$

    a. Find the system differential equation.

    b. Find the system impulse response $h(t)$.

    c. What must be true of the constant a in order for this system to be BIBO stable?

1–13. Consider a causal system having the transfer function

$$H(s) = \frac{s}{s + a}$$

where $a > 0$.

a. Find the impulse response $h(t)$ of this system.

HINT: $h(t)$ includes an impulse.

b. Find the system differential equation.

1–14. Consider a system functioning as an *integrator*:

$$y(t) = T[x(t)] = \int_{-\infty}^{t} x(\tau)d\tau$$

a. Show that $H(s) = {}^1/s$.

b. The impulse response of this system is $h(t) = u(t)$. Suppose the input to this system is $x(t) = u(t)$. Then the output is $y(t) = h(t) * x(t) = u(t) * u(t)$, where $*$ denotes convolution.

　　i. Use the Laplace transform and Table 1–1 to show that $y(t) = tu(t)$.

　　ii. Use the convolution integral (Equation [1–45]) to show that $y(t) = tu(t)$.

c. Based on part (b), would you classify this system as BIBO stable? Why or why not?

d. This system has one pole. Is this pole in the left half plane, the right half plane, or on the imaginary axis?

1–15. Consider a system having the impulse response

$$h(t) = e^{-10t}u(t)$$

Suppose the input to this system is $x(t) = u(t)$.

a. Use the Laplace transform to evaluate $y(t) = h(t) * x(t)$. ($*$ denotes convolution.)

b. Is this system causal? Why or why not?

c. Is this system BIBO stable? Why or why not?

1–16. Find the Laplace transform of the following time function:

$$x(t) = e^{-a|t|}$$

This function is defined for all values of t, both positive and negative. The constant a is greater than zero.

HINT: $\qquad \int_{-\infty}^{\infty} x(t)e^{-st}dt = \int_{-\infty}^{0} x(t)e^{-st}dt + \int_{0}^{\infty} x(t)e^{-st}dt$

1–17. Consider a system having the following impulse response:

$$h(t) = e^{-a|t|}$$

This impulse response is defined for all values of t, both positive and negative. The constant a is greater than zero.

a. Sketch $h(t)$.

b. Is this system causal or noncausal? Why?

1–18. Show that $A\delta(t - t_o) * x(t) = Ax(t - t_o)$ where $*$ denotes convolution.

HINTS: This can be shown using the Laplace transform. It can also be shown directly, using the convolution integral and the sifting property of the impulse:

$$A\delta(t - t_o) * x(t) = A\int_{-\infty}^{\infty} \delta(t - \tau)x(\tau - t_o)d\tau$$

1–19. Consider a system where $x(t)$ is the input, $y(t)$ is the output, and the input/output relationship is

$$y(t) = x(t) + ax(t - t_o)$$

where t_o is positive. (The output is the input signal plus a time-delayed version of the input signal. If $x(t)$ is a speech signal, the effect would be to create an echo.) Find the transfer function $H(s)$ of this system.

HINT: take the Laplace transform of both sides of the system equation. You will have to use the time shifting property. Solve for $Y(s)/X(s)$.

1–20. Consider a system with the following impulse response:

$$h(t) = \delta(t) + \delta(t - 10) - 3\delta(t - 20)$$

   a. Find the system equation. That is, $y(t) = T[x(t)]$: describe $T[x(t)]$.
   b. Find the transfer function $H(s)$ of this system.
   c. Find and sketch the output of this system if the input is $x(t) = u(t)$.

1–21. Suppose a system has the impulse response $h(t)$ shown in Figure P1–21. Suppose the input to this system is

$$x(t) = \delta(t) + 2\delta(t - T) + 3\delta(t - 2T) - 2\delta(t - 3T)$$

   a. Sketch $x(t)$.
   b. Let $y(t) - h(t) * x(t)$, where $*$ denotes convolution. Sketch $y(t)$.
      HINT: $h(t) * A\delta(t - t_o) = Ah(t - t_o)$

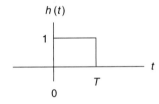

FIGURE P1–21

1–22. Suppose $x(t)$ is the pulse shown in Figure P1–22(a). Use the Laplace transform to show that if this pulse is convolved with itself; that is, $x(t) * x(t)$, then the result is the triangle function shown in Figure P1–22(b).

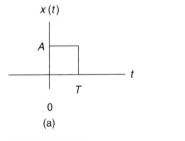
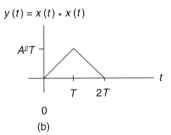

       (a)   (b)

FIGURE P1–22

1–23. Prove that a *sufficient* condition for the BIBO stability of an LSI system is that its impulse response $h(t)$ satisfies

$$\int_{-\infty}^{\infty} |h(t)| dt < \infty$$

(i.e., this definite integral has a finite value)

HINT: $\left|y(t)\right| = \left|\int_{-\infty}^{\infty} h(\tau)x(t-\tau)d\tau\right| \leq \int_{-\infty}^{\infty} \left|h(\tau)\right|\left|x(t-\tau)\right|d\tau$

if $\left|x(t)\right|$ is bounded, so is $\left|x(t-\tau)\right|$

1–24. Are the systems described in Problems 1–19 and 1–20 BIBO stable? Why or why not?

HINT: The sufficient condition for BIBO stability described in Problem 1–23 *can* be used to answer this question, but an answer can also be obtained by inspecting the system equation and determining whether or not "$x(t)$ is bounded" implies "$y(t)$ is bounded."

1–25. Evaluate the following integrals using the sifting property of the impulse:

a. $\int_{-\infty}^{\infty} \delta(t)\cos\left(2\pi f_o t\right)dt$

b. $\int_{-5}^{10} \delta(t-3)e^{-2t}dt$

c. $\int_{-\infty}^{\infty} e^{-2(\alpha-1)}\delta(\alpha+3)d\alpha$

d. $\int_{0}^{20} \delta(t-10)dt$

e. $\int_{0}^{5} \delta(t-10)dt$

2

FREQUENCY DOMAIN ANALYSIS FOR LSI CONTINUOUS TIME SYSTEMS

2–1 COMPLEX NUMBERS AND COMPLEX FUNCTIONS OF A REAL VARIABLE

In this chapter we will consider the frequency response function of an LSI continuous time system. (In Chapter 3 we will extend the discussion to the Fourier transform.) Since frequency domain analysis involves the manipulation of complex functions of a real variable, let's begin with a brief review of complex numbers.

Let z be a complex number. When expressed in rectangular form, we have

$$z = a + jb \tag{2–1}$$

where a and b are real numbers constituting the real part of z and the imaginary part of z, respectively. This relationship can be expressed as follows:

$$a = \text{Re}\{z\}$$
$$b = \text{Im}\{z\} \tag{2–2}$$

Figure 2–1 shows $z = a + jb$ as a point on the complex z plane. This can also be thought of as a vector with its tail at the origin and head at point z.

The magnitude (or *modulus* or *absolute value*) of z, denoted $|z|$, and the *phase* of z, denoted $\theta = \angle\{z\}$, are given by

$$|z| = \sqrt{a^2 + b^2}$$
$$\theta = \angle\{z\} = \tan^{-1}\left(\frac{b}{a}\right) \tag{2–3}$$

23

FIGURE 2–1

Geometric interpretation of the complex number z.

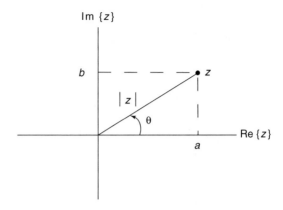

Euler's theorem allows us to express the polar form of z in a convenient manner. Euler's theorem is

$$e^{j\theta} = \cos(\theta) + j\sin(\theta)$$
$$e^{-j\theta} = \cos(\theta) - j\sin(\theta)$$

(2–4)

From trigonometry, we have $a = |z|\cos(\theta)$ and $b = |z|\sin(\theta)$. Therefore, z can be expressed as:

$$z = a + jb = |z|\cos(\theta) + j|z|\sin(\theta)$$
$$\therefore z = |z|[\cos(\theta) + j\sin(\theta)]$$

(2–5)

Applying Euler's theorem to Equation (2–5) gives us the polar form of z:

$$z = |z|e^{j\theta}$$

(2–6)

When two complex numbers are added together, the rectangular form is most convenient. Suppose $z_1 = a_1 + jb_1$ and $z_2 = a_2 + jb_2$. Then

$$z_1 + z_2 = (a_1 + a_2) + j(b_1 + b_2)$$

(2–7)

On the other hand, when two complex numbers are multiplied, the polar form is more convenient:

$$z_1 z_2 = |z_1|e^{j\theta_1} \cdot |z_2|e^{j\theta_2} = |z_1||z_2|e^{j(\theta_1 + \theta_2)}$$

(2–8)

The polar form is also convenient for division:

$$\frac{z_1}{z_2} = \frac{|z_1|e^{j\theta_1}}{|z_2|e^{j\theta_2}} = \frac{|z_1|}{|z_2|}e^{j(\theta_1 - \theta_2)}$$

(2–9)

The complex conjugate of z, denoted z^*, is found by changing the sign of the imaginary part of z:

$$z = a + jb = |z|e^{j\theta}$$
$$z^* = a - jb = |z|e^{-j\theta}$$

(2–10)

TABLE 2–1
Properties of Complex Numbers

1. $\|z_1 z_2\| = \|z_1 z_2\|$	6. $z + z^* = 2\text{Re}\{z\}$
2. $\left\|\dfrac{z_1}{z_2}\right\| = \left\|\dfrac{z_1}{z_2}\right\|$	7. $\left(z_1 + z_2\right)^* = z_1^* + z_2$
	8. $\left(z_1 z_2\right)^* = z_1^* z_2^*$
3. $\angle\{z_1 z_2\} = \angle\{z_1\} + \angle\{z_2\}$	
4. $\angle\left\{\dfrac{z_1}{z_2}\right\} = \angle\{z_1\} - \angle\{z_2\}$	9. $\left(\dfrac{z_1}{z_2}\right)^* = \dfrac{z_1^*}{z_2^*}$
	10. $\|z_1 + z_2\| \le \|z_1\| + \|z_2\|$
5. $\|z\|^2 = \|z^*\|^2 = zz^*$	

Table 2–1 contains a list of useful properties of complex numbers.

NOTE: The last entry in Table 2–1 has a less-than-or-equal-to sign; this relationship is sometimes called the *triangle inequality.*

In general, z could be a function of some real variable, say f; in this case, we have the complex function $z(f)$:

$$z(f) = \text{Re}\{z(f)\} + j\,\text{Im}\{z(f)\} \tag{2–11}$$

or

$$z(f) = \|z(f)\|e^{j\theta(f)} \tag{2–12}$$

where

$$\theta(f) = \angle\{z(f)\} = \tan^{-1}\left(\frac{\text{Im}\{z(f)\}}{\text{Re}\{z(f)\}}\right) \tag{2–13}$$

$$\|z(f)\| = \sqrt{\text{Re}^2\{z(f)\} + \text{Im}^2\{z(f)\}}$$

If $z(f)$ is the ratio of two functions, say $z(f) = N(f)/D(f)$, then the phase function can also be calculated as $\theta(f) = \angle\{N(f)\} - \angle\{D(f)\}$, and the amplitude function can be calculated as $\|z(f)\| = \|N(f)\|/\|D(f)\|$.

All of the properties of z discussed earlier still apply; just replace z with $z(f)$. A simple example should help to clarify this point. Suppose

$$z(f) = 10f + j\cos(2\pi f) \tag{2–14a}$$

Then

$$\text{Re}\{z(f)\} = 10f$$
$$\text{Im}\{z(f)\} = \cos(2\pi f) \tag{2–14b}$$

$$\|z(f)\| = \sqrt{(10f)^2 + \left[\cos(2\pi f)\right]^2}$$

$$\theta(f) = \tan^{-1}\left(\frac{\cos(2\pi f)}{10f}\right) \tag{2–15}$$

Observe that while the complex function $z(f)$ cannot be plotted in two dimensions, the functions $|z(f)|$, $\theta(f)$, $\text{Re}\{z(f)\}$, and $\text{Im}\{z(f)\}$ are real functions that can be plotted in two dimensions, with f as the independent variable.

2-2 INTRODUCTION TO THE FREQUENCY RESPONSE FUNCTION

We now turn our attention to the frequency response function of a BIBO stable LSI system. As an introduction, let us consider a familiar problem. In Figure 2–2, the input voltage $v_1(t)$ is a sinusoid, that is,

$$v_i(t) = A\cos(2\pi ft + \phi) \tag{2-16}$$

The problem is to find the *steady-state* output voltage $v_o(t)$. The exact value of frequency (f) is unspecified at this point. (The reader presumably recognizes the circuit of Figure 2–2 as a simple lowpass filter.)

For steady-state analysis, the input voltage $v_1(t)$ can be represented by the phasor $\underline{V_i}$, which carries the amplitude and phase information:

$$\underline{V_i} = |V_i|e^{j\angle\{V_i\}} = Ae^{j\phi} \tag{2-17}$$

Note that we are using the peak value of the sinusoid as the magnitude of the phasor. The reader can convert to an RMS value if he insists.

The output voltage phasor $\underline{V_o}$ can be found by using the voltage divider formula (where $1/(j2\pi fC)$ is the impedance of the capacitor):

$$\underline{V_o} = \underline{V_i}\left(\frac{\dfrac{1}{j2\pi fC}}{R + \dfrac{1}{j2\pi fC}}\right) \tag{2-18}$$

The quantity in parentheses in Equation (2–18) is a complex function of the real frequency variable f that relates the input and output phasors. This function is the *frequency*

FIGURE 2–2
Simple lowpass filter circuit.

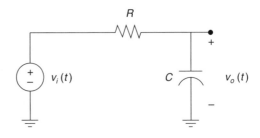

response function of the LSI system of Figure 2–2. The frequency response function of an LSI system will be denoted as $H_a(f)$ in this book. That is, for the LSI system of Figure 2–2,

$$H_a(f) = \frac{\dfrac{1}{j2\pi f C}}{R + \dfrac{1}{j2\pi f C}} \tag{2–19}$$

As a complex function, $H_a(f)$ has a magnitude, a phase, a real part, and an imaginary part, and can therefore be represented as

$$H_a(f) = \text{Re}\{H_a(f)\} + j\,\text{Im}\{H_a(f)\} \tag{2–20}$$

or as

$$H_a(f) = |H_a(f)|e^{j\theta_H(f)} \tag{2–21}$$

where

$$|H_a(f)| = \sqrt{\text{Re}^2\{H_a(f)\} + \text{Im}^2\{H_a(f)\}} \tag{2–22}$$

$$\theta_H(f) = \angle\{H_a(f)\} = \tan^{-1}\left[\frac{\text{Im}\{H_a(f)\}}{\text{Re}\{H_a(f)\}}\right] \tag{2–23}$$

$|H_a(f)|$ is the *amplitude response function* of the system, and $\theta_H(f)$ is the *phase response function* of the system. If the frequency response function is the ratio of two functions, say $H_a(f) = N_a(f)/D_a(f)$, then the phase response function can also be calculated as $\theta_H(f) = \angle\{N_a(f)\} - \angle\{D_a(f)\}$ and the amplitude response function can be calculated as $|H_a(f)| = |N_a(f)|/|D_a(f)|$. Both of these functions can be plotted in two dimensions, with f as the independent variable (an example of which will be presented shortly).

The output phasor can be found by multiplying the input phasor by the frequency response function, that is,

$$\begin{aligned} \underline{V_o} &= \underline{V_i} H_a(f) \\ &= Ae^{j\phi}|H_a(f)|e^{j\theta_H(f)} \\ &= A|H_a(f)|e^{j(\phi+\theta_H(f))} \end{aligned} \tag{2–24}$$

Translating to the time domain results in

$$v_o(t) = A|H_a(f)|\cos[2\pi ft + \phi + \theta_H(f)] \tag{2–25}$$

For a specific frequency f, $|H_a(f)|$ and $\theta_H(f)$ are evaluated at that specific frequency.

[NOTE: If the input is $v_i(t) = A\sin(2\pi ft + \phi)$, the output is $v_o(t) = A\left|H_a(f)\right|\sin(2\pi ft + \phi + \theta_H(f))$].

Now let's find expressions for the amplitude response function $\left|H_a(f)\right|$ and phase response function $\theta_H(f)$ for the system of Figure 2–2. The frequency response function of that system is given in Equation (2–19); it can be rewritten as follows:

$$H_a(f) = \frac{1}{1 + j2\pi fRC} \tag{2-26}$$

Since the frequency response function is the ratio of two functions, that is,

$$H_a(f) = N_a(f)/D_a(f) \tag{2-27}$$

where $N_a(f) = 1$ and $D_a(f) = (1 + j2\pi fRC)$, the amplitude and phase response functions can be determined as follows:

$$\left|H_a(f)\right| = \frac{\left|N_a(f)\right|}{\left|D_a(f)\right|} = \frac{\sqrt{1^2 + 0^2}}{\sqrt{1^2 + (2\pi fRC)^2}} = \frac{1}{\sqrt{1 + (2\pi fRC)^2}} \tag{2-28}$$

$$\theta_H(f) = \angle\{N_a(f)\} - \angle\{D_a(f)\} = \angle\{1 + j0\} - \angle\{1 + j2\pi fRC\}$$
$$= 0 - \tan^{-1}\left(\frac{2\pi fRC}{1}\right) = \tan^{-1}(-2\pi fRC) \tag{2-29}$$

Therefore, with respect to Figure 2–2 and Equation (2–25), we have

$$v_o(t) = \frac{A}{\sqrt{1 + (2\pi fRC)^2}} \cos\left[2\pi ft + \phi + \tan^{-1}(-2\pi fRC)\right] \tag{2-30}$$

Let's consider a specific example: Suppose $R = 100\ \Omega$, $C = 15.9\ \mu F$, and $f = 100$Hz. Then $\left|H_a(100)\right| = 0.7074$, $\theta_H(100) = -0.7874$ radians, and the output voltage is

$$v_o(t) = 0.7074A\,\cos\left[2\pi(100)t + \phi - 0.7849\right] \tag{2-31}$$

Figure 2–3 shows the amplitude and phase response functions for the circuit of Figure 2–2 when $R = 100\ \Omega$ and $C = 15.9\ \mu F$. The cutoff frequency of this simple lowpass filter is usually defined as $f_c = 1/(2\pi RC)$, which is 100Hz in this example. The cutoff frequency can also be thought of as the *bandwidth* of this system.

The fact that Figure 2–3 shows these functions plotted over a range of both positive and negative frequency may strike the reader as unusual (how can the frequency be negative?). The significance of negative frequency will be explored in Section 2–3.

The example above serves to illustrate a well-known point: In an LSI system, if the input to the system is a sinusoid with frequency f, then the steady-state output is also a sinusoid with the same frequency f. However, the amplitude and phase of the sinusoid are altered by the frequency response function of the system.

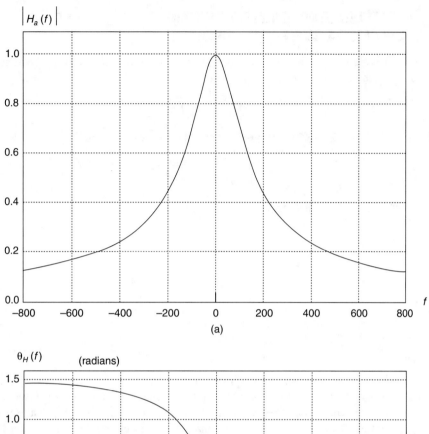

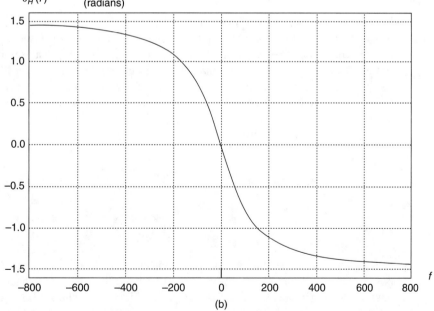

FIGURE 2–3

Frequency response plots. (a) Amplitude response function. (b) Phase response function.

2–3   COMPLEX EXPONENTIAL SIGNALS AND NEGATIVE FREQUENCY

Before considering a more general view of the frequency response function of an LSI system, let us consider a *complex exponential signal* given by

$$x(t) = e^{j2\pi ft} \tag{2-32}$$

By Euler's theorem, $x(t)$ can also be expressed as follows:

$$x(t) = e^{j2\pi ft} = \cos(2\pi ft) + j\sin(2\pi ft) \tag{2-33}$$

Thus the real-world sinusoid $\cos(2\pi ft)$ is the *real part* of the complex exponential. Note also that by Euler's theorem,

$$e^{-j2\pi ft} = \cos(2\pi ft) - j\sin(2\pi ft) \tag{2-34}$$

By adding Equations (2–33) and (2–34), we obtain

$$\cos(2\pi ft) = \tfrac{1}{2}e^{j2\pi ft} + \tfrac{1}{2}e^{-j2\pi ft} \tag{2-35}$$

which can also be expressed as follows:

$$\cos(2\pi ft) = \tfrac{1}{2}e^{j2\pi ft} + \tfrac{1}{2}e^{j2\pi(-f)t} \tag{2-36}$$

Equation (2–36) shows that mathematically, the real-world sinusoid $\cos(2\pi ft)$ is composed of a pair of these complex exponentials, one having a positive frequency (f) and the other having a negative frequency ($-f$). This negative frequency is a useful mathematical fiction. If we are going to use this mathematical fiction in our analysis of LSI systems, we will need to allow the frequency response function $H_a(f)$ to be defined also for negative f (which is why the amplitude and phase response functions in Figure 2–3 are shown for both positive and negative frequency).

The reader should also note that Euler's theorem can handle an arbitrary gain constant and phase shift:

$$A\cos(2\pi ft + \phi) = \tfrac{A}{2}e^{j(2\pi ft + \phi)} + \tfrac{A}{2}e^{-j(2\pi ft + \phi)}$$

$$= \left(\tfrac{A}{2}e^{j\phi}\right)e^{j2\pi ft} + \left(\tfrac{A}{2}e^{-j\phi}\right)e^{j2\pi(-f)t} \tag{2-37}$$

$$= ce^{j2\pi ft} + c^*e^{j2\pi(-f)t}$$

Here again we observe that the real-world sinusoid $A\cos(2\pi ft + \phi)$ is mathematically composed of a pair of complex exponentials, one having positive frequency and the other having negative frequency. Note that the phase shift ϕ causes the gain constants in front of the complex exponentials to be complex (one is the complex conjugate of the other).

Let us now consider the following problem: Suppose the steady-state input to an LSI system with frequency response function $H_a(f)$ is a complex exponential, that is,

$$x(t) = ce^{j2\pi ft} \tag{2-38}$$

in which the gain constant c is complex ($c = A\exp(j\phi)$). (It is true that there is no signal generator in your lab that creates such a signal but from a strictly mathematical perspective there is no problem in considering it.) Let $y(t)$ be the steady-state output. It will now be shown that

$$y(t) = cH_a(f)e^{j2\pi ft} \tag{2-39}$$

In other words, in this very special case the output is just the input multiplied by $H_a(f)$. To show this, let's use Euler's theorem to express the input as follows:

$$x(t) = c\cos(2\pi ft) + cj\sin(2\pi ft) \tag{2-40}$$

NOTE: The constants c and cj are treated mathematically like any other constant multipliers. Since the system is *linear*, we can use *superposition* to determine $y(t)$. First, the steady-state output due to $c\cos(2\pi ft)$ acting alone is

$$c|H_a(f)|\cos[2\pi ft + \theta_H(f)] \tag{2-41}$$

Second, the steady-state output due to $cj\sin(2\pi ft)$ acting alone is

$$cj|H_a(f)|\sin[2\pi ft + \theta_H(f)] \tag{2-42}$$

The steady-state output $y(t)$ is the sum of these two results:

$$y(t) = c|H_a(f)|\cos[2\pi ft + \theta_H(f)] + cj|H_a(f)|\sin[2\pi ft + \theta_H(f)] \tag{2-43}$$

Now factor out $c|H_a(f)|$ and apply Euler's theorem:

$$\begin{aligned} y(t) &= c|H_a(f)|\{\cos[2\pi ft + \theta_H(f)] + j\sin[2\pi ft + \theta_H(f)]\} \\ &= c|H_a(f)|e^{j[2\pi ft + \theta_H(f)]} \end{aligned} \tag{2-44}$$

The result can also be expressed as:

$$y(t) = c\left[|H_a(f)|e^{j\theta_H(f)}\right]e^{j2\pi ft} \tag{2-45}$$

The expression inside the brackets in Equation (2–45) is the polar form of $H_a(f)$. Therefore,

$$y(t) = cH_a(f)e^{j2\pi ft} \tag{2-46}$$

which is what we set out to demonstrate.

Let us now consider how this result is related to the phasors the reader is accustomed to using. Recall that the constant c is complex: $c = A\exp(j\phi)$. If this is substituted into Equations (2–38) and (2–39), the following is obtained:

$$x(t) = \left(Ae^{j\phi}\right)e^{j2\pi ft} = Ae^{j(2\pi t + \phi)} \tag{2-47}$$

and

$$y(t) = Ae^{j\phi}H_a(f)e^{j2\pi ft} = A|H_a(f)|e^{j[2\pi ft + \phi + \theta_H(f)]} \tag{2-48}$$

Observe that the *real part* of $x(t)$ is

$$\text{Re}\{x(t)\} = A\cos(2\pi ft + \phi) \tag{2-49}$$

And the *real part* of $y(t)$ is

$$\text{Re}\{y(t)\} = A|H_a(f)|\cos(2\pi ft + \phi + \theta_H(f)) \tag{2-50}$$

Thus the real part of this relationship corresponds to the real-world steady-state input-output relationship described in Section 2–2. Note also that the complex constant $A\exp(j\phi)$ is exactly the same as the phasor that would be used to describe $A\cos(2\pi ft + \phi)$. In other words, the phasor for $A\cos(2\pi ft + \phi)$ can be obtained by considering $A\cos(2\pi ft + \phi)$ to be the real part of the complex exponential signal $c\exp(2\pi ft)$, where $c = A\exp(j\phi)$ is the phasor.

2–4 FREQUENCY RESPONSE FUNCTION: RELATIONSHIP TO SYSTEM IMPULSE RESPONSE

Let $x(t) = c\exp(j2\pi ft)$ be the input applied to the LSI system under consideration (c is some constant that could be complex, as in the last section). Assume furthermore that this input is applied at $t = -\infty$ (meaning a long time ago: we are after a steady-state result here.) Let $h(t)$ be the system impulse response. Let $y(t)$ be the system output. Using the convolution integral (see Chapter 1), we have

$$y(t) = \int_{-\infty}^{\infty} h(\tau)x(t - \tau)d\tau = \int_{-\infty}^{\infty} h(\tau)ce^{j2\pi f(t-\tau)}d\tau$$

$$= ce^{j2\pi ft}\int_{-\infty}^{\infty} h(\tau)e^{-j2\pi f\tau}d\tau \tag{2-51}$$

Changing the dummy variable of integration back to t gives us

$$y(t) = ce^{j2\pi ft}\left[\int_{-\infty}^{\infty} h(t)e^{-j2\pi ft}dt\right] \tag{2-52}$$

Considering the results from the previous section, in which a steady-state input $x(t) = c\exp(j2\pi ft)$ was shown to result in a steady-state output $y(t) = cH_a(f)\exp(j2\pi ft)$, we must conclude that the complex function of f enclosed in the brackets in the above Equation (2–52) is in fact the frequency response function $H_a(f)$. In other words, we conclude that

$$H_a(f) = \int_{-\infty}^{\infty} h(t)e^{-j2\pi ft}dt \tag{2-53}$$

The integral on the right side of Equation (2–53) is the *Fourier transform* of $h(t)$. Thus the system frequency response function is the Fourier transform of the system impulse response. (The Fourier transform is covered in more detail in Chapter 3.)

Let us verify with a simple example that the frequency response function given by the integral of Equation (2–53) is the same thing that we would obtain via circuit analysis.

Consider once again the circuit of Figure 2–2. Let $X(s)$ and $Y(s)$ be the input and output, respectively, in the Laplace domain. The system transfer function can be found using the voltage divider formula:

$$\frac{Y(s)}{X(s)} = H(s) = \frac{\dfrac{1}{sC}}{R + \dfrac{1}{sC}} = \frac{\dfrac{1}{RC}}{s + \dfrac{1}{RC}} \tag{2–54}$$

Taking the inverse Laplace transform of $H(s)$ gives us the system impulse response

$$h(t) = \frac{1}{RC} e^{-\left(\frac{1}{RC}\right)t} u(t) \tag{2–55}$$

The next step is to use Equation (2–53) to find the frequency response function:

$$H_a(f) = \int_0^\infty \frac{1}{RC} e^{-\left(\frac{1}{RC}\right)t} e^{-j2\pi ft}\, dt \tag{2–56}$$

The lower limit of integration is 0 because $h(t)$ starts at $t = 0$. Continuing

$$H_a(f) = \frac{1}{RC} \int_0^\infty e^{-\left(j2\pi f + \frac{1}{RC}\right)t}\, dt$$

$$= \frac{-\dfrac{1}{RC}}{j2\pi f + \dfrac{1}{RC}} e^{-\left(j2\pi f + \frac{1}{RC}\right)t}\bigg|_{t=0}^{\infty} \tag{2–57}$$

$$= \frac{\dfrac{1}{RC}}{j2\pi f + \dfrac{1}{RC}}$$

Multiplying the numerator and denominator by RC results in

$$H_a(f) = \frac{1}{j2\pi fRC + 1} \tag{2–58}$$

which is the same result obtained by using circuit analysis directly (see Equations [2–18] through [2–26]).

2–5   STEADY-STATE RESPONSE TO PERIODIC SIGNALS

In the last section we considered the steady-state response of an LSI system to a complex exponential input. Using this result in conjunction with superposition and the Fourier series, we can now consider the steady-state response of an LSI system to an arbitrary periodic input signal.

Let $x(t)$ be periodic, that is, $x(t) = x(t + T)$, where T is the period. Since $x(t)$ is periodic, it can be expressed as a Fourier series. The complex exponential form of the Fourier series is

$$x(t) = \sum_{n=-\infty}^{\infty} c_n e^{j2\pi nf_o t} \tag{2-59}$$

where $f_o = 1/T$ is the *fundamental frequency* and the coefficients c_n are the Fourier series coefficients, given by

$$c_n = \frac{1}{T} \int_T x(t)e^{-j2\pi nf_o t} dt \tag{2-60}$$

The Fourier series coefficients are in general complex. Equation (2–59) says that a periodic signal is mathematically composed of a sum of complex exponentials with frequencies f_o, $2f_o$, $3f_o$, . . . and negative frequencies $-f_o$, $-2f_o$, $-3f_o$, . . . (The $n = 0$ term reduces to a constant.) The Fourier series can be expressed in other forms, such as

$$x(t) = a_0 + \sum_{n=1}^{\infty} a_n \cos(2\pi nf_o t) + \sum_{n=1}^{\infty} b_n \sin(2\pi nf_o t) \tag{2-61}$$

where a_0, a_n, and b_n are real numbers that can be derived from the complex c_n's in Equation (2–59). The cosines and sines in Equation (2–61) can be combined to obtain another form of the Fourier series:

$$x(t) = a_0 + \sum_{n=1}^{\infty} d_n \cos(2\pi nf_o t + \phi_n) \tag{2-62}$$

where a_0, d_n, and ϕ_n are also real numbers which can be derived from the c_ns. In this form (Equation [2–62]) we see the essential idea: Any periodic waveform (with fundamental frequency $f_o = 1/T$) can be expressed as a sum of sinusoids with frequencies f_o, $2f_o$, $3f_o$, . . . , with respective amplitudes d_1, d_2, d_3, . . . , and respective phase shifts ϕ_1, ϕ_2, ϕ_3, . . . , and a possible DC offset term a_0. The component with frequency f_o is called the fundamental; the component with frequency $2f_o$ is called the second harmonic; the component with frequency $3f_o$ is called the third harmonic; etc. Although the Fourier series as expressed in Equation (2–62) is easier to visualize, the complex exponential form of the Fourier series is often easier to deal with mathematically. Therefore, the complex exponential form is the "weapon of choice" here.

Suppose a periodic signal $x(t)$, which can be represented by a Fourier series in complex exponential form, is the steady-state input to a system with a frequency response function $H_a(f)$. Using the superposition principle, each individual component $c_n \exp(j2\pi nf_o t)$ can be dealt with separately; the resulting individual output (as explained in Section 2–3) is $c_n H_a(nf_o) \exp(j2\pi nf_o t)$.

NOTE: The notation $H_a(nf_o)$ means the frequency response function evaluated at the particular frequency $f = nf_o$. When the frequency response function is evaluated at a particular frequency, the result is a complex number. The results are then added together to obtain

$$y(t) = \sum_{n=-\infty}^{\infty} c_n e^{j2\pi nf_o t} H_a(nf_o) \tag{2-63}$$

In other words, the steady-state output $y(t)$ is also a Fourier series, but the new Fourier series coefficients are $c_n H_a(nf_o)$.

If the input periodic signal is expressed as in Equation (2–62), then the steady-state output is

$$y(t) = a_0 H_a(0) + \sum_{n=1}^{\infty} d_n |H_a(nf_o)| \cos[2\pi nf_o t + \phi_n + \theta_H(nf_o)] \qquad (2\text{--}64)$$

In this form we see that each individual sinusoidal component has its amplitude and phase changed by the frequency response function of the system, according to the principles discussed in Section 2–2. ($H_a(0)$ is the DC gain of the system.)

The steady-state sinusoidal input-output relationships considered in this chapter are only valid if the system under consideration is BIBO stable. When a sinusoid is first applied to the input of a system, the output will have two components: the steady-state component considered in this chapter, and a *natural response* component. The form of the natural response component is determined by the location of the system poles. If the system is BIBO stable (all poles are in the left half of the s plane), the natural response component is a *transient* component that will die out as time goes on, leaving us with the steady-state component only. However, if the system is not BIBO stable, there will be parts of the natural response that will either increase without bound or will be bounded but never die out.

In Chapter 3 we will consider signals that are more general, that is, not restricted to sinusoids or sums of sinusoids, and consider frequency domain analysis in a much more general sense. Most signals of interest, even if they are not periodic (a single pulse, for example), can be thought of as having a *spectrum*, or *frequency content*, in some sense. If such a signal is applied to the input of a BIBO stable LSI system, then the resulting output signal has a spectrum that can be found by multiplying the input signal spectrum by the frequency response function of the system. The mathematical tool used to explain this concept is the *Fourier transform*. The Fourier transform can be used to obtain straightforward explanations for concepts such as signal bandwidth, system bandwidth, filtering, and modulation (frequency shifting). Furthermore, the relationship between *continuous time* and *discrete time* systems is best understood in the Fourier transform domain.

PROBLEMS

2–1. Let z be a complex number.
    a. Show that $z + z^* = 2\text{Re}\{z\}$.
    b. Show that $zz^* = |z|^2$.

2–2. Let

$$z = \frac{1}{2} + j\frac{1}{4}$$

    Find the following: $|z|$, $\angle\{z\}$, $\text{Re}\{z\}$, $\text{Im}\{z\}$, and z^*.

2–3. Let

$$z(f) = \frac{1 + j0.5f}{1 + j0.25f}$$

    Find: $|z(f)|$, $\theta(f)$, $\text{Re}\{z(f)\}$, $\text{Im}\{z(f)\}$, and $z^*(f)$.

2–4. Show that

$$\left|e^{j2\pi ft}\right| = 1$$

2–5. Show that

$$\sin(2\pi ft) = \frac{1}{2j}e^{j2\pi ft} - \frac{1}{2j}e^{-j2\pi ft}$$

2–6. Prove the following:
   a. $-1 = e^{\pm j\pi}$
   b. $j = e^{j\frac{\pi}{2}}$
   c. $-j = e^{-j\frac{\pi}{2}}$

2–7. Let

$$H_a(f) = \frac{1}{1 + j\left(\frac{1}{10}\right)f}$$

be the system frequency response function.
Let $x(t) = 10\cos[2\pi(20)t]$ be the steady-state input to this system.
Find $y(t)$ (the steady-state output signal).

2–8. Let

$$H_a(f) = \frac{j\left(\frac{1}{10}\right)f}{1 + j\left(\frac{1}{10}\right)f}$$

be the system frequency response function.
   a. Find $\left|H_a(f)\right|$ and $\theta_H(f)$.
   b. Sketch $\left|H_a(f)\right|$ and $\theta_H(f)$, showing both positive and negative frequency.
   c. Let

$$x(t) = 2\exp\left\{j\left[2\pi(10)t + \frac{\pi}{3}\right]\right\}$$

   be the steady-state input to this system.
   Find $y(t)$ (the steady-state output signal).

2–9. Given the following signal:

$$x(t) = \sum_{n=-2}^{2} \frac{1}{2}e^{j2\pi(10n)t}$$

This signal can also be expressed as

$$x(t) = a_0 + a_1\cos(2\pi f_1 t) + a_2\cos(2\pi f_2 t)$$

Find $a_0, a_1, a_2, f_1,$ and f_2.

HINT: Write out the terms of the summation and look for ways to use Euler's theorem to combine complex exponentials into cosines.

2–10. Suppose the following signal is the steady-state input to the system having the frequency response function shown in Problem 2–8:

$$x(t) = \cos\left[2\pi(10)t\right] + \cos\left[2\pi(20)t - \frac{\pi}{4}\right]$$

Find $y(t)$ (the steady-state output signal).

3

FOURIER TRANSFORM ANALYSIS OF CONTINUOUS TIME LSI SYSTEMS

3–1 THE FOURIER TRANSFORM, SIGNAL SPECTRUM, AND FREQUENCY RESPONSE

The Fourier transform $X_a(f)$ of a time domain function $x(t)$ is given by

$$\mathbf{F}\{x(t)\} = X_a(f) = \int_{-\infty}^{\infty} x(t)e^{-j2\pi ft}dt \tag{3–1}$$

where $\mathbf{F}\{x(t)\}$ is shorthand for "the Fourier transform of $x(t)$." The transform pair relationship is indicated by

$$x(t) \leftrightarrow X_a(f) \tag{3–2}$$

The function $x(t)$ is said to be in the *time domain* and $X_a(f)$ is said to be in the *frequency domain*. The time domain is defined for both positive and negative t, and the frequency domain is defined for both positive and negative f. The Fourier transform of a signal $x(t)$ is interpreted as the *spectrum*, or *spectral content*, or *frequency content* of the signal.

If $X_a(f)$ is known, the corresponding time domain signal $x(t)$ can be found using the inverse Fourier transform, given by

$$x(t) = \int_{-\infty}^{\infty} X_a(f)e^{j2\pi ft}df \tag{3–3}$$

$X_a(f)$ is in general a complex function of frequency variable f. It can therefore be represented in rectangular or polar form as follows:

$$X_a(f) = \text{Re}\{X_a(f)\} + j\text{Im}\{X_a(f)\} \tag{3–4}$$

or

$$X_a(f) = |X_a(f)|e^{j\theta x(f)} \tag{3–5}$$

39

where

$$\left|X_a(f)\right| = \sqrt{\text{Re}^2\{X_a(f)\} + \text{Im}^2\{X_a(f)\}}$$

$$\theta_X(f) = \angle\{X_a(f)\} = \tan^{-1}\left(\frac{\text{Im}\{X_a(f)\}}{\text{Re}\{X_a(f)\}}\right) \quad\quad (3\text{–}6)$$

The function $\left|X_a(f)\right|$ could be defined as the *spectral amplitude function* and the function $\theta_X(f)$ as the *phase function*.

From Equation (2–53) (in Chapter 2), we can see that *the frequency response function of an LSI system is the Fourier transform of the system impulse response*:

$$h(t) \leftrightarrow H_a(f) \quad\quad (3\text{–}7)$$

The Fourier transform is a linear operator. That is,

$$ax(t) + bg(t) \leftrightarrow aX_a(f) + bG_a(f) \qu\quad (3\text{–}8)$$

The proof of this *linearity property* follows directly from the definition of the Fourier transform:

$$\mathbf{F}\{ax(t) + bg(t)\} = \int_{-\infty}^{\infty}\left[ax(t) + bg(t)\right]e^{j2\pi ft}\,dt$$

$$= a\int_{-\infty}^{\infty} x(t)e^{j2\pi f}\,dt + b\int_{-\infty}^{\infty} g(t)e^{j2\pi f}\,dt = a\mathbf{F}\{x(t)\} + b\mathbf{F}\{g(t)\} \quad\quad (3\text{–}9)$$

3–2 RELATIONSHIP OF FOURIER AND LAPLACE TRANSFORMS

The reader may have already noticed the similarity between the Fourier transform integral and the Laplace transform integral, which are shown together here for convenience:

$$X(s) = \int_{-\infty}^{\infty} x(t)e^{-st}\,dt \quad\quad (3\text{–}10)$$

$$X_a(f) = \int_{-\infty}^{\infty} x(t)e^{-j2\pi ft}\,dt \quad\quad (3\text{–}11)$$

The Fourier transform of a signal $x(t)$ is the Laplace transform of the signal evaluated on the imaginary axis of the s plane, that is, at $s = j2\pi f$. There is a catch, however: We can obtain $X_a(f)$ from $X(s)$ by replacing s with $j2\pi f$ provided $X(s)$ is actually defined on the imaginary axis in the s plane. In other words, the Laplace transform integral must converge when $s = j2\pi f$; another way of stating the condition is that the region of convergence for $X(s)$ must include the imaginary axis of the s plane. For almost all practical signals and systems of interest, this is not a problem. In fact, it can be shown that as long as an LSI system is BIBO stable, we can obtain its frequency response function directly from its transfer function:

$$H_a(f) = H(s)\big|_{s=j2\pi f} = H(j2\pi f) \quad\quad (3\text{–}12)$$

NOTE: An example of a theoretical signal in which $s = j2\pi f$ does *not* work is the unit step function $u(t)$. The Laplace transform of $u(t)$ is $1/s$ but the Fourier transform of $u(t)$ is *not* $1/j2\pi f$. It can also be shown that a system having $u(t)$ as its impulse response is *not* BIBO stable.

Given the close relationship between the Laplace transform and the Fourier transform, the following relationship should come as no surprise. Let $X_a(f)$ and $Y_a(f)$ be the Fourier transforms of the LSI system input and output signals, respectively. Let $H_a(f)$ be the frequency response function of this system. It can be shown that the input and output spectra are related by

$$Y_a(f) = H_a(f)X_a(f) \tag{3-13}$$

That is, *the spectrum of the output signal is found by multiplying the spectrum of the input signal by the frequency response function of the system.* Thus the system alters the frequency content of the input signal. We see that the frequency response function relates the input and output spectra in a manner analogous to the way in which the transfer function relates the input and output Laplace transforms $[Y(s) = H(s)X(s)]$.

3-3   RELATIONSHIP BETWEEN FREQUENCY RESPONSE FUNCTION AND SYSTEM POLES AND ZEROS

Recall from Chapter 1 that the transfer function of a continuous time LSI system can be expressed as:

$$H(s) = \frac{G\prod_{i=1}^{M}(s - z_i)}{\prod_{i=1}^{N}(s - p_i)} \tag{3-14}$$

where p_i and z_i are the system poles and zeros, respectively. The system poles and zeros can be plotted as points on the s plane, as shown in Figure 1–3 (Page 11, Chapter 1). These points can also be thought of as *vectors*. For example, if the system has poles at $s = -1 \pm j$ and a zero at $s = -1$, the location of these vectors is as depicted in Figure 3–1(a).

As noted in Section 3–2, the frequency response function of a BIBO stable continuous time LSI system can be found by evaluating the transfer function at $s = j2\pi f$ (Equation [3–12]). For any particular frequency f, $j2\pi f$ is a point on the imaginary axis of the s plane; it is also a vector, as shown in Figure 3–1(b). The length of this vector changes as the frequency (f) changes.

The amplitude response function for a system having the transfer function of Equation (3–14) is:

$$\left| H_a(f) \right| = \left| H(j2\pi f) \right| = \frac{|G|\prod_{i=1}^{M}\left| j2\pi f - z_i \right|}{\prod_{i=1}^{N}\left| j2\pi f - p_i \right|} \tag{3-15}$$

FIGURE 3–1
(a) Vector interpretation of poles and zeros on the s plane. (b) The vector $s = j2\pi f$. (c) Tip-to-tail vector addition.

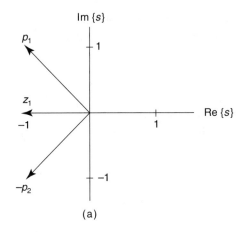

(a)

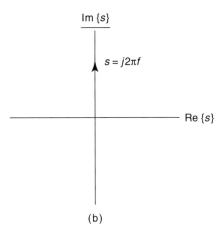

(b)

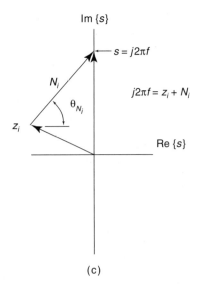

(c)

Let us consider the ith factor in the numerator of Equation (3–15): $|j2\pi f - z_i|$, evaluated at a particular frequency (f). The vectors z_i and $j2\pi f$ are shown in Figure 3–1(c) along with a third vector labeled N_i. Using the tip-to-tail method of vector addition, we see that the relationship between these three vectors is

$$j2\pi f = z_i + N_i \qquad (3\text{–}16)$$

Therefore,

$$N_i = j2\pi f - z_i \qquad (3\text{–}17)$$

It follows that the *length* of vector N_i is $|N_i| = |j2\pi f - z_i|$. That is, $|j2\pi f - z_i|$ is the length of the line segment from the point $s = z_i$ to the point $s = j2\pi f$. As the frequency (f) changes, the length of this line segment changes.

Similarly, the denominator factor $|j2\pi f - p_i|$ is the length of the line segment from the point $s = p_i$ to the point $s = j2\pi f$. As the frequency changes, the length of this line segment changes. Let us define the following:

$$D_i = j2\pi f - p_i$$
$$|D_i| = |j2\pi f - p_i| \qquad (3\text{–}18)$$

Using the above definitions of lengths $|N_i|$ and $|D_i|$, we see that the amplitude response function can be expressed in terms of these lengths:

$$|H_a(f)| = \frac{|G|\prod_{i-1}^{M}|N_i|}{\prod_{i=1}^{N}|Di|} \qquad (3\text{–}19)$$

If there happens to be a system zero z_i located on the imaginary axis of the s plane, then at the particular frequency f where $z_i = j2\pi f$, the length $|N_i|$ is equal to zero, and therefore the amplitude response function is equal to zero at that frequency. Since all of the poles of a BIBO stable system must be in the left half of the s plane, such a system will not have any poles on the imaginary axis. Therefore, there will be no frequencies where $|D_i| = 0$. However, if there is a pole p_i located very close to the imaginary axis, $|D_i|$ will be very small when the frequency (f) is such that $j2\pi f$ is a point near this pole.

Let us consider an example. Suppose we have a system with a zero at $s = -10$ and a pair of complex conjugate poles at $s = -5 \pm j20$. Figure 3–2(a) shows the lengths of $N_1, D_1,$ and D_2 when the frequency is $f = 10/(2\pi)$ Hz. Figure 3–2(b) shows these lengths when the frequency is $f = 20/(2\pi)$ Hz. Figure 3–2(c) shows these lengths when the frequency is $f = 30/(2\pi)$Hz. The reader is invited to use a ruler to measure the lengths of these line segments to determine which of these three frequencies results in the largest value of $|H_a(f)|$.

Using complex number properties 3 and 4 from Table 2–1, on page 25, it can be shown that the phase response function for a system having the transfer function of Equation (3–14) (assuming G is positive) is

$$\theta_H(f) = \prod_{i=1}^{M}\theta_{N_i} - \prod_{i=1}^{N}\theta_{D_i} \qquad (3\text{–}20)$$

FIGURE 3–2
Change in lengths D_1, D_2, and N_1 as frequency changes.

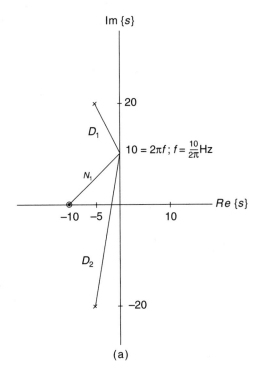

(a)

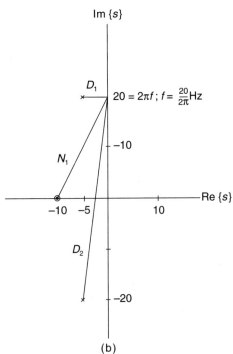

(b)

FIGURE 3–2
(continued)

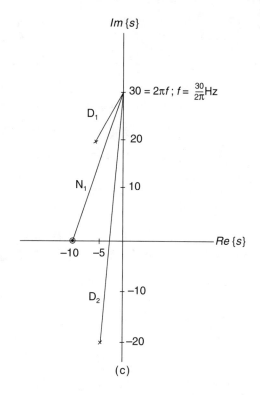

(c)

where

$$\theta_{N_i} = \angle\{N_i\} = \angle\{j2\pi f - z_i\}$$
$$\theta_{D_i} = \angle\{D_i\} = \angle\{j2\pi f - p_i\}$$

$$(3\text{–}21)$$

These angles are defined as usual for vectors. (Figure 3–1[c] shows θ_{N_i}). If G is negative, a constant phase shift of $\pm\pi$ radians must be added to the right side of Equation (3–20).

3–4 FOURIER TRANSFORM CONVOLUTION THEOREMS

In Chapter 1 we observed that the output signal from an LSI system can be found by convolving the input signal with the system impulse response, that is, $y(t) = x(t) * h(t)$. This fact, combined with Equation (3–13) above, implies the following Fourier transform pair:

$$x(t) * h(t) \leftrightarrow X_a(f)H_a(f)$$

$$(3\text{–}22)$$

That is, *convolution* in the time domain corresponds to *multiplication* in the frequency domain. It can be shown that the reverse is also true: Multiplication in the time domain corresponds to convolution in the frequency domain. In other words, suppose two signals $x(t)$ and

$g(t)$ are multiplied together. The resulting spectrum is the convolution of the spectra of $x(t)$ and $g(t)$:

$$x(t)g(t) \leftrightarrow X_a(f) * G_a(f) \tag{3-23}$$

Convolution in the frequency domain is defined in a manner analogous to that in the time domain (compare Equation [3–24] with Equation [1–42]):

$$X_a(f) * G_a(f) = \int_{-\infty}^{\infty} X_a(\alpha) G_a(f - \alpha) d\alpha \tag{3-24}$$

where α is the *dummy variable* of integration. The convolution operation blends the two original spectra together in a special and complicated way. The result of this operation is a spectrum that won't look like either of its parents but will inherit characteristics from both of them.

3-5 FOURIER TRANSFORM TIME AND FREQUENCY SHIFTING THEOREMS

An important Fourier transform theorem is the time shifting theorem. If

$$x(t) \leftrightarrow X_a(f) \tag{3-25}$$

then

$$x(t - t_o) \leftrightarrow e^{-j2\pi f t_o} X_a(f) \tag{3-26}$$

where $x(t - t_o)$ is a time shifted version of $x(t)$. When t_o is a positive number, $x(t - t_o)$ is a time-delayed version of $x(t)$, as illustrated by Figure 3–3. The proof of the time shifting theorem is straightforward. We begin by taking the Fourier transform of $x(t - t_o)$:

$$\mathbf{F}\{x(t - t_o)\} = \int_{-\infty}^{\infty} x(t - t_o) e^{-j2\pi f t} dt \tag{3-27}$$

FIGURE 3-3
Shifting in the time domain.

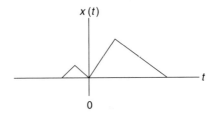

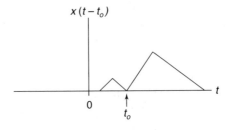

Next we change variables in the integral: Let $(t - t_o) = \alpha$. Then $t = \alpha + t_o$, and $dt = d\alpha$. We now have

$$\mathbf{F}\{x(t - t_o)\} = \int_{-\infty}^{\infty} x(\alpha) e^{-j2\pi f(\alpha + t_o)} d\alpha \qquad (3\text{--}28)$$

But the integral can now be expressed as follows:

$$\mathbf{F}\{x(t - t_o)\} = e^{-j2\pi f t_o} \left[\int_{-\infty}^{\infty} x(\alpha) e^{-j2\pi f \alpha} d\alpha \right] \qquad (3\text{--}29)$$

The expression in brackets is the Fourier transform of $x(t)$ (with α used as a dummy variable of integration). Thus the time shifting theorem is demonstrated.

NOTE: Since the magnitude of $\exp(-j2\pi f t_o)$ is 1, a time delay does not alter the spectral amplitude function $|X_a(f)|$.

The frequency shifting (or *modulation*) theorem is the *dual* of the time shifting theorem. If $x(t) \leftrightarrow X_a(f)$, then

$$e^{j2\pi f_o t} x(t) \leftrightarrow X_a(f - f_o) \qquad (3\text{--}30)$$

where $X_a(f - f_o)$ is a frequency-shifted version of $X_a(f)$. When f_o is a positive number, the frequency shift is to the right, as depicted in Figure 3–4 .

The proof of the frequency shifting theorem is relatively straightforward:

$$\mathbf{F}\{e^{j2\pi f_o t} x(t)\} = \int_{-\infty}^{\infty} e^{j2\pi f_o t} x(t) e^{-j2\pi f t} dt$$

$$= \int_{-\infty}^{\infty} x(t) e^{-j2\pi (f - f_o) t} dt \qquad (3\text{--}31)$$

$$= X_a(f - f_o)$$

An important extension of the frequency shifting property (obtained using Euler's theorem) is

$$x(t)\cos(2\pi f_o t) \leftrightarrow \frac{1}{2} X_a(f - f_o) + \frac{1}{2} X_a(f + f_o) \qquad (3\text{--}32)$$

The positive and negative frequency shifts are shown in Figure 3–5.

FIGURE 3–4
Shifting in the frequency domain.

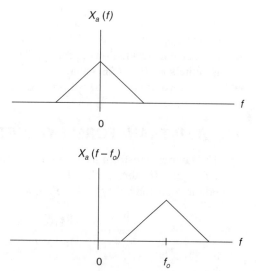

FIGURE 3–5
(a) Spectrum of $x(t)$. (b)
Spectum of $x(t)\cos(2\pi f_o t)$

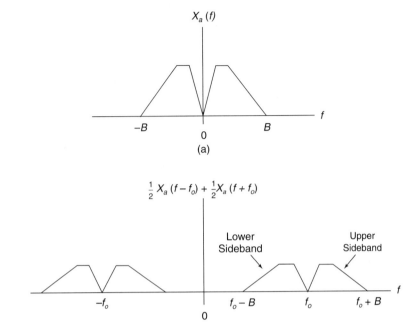

NOTE: Figures 3–4 and 3–5 both suggest that $X_a(f)$ has certain symmetries about the point $f = 0$. The symmetry properties of the Fourier transform will be considered in Section 3–6.

In communication system theory, the multiplication of a *message signal* $x(t)$ (say, a speech signal) by a *carrier* $A\cos(2\pi f_o t)$ is accomplished by a *balanced modulator* circuit. The resulting signal is a *double-sideband* signal. Note that before $x(t)$ is multiplied by the carrier signal, there is a negative frequency part of the spectrum $X_a(f)$ that is arguably a mathematical fiction. However, after being multiplied by the carrier signal, the negative frequency part of the original spectrum is *what becomes the lower sideband*. The existence of both upper and lower sidebands can be verified by physical measurements. In this case, the mathematical fiction of negative frequency ends up manifesting itself as something that is entirely *real world* in nature.

A list of Fourier transform theorems is shown in Table 3–1.

3–6 FOURIER TRANSFORM SYMMETRY PROPERTIES

The Fourier transform has certain symmetry properties that are worth noting. Assuming $x(t)$ is real (in this book, all time domain signals are real unless stated otherwise), its Fourier transform $X_a(f)$ has the following properties (it can be shown):

$$\mathrm{Re}\{X_a(f)\} = \mathrm{Re}\{X_a(-f)\}$$
$$\mathrm{Im}\{X_a(f)\} = -\mathrm{Im}\{X_a(-f)\}$$

(3–33)

TABLE 3–1
Fourier Transform Theorems

$$x(t) \leftrightarrow X_a(f)$$

$$ax(t) + bg(t) \leftrightarrow aX_a(f) + bG_a(f)$$

$$x(t - t_o) \leftrightarrow e^{-j2\pi f t_o} X_a(f)$$

$$x(-t) \leftrightarrow X_a(-f)$$

$$x(t)e^{j2\pi f_o t} \leftrightarrow X_a(f - f_o)$$

$$x(t)\cos(2\pi f_o t) \leftrightarrow \frac{1}{2}X_a(f - f_o) - \frac{1}{2}X_a(f + f_o)$$

$$x(t)\sin(2\pi f_o t) \leftrightarrow \frac{1}{2j}X_a(f - f_o) - \frac{1}{2j}X_a(f + f_o)$$

$$x(t) * g(t) \leftrightarrow X_a(f)G_a(f)$$

$$x(t)g(t) \leftrightarrow X_a(f) * G_a(f)$$

In other words, the real part of the spectrum is an *even* function (i.e., has even symmetry about the origin), and the imaginary part of the spectrum is an *odd* function (i.e., has odd symmetry, or *antisymmetry*, about the origin). Consequently, the amplitude spectrum and phase spectrum have the following symmetries:

$$|X_a(f)| = |X_a(-f)|$$
$$\theta_X(f) = -\theta_X(-f)$$

(3–34)

That is, the amplitude spectrum is an *even* function, and the phase spectrum is an *odd* function. These symmetries are illustrated by Figure 2–3 (page 29, Chapter 2), in which the case at hand is the Fourier transform of the impulse response, $h(t)$, of a system. The Fourier transform examples in Section 3–7 will also illustrate the Fourier transform symmetry properties.

Another symmetry property of interest is the following: If the time domain signal $x(t)$ has even symmetry about the point $t = 0$, then its Fourier transform is a *real* function of frequency (f). In other words, if $x(t) = x(-t)$, then $X_a(f)$ is real. To show this, we begin by observing that the Fourier transform of $x(t)$ can be expressed as follows:

$$\int_{-\infty}^{\infty} x(t)e^{j2\pi f t}\,dt = \int_{-\infty}^{\infty} x(t)\cos(2\pi f t)\,dt - j\int_{-\infty}^{\infty} x(t)\sin(2\pi f t)\,dt$$

(3–35)

Now observe the following:

1. $\cos(2\pi f t)$ is an *even* function of t.
2. $\sin(2\pi f t)$ is an *odd* function of t.
3. The product of two *even* functions is an *even* function.
4. The product of an *even* function and an *odd* function is an *odd* function.
5. If $r_{odd}(t)$ is an *odd* function of t, then

$$\int_{-\infty}^{\infty} r_{odd}(t)\,dt = 0$$

(3–36)

Putting all of this information together leads to the conclusion that if $x(t) = x(-t)$, then

$$\int_{-\infty}^{\infty} x(t)\sin(2\pi ft)dt = 0 \qquad (3\text{--}37)$$

Therefore, if $x(t) = x(-t)$, then

$$X_a(f) = \int_{-\infty}^{\infty} x(t)\cos(2\pi ft)dt \qquad (3\text{--}38)$$

Since there are no js in Equation (3–38), $X_a(f)$ is a real function of the variable f.

3-7   FOURIER TRANSFORM EXAMPLES

EXAMPLE 3–7.1

Consider the rectangular pulse depicted in Figure 3–6(a). Let us give this pulse the name $r(t)$ instead of the generic name $x(t)$, since it will play a special role in some of the examples that follow this one. The spectrum of the pulse $r(t)$ is given by

$$
\begin{aligned}
R_a(f) &= \int_{-T/2}^{T/2} Ae^{-j2\pi ft}\,dt \\[1mm]
&= \frac{A}{-j2\pi f}e^{-j2\pi ft}\Big|_{t=-T/2}^{T/2} \\[1mm]
&= \frac{A}{-j2\pi ft}\left[e^{-j2\pi f\frac{T}{2}} - e^{j2\pi f\frac{T}{2}}\right]
\end{aligned}
\qquad (3\text{--}39)
$$

Using Euler's theorem, this reduces to

$$R_a(f) = \frac{A\sin(\pi fT)}{\pi f} \qquad (3\text{--}40)$$

$R_a(f)$ is shown in Figure 3–6(b). Observe that the spectrum is real in this example. As noted in Section 3–6, if a signal $x(t)$ is even symmetric about the origin, that is, $x(t) = x(-t)$

FIGURE 3–6
(a) Rectangular pulse.
(b) Spectrum of a rectangular pulse.

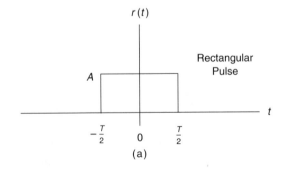

(a)

FIGURE 3–6
(continued)

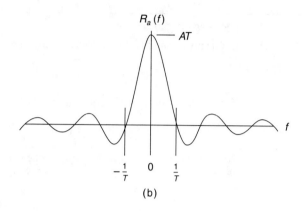

(b)

(as is the case with this example), then $X_a(f)$ is real. (However, the reader should keep in mind that in general the spectrum of a signal is complex.)

Example 3–7.1 also serves to introduce the concept of *bandwidth*. There are several ways in which bandwidth can be defined mathematically, but the basic idea behind all of these is that *the bandwidth of a signal x(t) is defined as that range of frequencies in which the amplitude spectrum* $|X_a(f)|$ *is significant.* In Figure 3–6(b), observe that roughly speaking, $|R_a(f)|$ is significant in the frequency range $-1/T \le f \le 1/T$ (the location of the main lobe of the spectrum). In terms of positive frequency only, the spectrum is significant between 0 and $1/T$ Hz; thus the bandwidth *could* be defined as $B = 1/T$ Hz for this pulse signal. There is no single formula for bandwidth that is applicable to all cases. (What is important here is the *concept* of bandwidth.)

In the strictest sense, a signal is bandlimited to B Hz if $X_a(f) = 0$ for all frequencies greater than B. Figure 3–7 shows a hypothetical spectrum that is strictly bandlimited to B Hz. *Real-world signals are never strictly bandlimited in this sense.*

Returning to Example 3–7.1 (rectangular pulse and its spectrum), observe that if the pulse is made wider (i.e., if T is increased), the bandwidth ($1/T$) decreases. This serves to illustrate a general rule of thumb: *The bandwidth of a pulse is inversely proportional to its time-width.* (This is why higher data transmission rates require a communications channel with a wider bandwidth.)

FIGURE 3–7
Spectrum of a hypothetical
signal strictly bandlimited to
B Hz.

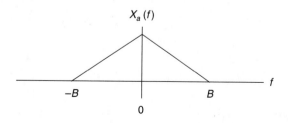

EXAMPLE 3–7.2

Let us now consider the Fourier transform of an impulse function, that is,

$$x(t) = A\delta(t - t_o) \tag{3–41}$$

This signal is shown in Figure 3–8(a). The Fourier transform of this signal is

$$X_a(f) = \int_{-\infty}^{\infty} A\delta(t - t_o) e^{-j2\pi ft} dt \tag{3–42}$$

To evaluate this integral, we must appeal to the sifting property of the impulse, as described in Chapter 1. The result is the Fourier transform pair

$$A\delta(t - t_o) \leftrightarrow A e^{-j2\pi ft_o} \tag{3–43}$$

A special case is one in which $t_o = 0$:

$$A\delta(t) \leftrightarrow A \tag{3–44}$$

It is interesting to note that regardless of the value the time shift t_o, the spectral amplitude function is

$$|X_a(f)| = |A| \tag{3–45}$$

Since the spectral amplitude function is constant for all frequencies, as shown in Figure 3–8(b), an impulse signal has an infinite bandwidth.

FIGURE 3–8
(a) Impulse function $A\delta(t - t_o)$.
(b) Amplitude spectrum of this impulse.

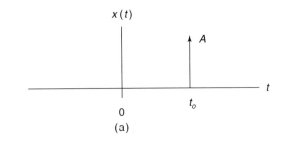

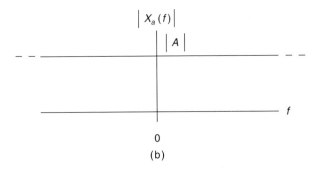

EXAMPLE 3–7.3

Consider an *ideal lowpass filter* with cutoff frequency f_c, as depicted in the frequency domain in Figure 3–9 . This filter will have no effect on spectral components lower than f_c Hz, but will completely eliminate spectral components above f_c Hz.

Use the inverse Fourier transform to find the impulse response $h(t)$ of the ideal lowpass filter (see Equation [3–3]):

$$h(t) = \int_{-f_c}^{f_c} (1)e^{j2\pi ft} df$$

$$= \left. \frac{e^{j2\pi ft}}{j2\pi t} \right|_{f=-f_c}^{f_c} \qquad\qquad (3\text{–}46)$$

$$= \frac{e^{j2\pi f_c t} - e^{-j2\pi f_c t}}{j2\pi t}$$

Using Euler's theorem, this cleans up to

$$h(t) = \frac{\sin(2\pi f_c t)}{\pi t} \qquad\qquad (3\text{–}47)$$

The ideal lowpass filter impulse response $h(t)$ is shown in Figure 3–10.

FIGURE 3–9
Ideal lowpass filter.

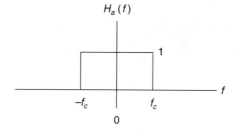

FIGURE 3–10
Impulse response of the ideal lowpass filter depicted in Figure 3–9.

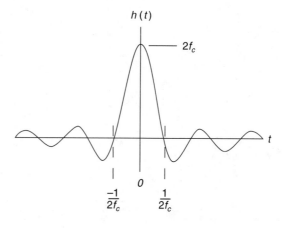

There are two important things to observe about the impulse response of an ideal low-pass filter. First of all, notice that the impulse response starts *before* the impulse is applied to the system (at $t = 0$). This system is therefore *noncausal*. (See Chapter 1 for a discussion of causal versus noncausal systems.) The second thing to observe is that the impulse response of an ideal lowpass filter exists for all time (it starts at $t = -\infty$); therefore, there is no possibility of shifting this impulse response to the right in order to make it strictly causal. For these (and other) reasons, *an ideal lowpass filter can not be realized in the real world.* However, the ideal lowpass filter is often a theoretical starting point in the filter design process.

NOTE: Think of $h(t)$ as a pulse with a width of $1/f_c$, since most of the pulse energy lies in the interval $-1/(2f_c) \le f \le 1/(2f_c)$; we therefore have another illustration of the "pulsewidth is inversely proportional to bandwidth" rule of thumb.

EXAMPLE 3–7.4
In order to illustrate the concept of filtering with respect to the Fourier transform, in particular the concept of how a filter alters the spectrum of its input signal, consider a lowpass filter with input and output signals $x(t)$ and $y(t)$, respectively. In order to simplify the picture in the frequency domain and thus simplify the concept, let us suppose that the input spectrum $X_a(f)$ is as shown in Figure 3–11(a), and that the lowpass filter is ideal, $H_a(f)$, as

FIGURE 3–11
Filtering with an ideal lowpass filter. (a) Spectrum of the input signal. (b) Ideal lowpass filter frequency response function. (c) Spectrum of the output signal.

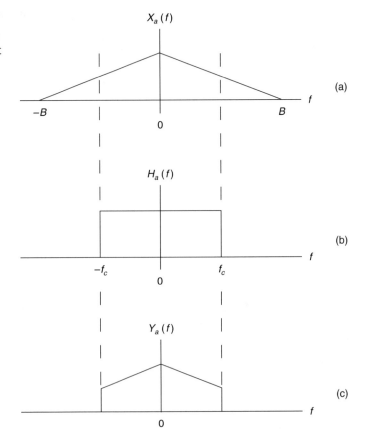

in Figure 3–11(b). The output spectrum is given by the product of the input spectrum and the filter frequency response function:

$$Y_a(f) = X_a(f)H_a(f) \tag{3-48}$$

$Y_a(f)$ is shown in Figure 3–11(c). The cutoff frequency of the lowpass filter is f_c Hz, which causes the frequency content of the input signal above f_c Hz to be eliminated; the frequency content below f_c Hz is preserved.

EXAMPLE 3–7.5

Let us take the ideal lowpass filter impulse response (Equation [3–47] and Figure 3–10) and *truncate* it to create a new impulse response $h(t)$ of duration T seconds:

$$h(t) = \begin{cases} \dfrac{\sin(2\pi f_c t)}{\pi t}, & \dfrac{-T}{2} \le t \le \dfrac{T}{2} \\ 0, & otherwise \end{cases} \tag{3-49}$$

This result can also be represented as the multiplication of an ideal lowpass filter impulse response by the rectangular window function $r(t)$:

$$h(t) = r(t)\left[\frac{\sin(2\pi f_c t)}{\pi t}\right] \tag{3-50}$$

where

$$r(t) = \begin{cases} 1, & \dfrac{-T}{2} \le t \le \dfrac{T}{2} \\ 0, & otherwise \end{cases} \tag{3-51}$$

The rectangular window function described here is the same as the rectangular pulse depicted in Example 3–7.1; thus the spectrum of this rectangular window is the same as that shown for $R_a(f)$ in Figure 3–6(b) and Equation (3–40).

Let us now consider the frequency response function of the filter with the truncated impulse response $h(t)$ (Equation [3.50]). As noted in Section 3–4, multiplication of two functions in the time domain transforms to the convolution of their respective spectra in the frequency domain. Thus the frequency response function of the system under consideration here can be obtained by convolving the spectra depicted in Figures 3–6(b) and 3–9. The result is that the ideal lowpass filter spectrum (Figure 3–9) will be altered in the following manner (it can be shown):

1. The discontinuities in the ideal lowpass filter frequency response function (at f_c and $-f_c$) will be smoothed off. The transition from passband to stopband will not be as abrupt in the resulting filter.

2. The ripples in the spectrum of the rectangular window will manifest themselves as ripples in the frequency response function of the resulting filter.

Figure 3–12 shows how the amplitude response function of the resulting filter compares with that of the ideal lowpass filter.

FIGURE 3–12
Amplitude response function
of a lowpass filter created
by truncating the impulse
response of an ideal lowpass
filter.

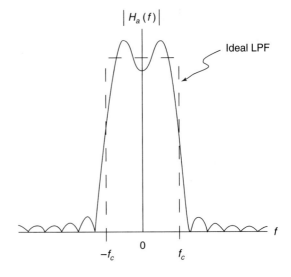

As one might intuitively expect, as more of the original ideal lowpass filter impulse response is preserved (i.e., the larger T in Equations [3–49] and [3–51]), the more closely the resulting amplitude response function will resemble the ideal lowpass filter. Figure 3–13 shows the amplitude response function of a truncated ideal lowpass filter, in which T is approximately five times greater than that corresponding to Figure 3–12. The transition from passband to stopband is more abrupt; however, the magnitude of the overshoot, or ripple, in the vicinity of frequencies f_c and $–f_c$ is approximately the same in both Figures 3–12 and 3–13.

The impulse response $h(t)$ described by Equation (3–49) is still noncausal. A causal version can be created by shifting $h(t)$ to the right on the time axis such that the starting

FIGURE 3–13
Amplitude response function
of a lowpass filter created
by truncating the impulse
response of an ideal lowpass
filter. (The truncation is not as
severe as that corresponding
to Figure 3–12.)

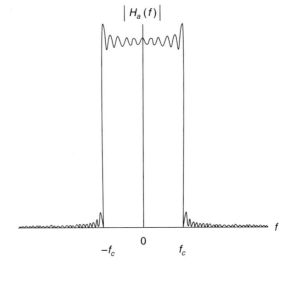

point is $t = 0$ instead of $t = -T/2$. Let us call this new causal impulse response $h_c(t)$, which is related to $h(t)$ as follows:

$$h_c(t) = h\left(t - \frac{T}{2}\right) \tag{3-52}$$

That is,

$$h_c(t) = \begin{cases} \dfrac{\left|\sin\left[2\pi f_c\left(t - \dfrac{T}{2}\right)\right]\right|}{\pi\left(t - \dfrac{T}{2}\right)}, & 0 \leq t \leq T \\ 0, & otherwise \end{cases} \tag{3-53}$$

Let us denote the frequency response function of this new filter as $H_{ac}(f)$. Using the time shifting theorem (Section 3–5), we obtain the following:

$$H_{ac}(f) = e^{-j2\pi f \frac{T}{2}} H_a(f) \tag{3-54}$$

where $H_a(f)$ is the frequency response function associated with $h(t)$.

Observe that the time shift applied to $h(t)$ does not alter the amplitude response function of the resulting filter, since

$$\left|H_{ac}(f)\right| = \left|e^{-j2\pi f \frac{T}{2}}\right|\left|H_a(f)\right| = \left|H_a(f)\right| \tag{3-55}$$

It is not actually possible to realize a continuous time filter with the impulse response of Equation (3–53), even though it is causal. However, it is possible to realize a digital (i.e., discrete time) filter with an impulse response that is a truncated and time-shifted (causal) version of the impulse response of an ideal discrete time lowpass filter. This will be discussed in later chapters.

EXAMPLE 3–7.6

We next consider the Fourier transform of the complex exponential signal given by

$$x(t) = ce^{j2\pi f_o t} \tag{3-56}$$

The Fourier transform pair turns out to be

$$ce^{j2\pi f_o t} \leftrightarrow c\delta(f - f_o) \tag{3-57}$$

where $c\delta(f - f_o)$ is an impulse located at $f = f_o$, as shown in Figure 3–14. This impulse is a Dirac delta function, that is, its amplitude approaches infinity, its width approaches zero, and its area (or weight) is equal to the constant c. (The constant c can be a complex number, in which case the area has a real part and an imaginary part.)

FIGURE 3–14
The Fourier transform of the complex exponential function $x(t) = c\exp(j2\pi f_o t)$ is an impulse at frequency $f = f_o$.

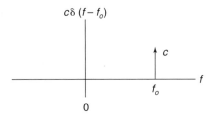

It is easy to demonstrate the validity of this Fourier transform pair by using the inverse Fourier transform integral along with the sifting property of the impulse:

$$\int_{-\infty}^{\infty} c\delta(f - f_o)e^{j2\pi ft}\,df = ce^{j2\pi f_o t} \tag{3–58}$$

One immediate result is that we can now find the Fourier transform of a sinusoid. To do this, we start by expressing the sinusoid in terms of complex exponentials using Euler's theorem:

$$x(t) = A\cos(2\pi f_o t) = \frac{A}{2}e^{j2\pi f_o t} + \frac{A}{2}e^{-j2\pi f_o t} \tag{3–59}$$

Applying the linearity property, we obtain the following Fourier transform pair:

$$A\cos(2\pi f_o t) \leftrightarrow \frac{A}{2}\delta(f - f_o) + \frac{A}{2}\delta(f + f_o) \tag{3–60}$$

Thus the Fourier transform of a sinusoid of frequency f_o consists of a pair of impulses in the frequency domain, as shown in Figure 3–15. One impulse is located at $f = f_o$, and the other is located at $f = -f_o$.

The results presented in this section allow us to consider the Fourier transform of a periodic signal. If $x(t)$ is periodic with period T and fundamental frequency $f_o = 1/T$, then $x(t)$ can be expressed as a Fourier series (see Section 2–5):

$$x(t) = \sum_{n=-\infty}^{\infty} c_n e^{j2\pi n f_o t} \tag{3–61}$$

where the c_ns are the Fourier series coefficients. To find the Fourier transform of this signal, we must use the linearity property along with Equation (3–57). The result is

$$X_a(f) = \sum_{n=-\infty}^{\infty} \mathbf{F}\{c_n e^{j2\pi n f_o t}\}$$

$$= \sum_{n=-\infty}^{\infty} c_n \delta(f - n f_o) \tag{3–62}$$

Thus the spectrum of a periodic signal is a so-called *line spectrum*: It consists of weighted impulses at discrete frequencies $n f_o$, where n ranges from minus to plus infinity. Figure 3–16 shows a hypothetical amplitude line spectrum $|X_a(f)|$. Note that the conventional way to illustrate the fact that the impulses have different weights $|c_n|$ is to make the height of the impulses in the illustration proportional to their weights.

FIGURE 3–15
The Fourier transform of $x(t)$ = $A\cos(2\pi f_o t)$ is a pair of impulses located at $f = \pm f_o$.

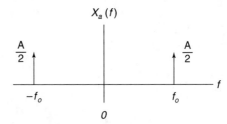

$X_a(f)$

FIGURE 3–16
Amplitude spectrum of a hypothetical periodic signal with fundamental frequency f_o.

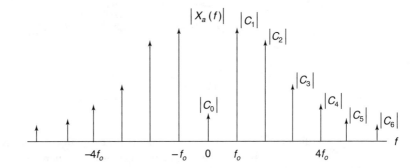

$|X_a(f)|$

EXAMPLE 3–7.7

The previous example demonstrated that the spectrum of a sinusoid given by $x(t) = A\cos(2\pi f_o t)$ is a pair of impulses located at $f = \pm f_o$. It is important to note that this sinusoid is defined for all time (t). If the sinusoid is truncated in the time domain, its spectrum will no longer consist of a pair of impulses.

Let $r(t)$ be the rectangular window truncation function defined by Equation (3–51). Let $x(t)$ be a truncated sinusoid, that is,

$$x(t) = r(t)A\cos(2\pi f_o t) \tag{3–63}$$

Using the frequency shifting theorem, the spectrum of $x(t)$ is

$$X_a(f) = \frac{A}{2} R_a(f - f_o) + \frac{A}{2} R_a(f + f_o) \tag{3–64}$$

where $R_a(f)$ is the Fourier transform of $r(t)$ (see Equation [3–40] and Figure 3–6[b]). Instead of impulses at $f = \pm f_o$, there are copies of $R_a(f)$ centered at $f = \pm f_o$.

NOTE: These copies will overlap to a certain extent, and are added together.

The approximate bandwidth of these copies of $R_a(f)$ is $2/T$ Hz, where T is the time duration of $r(t)$. Figure 3–17 shows $X_a(f)$ for the case $A = 1$, $f_o = 307.5$ Hz, and $T = 0.02$ second, plotted for both positive and negative frequency.

Somewhat more interesting is the case of a signal consisting of the sum of two sinusoids at different frequencies, which is then truncated:

$$x(t) = r(t)\left[A_1\cos(2\pi f_1 t) + A_2\cos(2\pi f_2 t)\right] \tag{3–65}$$

FIGURE 3–17
Fourier transform of a
truncated sinusoid.

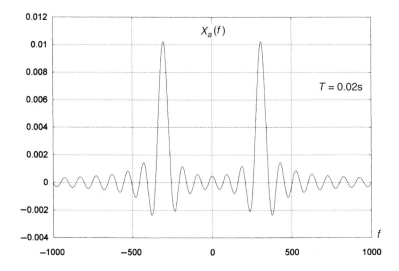

With no truncation (i.e., if $r(t)$ has the time duration $T = \infty$), $X_a(f)$ would be characterized by impulses at frequencies $\pm f_1$ and $\pm f_2$. However, when T is finite, a copy of $R_a(f)$ shows up at these frequencies instead of the impulses:

$$X_a(f) = \frac{A_1}{2} R_a(f - f_1) + \frac{A_1}{2} R_a(f + f_1) + \frac{A_2}{2} R_a(f - f_2) + \frac{A_2}{2} R_a(f + f_2) \qquad \textbf{(3–66)}$$

Figures 3–18(a) through 3–18(d) show $X_a(f)$ for the case $A_1 = A_2 = 1, f_1 = 307.5$ Hz, and $f_2 = 503.5$ Hz, at four different values of T. Figure 3–18(a) shows the spectrum when $T = 0.1$ second, which is relatively long compared to the periods of the sinusoids themselves. This spectrum is characterized by tall, narrow spikes at the frequencies $\pm f_1$ and $\pm f_2$. Figures 3–18(b) and 3–18(c) show what happens to the spectrum if T is reduced to 0.02 second and 0.01 second, respectively. As T is reduced, the spikes at $\pm f_1$ and $\pm f_2$ become shorter and wider. Figure 3–18(d) shows that if T is reduced to 0.005 second, the copies of $R_a(f)$ overlap to such an extent that when they are added together, the existence of two sinusoidal components is actually obscured.

This result can be applied to periodic signals in general. If $x(t)$ is periodic with fundamental frequency f_o and therefore has a Fourier series representation (Equation [3–61]), then the truncated version of this signal is

$$x(t) = r(t) \sum_{n=-\infty}^{\infty} c_n e^{j2\pi n f_o t} = \sum_{n=-\infty}^{\infty} c_n r(t) e^{j2\pi n f_o t} \qquad \textbf{(3–67)}$$

Using the linearity and frequency shifting properties, we can obtain the following expression for the spectrum of this signal:

$$X_a(f) = \sum_{n=-\infty}^{\infty} c_n R_a(f - n f_o) \qquad \textbf{(3–68)}$$

That is, instead of the weighted (c_n) impulses shown in Equation (3–62), weighted copies of $R_a(f)$ appear at all integer multiples of the fundamental frequency. If the truncation length T is large relative to the fundamental period $1/f_o$, distinct peaks will appear in the plot of the amplitude spectrum at the significant frequencies (perhaps the fundamental frequency and the first few harmonic frequencies). On the other hand, if the truncation length T is too short, the copies of $R_a(f)$ will overlap to such an extent that a plot of $|X_a(f)|$ will

FIGURE 3–18
Fourier transform of a truncated sum of two sinusoids. The shape of this spectrum depends on the truncation length (T).
(a) $T = 0.1$ sec. (b) $T = 0.02$ sec. (c) $T = 0.01$ sec.
(d) $T = 0.005$ sec.

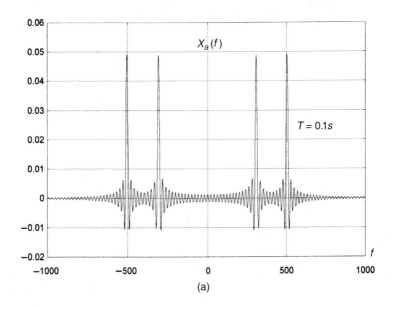

(a)

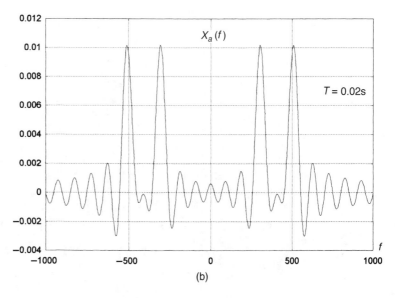

(b)

FIGURE 3–18
(continued)

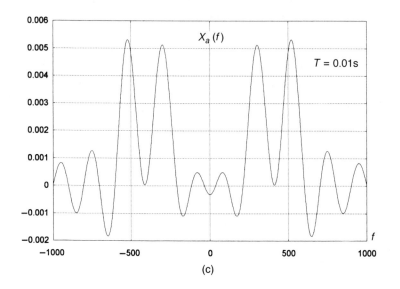

(c)

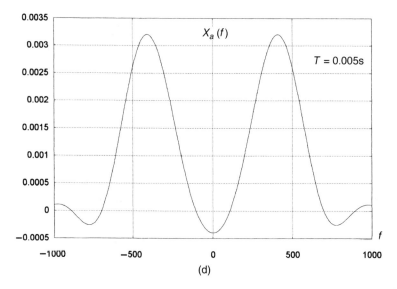

(d)

not show well-defined spectral peaks at these frequencies; however, it will still show the general range of frequencies at which the frequency content is significant.

The results shown in this example can also be obtained using the Fourier transform convolution property $g(t)x(t) \leftrightarrow G_a(f) * X_a(f)$. First, note the following:

$$c\delta(f - f_o) * R_a(f) = \int_{-\infty}^{\infty} c\delta(\lambda - f_o)R_a(f - \lambda)d\lambda$$
$$= cR_a(f - f_o)$$

(3–69)

This result is obtained by appealing to the sifting property of the impulse. It can easily be extended to a function consisting of multiple impulses, that is,

$$R_a(f) * \sum_n c_n \delta(f - f_n) = \sum_n c_n R_a(f - f_n) \tag{3-70}$$

This result can be applied to the problem of finding the spectrum of a truncated sinusoid or the truncated sum of several sinusoids. Prior to the truncation, the spectrum consists of a sum of weighted impulses located at various frequencies. After the time domain truncation, which is accomplished by *multiplying* the original signal by $r(t)$, the original impulsive spectrum is *convolved* with $R_a(f)$, as shown on the left side of Equation (3–70). The resulting spectrum (the right side of Equation [3–70]) has weighted copies of $R_a(f)$ located at the frequencies where there were originally impulses.

EXAMPLE 3–7.8

This example illustrates a voice-scrambling technique that is relatively easy to understand if the concepts presented in this chapter are employed. Suppose $x(t)$ is a voice signal that we want to transmit through a telephone channel, and we want to fix things such that anyone tapping into this channel to intercept our message will hear a scrambled, unintelligible signal. At the same time, the desired listener must have equipment to unscramble the signal.

The voice scrambler is depicted in Figure 3–19. The original voice signal $x(t)$ is presumed to be bandlimited to B Hz. (B is approximately 4000 Hz for a real-world communications-quality voice signal.) Suppose $x(t)$ has the spectrum $X_a(f)$ depicted in position 1, Figure 3–21.

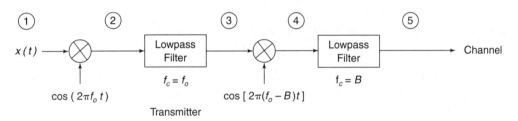

FIGURE 3–19
Voice-scrambling system based on sideband reversal.

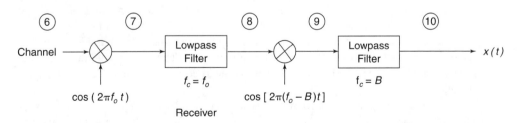

FIGURE 3–20
Receiver designed to unscramble the signal created by the system shown in Figure 3–19.

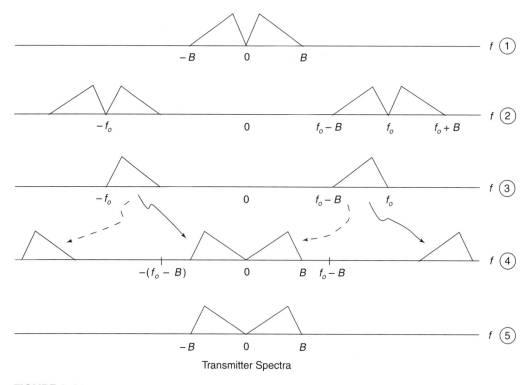

Transmitter Spectra

FIGURE 3–21
Spectrum at various points in the voice-scrambling system of Figure 3–19.

NOTE: The circled position numbers that appear in Figures 3–19 and 3–20 denote the location of the signals with the corresponding numbered spectra in Figures 3–21 and 3–22, respectively.

Input signal $x(t)$ is modulated by a carrier with frequency $f_o > 2B$; the resulting spectrum is shown for position 2. (Scaling factors have been neglected here.) This double-sideband signal is passed through a lowpass filter with a cutoff frequency f_c equal to carrier frequency f_o; the result is that the upper sideband is eliminated (position 3). The lower sideband signal is modulated by a carrier with frequency $(f_o - B)$ to create another signal (position 4), which is in turn bandlimited to B Hz by a lowpass filter (position 5).

Compare the spectra at positions 1 and 5; note how the sidebands have been reversed. The message is present, but it will be unintelligible. The listening effect is the same as receiving a lower-sideband signal on a communications receiver set to receive an upper-sideband signal.

A receiver designed to unscramble this signal is depicted in Figure 3–20. The corresponding numbered spectra are shown in Figure 3–22. (Note that the scrambling and unscrambling systems are the same.) The sideband-reversed spectrum at position 6 is unreversed at position 10. The intermediate results are shown for positions 7, 8, and 9.

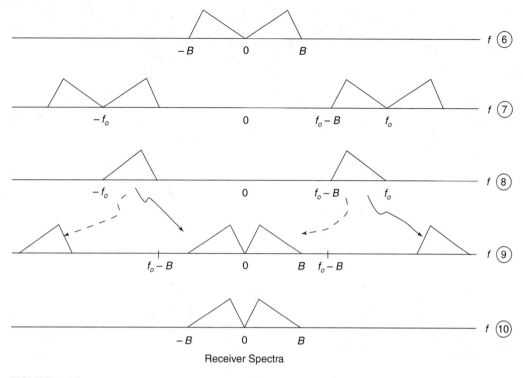

FIGURE 3–22
Spectrum at various points in the receiver shown in Figure 3–20.

PROBLEMS

3–1. Consider the rectangular pulse shown in Figure 3–6(a). Sketch the spectrum of this pulse—that is, sketch $R_a(f)$ for two cases and compare the bandwidth of these pulses:
   a. $A = 10, T = 0.01$ second
   b. $A = 100, T = 0.001$ second

3–2. Show that:
$$x(t)\sin(2\pi f_o t) \leftrightarrow \tfrac{1}{2j} X_a(f - f_o) - \tfrac{1}{2j} X_a(f + f_o)$$
HINT: Look back at Problem 2–5.

3–3. Show that $x(-t) \leftrightarrow X_a(-f)$.
   (This is the *time-reversal property* of the Fourier transform. See also Problem 3–20.)

3–4. Show that if $x(t) = -x(-t)$—that is, if $x(t)$ is an *odd* function—then $X_a(f)$ is imaginary—that is, it has no real component.

3–5. Consider the function $R_a(f)$ given by Equation (3–40) and plotted in Figure 3–6(b). When $f = 0$ is plugged into this function, the result is an indeterminant form (0/0). Use L'Hopital's rule to show that
$$\lim_{f \to 0} R_a(f) = AT$$

NOTE: L'Hopital's rule can be found in just about any undergraduate calculus textbook.

3–6. Consider the function $h(t)$, given in Equation (3–47) and plotted in Figure 3–10. (This is the impulse response of an ideal lowpass filter.) When $t = 0$ is plugged into this function, the result is an indeterminant form (0/0). Use L'Hopital's rule to show that

$$\lim_{t \to 0} h(t) = 2f_c$$

3–7. Let

$$x(t) = \begin{cases} A, & 0 \le t \le T \\ 0, & \text{otherwise} \end{cases}$$

    a. Find $X_a(f)$.

       HINT: Look back at Example 3–7.1 and the time shifting theorem.

    b. Sketch $|X_a(f)|$.

3–8. Consider the triangle pulse $y(t)$ shown in Figure P1–22(b) in Chapter 1. Find $Y_a(f)$ and sketch $|Y_a(f)|$.

       HINT: Look back at convolution theorems and Problem 3–7.

3–9. Let $x(t) = 10\cos(2\pi f_1 t) + 5\cos(2\pi f_2 t)$

    where $f_1 = 100$ Hz, and $f_2 = 300$ Hz. Let $g(t) = r(t)x(t)$, where $r(t)$ is the rectangular pulse (rectangular window) described in Example 3–7.1 and Figure 3–6(a), with a time duration of $T = 0.04$ second. Let $g(t) \leftrightarrow G_a(f)$. Sketch $G_a(f)$, showing positive and negative frequencies. (The sketch should be accurate with respect to main lobe widths. The main lobe heights should have the right proportions.)

3–10. Let

$$H(s) = \frac{1 - s}{1 + s}$$

    a. Find $H_a(f)$.

    b. Find and sketch $|H_a(f)|$.

       HINT: Consider on all-pass filter.

    c. Find $\theta_H(f)$.

3–11. Let $h_{LP}(t)$ be the impulse response of an ideal lowpass filter with a cutoff frequency of $f_c = 100$ Hz. Now define the following impulse response:

$$h(t) = 2h_{LP}(t)\cos\left[2\pi(300)t\right]$$

    Let $h(t) \leftrightarrow H_a(f)$.

    a. Find and sketch $H_a(f)$.

    b. What kind of ideal filter is this?

3–12. Consider an ideal filter with the following impulse response:

$$h(t) = \delta(t) - \frac{\sin(2\pi f_c t)}{\pi t}$$

    a. Find and sketch $H_a(f)$.

    b. What kind of ideal filter is this?

3–13. Consider an ideal filter with the following impulse response:

$$h(t) = \frac{\sin(2\pi f_2 t)}{\pi t} - \frac{\sin(2\pi f_1 t)}{\pi t}$$

where $f_1 = 300$ Hz, and $f_2 = 500$ Hz.
 a. Find and sketch $H_a(f)$.
 b. What kind of ideal filter is this?

3–14. Find the impulse response of an ideal bandstop filter (notch filter) with lower cutoff frequency f_1 and upper cutoff frequency f_2.

HINT: The solution involves combining two ideal filters.

3–15. Consider an ideal filter having the following frequency response function:

$$H_a(f) = \begin{cases} -j, & f > 0 \\ +j, & f < 0 \\ 0, & f = 0 \end{cases}$$

(This is an ideal *Hilbert transformer*. Hilbert transformers are useful in communications circuit theory. For example, the so-called phase shift method of generating single sideband signals uses a Hilbert transformer.)
 a. Find and sketch $|H_a(f)|$.
 b. Find and sketch $\theta_H(f)$.
 c. Let

$$x(t) = \sum_{n=1}^{M} a_n \cos\left(2\pi n f_o t + \phi_n\right)$$

be the steady-state input signal to the ideal Hilbert transformer. Find the steady-state output signal.

3–16. Consider a system having the impulse response $h(t) = \delta(t - t_o)$.
 a. Find $H_a(f)$.
 b. Show that this is a pure delay system: $y(t) = x(t - t_o)$.
 c. Find and sketch $|H_a(f)|$, showing positive and negative frequencies.
 d. Find and sketch $\theta_H(f)$, showing positive and negative frequencies.

3–17. Consider a system having the impulse response $h(t) = \delta(t) + \delta(t - t_o)$.
 a. Find $H_a(f)$, and put it in the form $H_a(f) = A(f)e^{-j()}$, where $A(f)$ is a real function.

HINT:

$$1 + e^{-jx} = e^{-j\frac{x}{2}}\left(e^{j\frac{x}{2}} + e^{-j\frac{x}{2}}\right)$$

 b. Show that this system creates an echo, that is, $y(t) = x(t) + x(t - t_o)$.
 c. Show that $|H_a(f)| = 2|\cos(\pi f t_o)|$.
 d. Suppose $t_o = 0.001$ second. Suppose the steady-state input to this system is $x(t) = \cos(2\pi f_1 t)$. For which frequencies (that is, which values of f_1) is the steady-state output $y(t) = 0$?
 e. Show that the phase response function of this system is

$$\theta_H(f) = \begin{cases} -\pi f t_o, & \text{when } \cos(\pi f t_o) > 0 \\ -\pi f t_o \pm \pi, & \text{when } \cos(\pi f t_o) < 0 \end{cases}$$

HINT: When $\cos(\pi f t_o) > 0$, $H_a(f)$ is already in polar form. When $\cos(\pi f t_o) < 0$, take advantage of Chapter 2, problem 6(a).

3–18. Consider a hypothetical system having a frequency response function of the form:

$$H_a(f) = A(f)e^{-j2\pi f t_o}$$

where $A(f)$ is a real function of f, but can take on positive, negative, or zero value, depending on f. This system is said to have generalized *linear phase*.

a. Show that $|H_a(f)| = |A(f)|$.

b. Show that the phase response function is

$$\theta_H(f) = \begin{cases} -2\pi f t_o, & A(f) > 0 \\ -2\pi f t_o \pm \pi, & A(f) < 0 \end{cases}$$

HINT: When $A(f) > 0$, $H_a(f)$ is already in polar form. When $A(f) < 0$, you can take advantage of Problem 2–6(a) in Chapter 2.

c. Suppose the steady-state input signal to this system is

$$x(t) = \cos(2\pi f_1 t)$$

i. Show that if $A(f_1)$ is positive, the steady-state output is:

$$y(t) = A(f_1)\cos(2\pi f_1(t - t_o))$$

ii. Show that if $A(f_1)$ is negative, the steady-state output is:

$$y(t) = -|A(f_1)| \cos(2\pi f_1(t - t_o))$$

Note that the input sinusoid is time delayed by the same amount (t_o seconds) *regardless* of what the frequency is (f_1). Therefore, if the input signal is periodic (and thus has a Fourier series), all of the frequency components of this signal will be time delayed by the same amount. The nature of the amplitude change depends on $A(f)$, which in general will vary according to frequency.

In order for a system to have generalized linear phase, it can be shown that its impulse response $h(t)$ must be symmetric about its midpoint. In the case of causal systems having poles (which means most continuous time systems of interest), it is *not* possible to obtain this symmetry, since these systems have impulse responses that start at $t = 0$ and asymptotically approach zero as $t \to \infty$. However, it is possible to design systems that come close to having this property; an example is a Bessel lowpass filter (also called a Thomson lowpass filter). Butterworth, Chebyshev, and Elliptic lowpass filters definitely *don't* have generalized linear phase.

3–19. Consider the system shown in Figure P3–19(a). Suppose the input signal $x(t)$ has the spectrum shown in Figure P3–19(b).

a. Find the carrier frequency (f_o) and filter type (including cutoff frequency) such that the spectrum of $y(t)$ is shown in Figure P3–19(c).

b. Find the carrier frequency (f_o) and filter type (including cutoff frequency) such that the spectrum of $y(t)$ is as shown in Figure P3–19(d).

3–20. Show that if $x(t)$ is real, then $X_a(-f) = X_a^*(f)$
where $X_a^*(f)$ is the complex conjugate of $X_a(f)$.

HINT:

$$X_a(f) - \int_{-\infty}^{\infty} x(t)\left[\cos(2\pi ft) - j\sin(2\pi ft)\right]dt$$

$$= \int_{-\infty}^{\infty} x(t)\cos(2\pi ft)dt - j\int_{-\infty}^{\infty} x(t)\sin(2\pi f)dt$$

NOTE: This means that if $x(t)$ is real, the time-reversal property (Problem 3–3) can be stated as

$$x(-t) \leftrightarrow X_a^*(f)$$

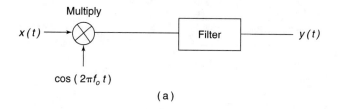

(a)

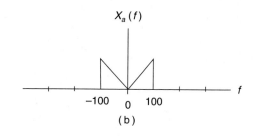

(b)

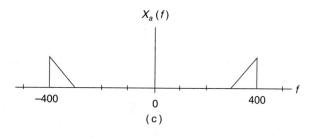

(c)

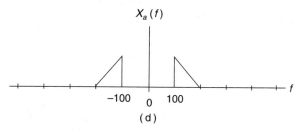

(d)

FIGURE P3–19

3–21. Let $x(t) = Ae^{-a|t|}$, defined for all time, where $a > 0$. Find $X_a(f)$.

HINT: $x(t)$ can be thought of as the sum of two functions, one of which is the time reversal of the other.

HINT: Look back at Equations (2–56) to (2–58) in Chapter 2.

3–22. Consider the pulse defined as follows:

$$x(t) = \begin{cases} +A, & 0 \le t \le T \\ -A, & -T \le t < \\ 0, & otherwise \end{cases}$$

Find $X_a(f)$.

HINT: $x(t)$ can be thought of as the sum of two functions, one of which is -1 times the time reversal of the other.

HINT: Look back at Problem 3–7.

4

BUTTERWORTH AND CHEBYSHEV LOWPASS AND HIGHPASS FILTERS

4–1   FREQUENCY RESPONSE FUNCTION: BUTTERWORTH LOWPASS FILTER

An Nth order Butterworth lowpass filter has an amplitude response function given by

$$|H_a(f)| = \frac{G}{\sqrt{1 + \left(\dfrac{f}{f_c}\right)^{2N}}} \tag{4–1}$$

where G is the passband gain (note that $|H_a(0)| = G$; for a Butterworth lowpass filter, the passband gain is the gain at 0 Hz), N is the filter order, and f_c is the *cutoff frequency*. Observe that $|H_a(f_c)| = G/\sqrt{2}$. In other words, for a Butterworth lowpass filter, the cutoff frequency is defined as that frequency at which the filter gain is $(1/\sqrt{2})$ times the passband gain. Figure 4–1 shows $|H_a(f)|$ for $N = 1, 2,$ and 3, with $f_c = 1000$ Hz in each case, plotted for positive f only. Note that the larger the filter order (N), the sharper the transition from passband to stopband. Note also that $|H_a(f)|$ is monotone decreasing: the curve has no ripples in either the passband or stopband.

In decibels (dB), we have

$$20 \log|H_a(f)| = 20 \log G - 10 \log\left[1 + \left(\frac{f}{f_c}\right)^{2N}\right] \tag{4–2}$$

FIGURE 4–1
Amplitude response function
of Butterworth lowpass
filters.

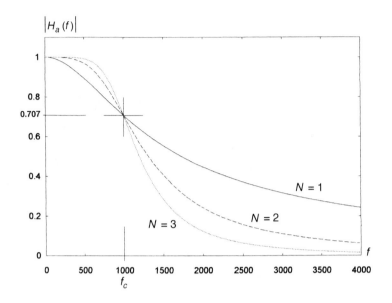

At frequencies much less than f_c, the term $(f/f_c)^{2N}$ is much less than 1; therefore, for frequencies much less than f_c, we have approximately:

$$20 \log|H_a(f)| \cong 20 \log G - 10 \log(1) = 20 \log G$$

$$f \ll f_c$$

(4–3)

At frequencies much greater than f_c, the term $(f/f_c)^{2N}$ is much greater than 1; therefore, for frequencies much greater than f_c, we have approximately

$$20 \log|H_a(f)| \cong 20 \log G - 20N \log\left(\frac{f}{f_c}\right)$$

$$= 20 \log G + 20N \log(f_c) - 20N \log(f)$$

$$f \gg f_c$$

(4–4)

At the cutoff frequency $f = f_c$, we have

$$20 \log|H_a(f_c)| = 20 \log G - 3.01$$

(4–5)

Let us define the 3 dB frequency, denoted f_{3dB}, as that frequency where the filter gain (in dB) is 3 dB below the passband gain (in dB). Equation (4–5) shows that for a Butterworth lowpass filter, the cutoff frequency and the 3 dB frequency are the same (that is, $f_c = f_{3dB}$). However, for a *Chebyshev* lowpass filter, f_c and f_{3dB} are *not* the same (as will be shown in Section 4–2).

Equation (4–3) shows that for frequencies below f_c, the gain of a Butterworth lowpass filter is approximately constant. Equation (4–4) shows that for frequencies above f_c the filter gain decreases at a rate of approximately $20N$ dB per decade (i.e., per tenfold increase in

FIGURE 4–2
Decibel amplitude response
function of Butterworth
lowpass filters.

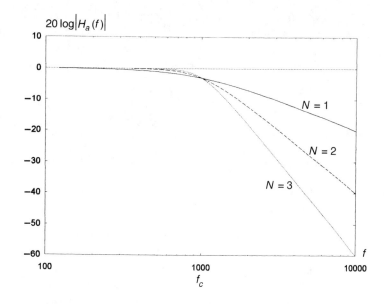

frequency). The Bode plots of the Butterworth lowpass filter amplitude response (dB) are
shown in Figure 4–2 for filter orders $N = 1$, 2, and 3.

4–2 FREQUENCY RESPONSE FUNCTION: CHEBYSHEV LOWPASS FILTER

An Nth order Chebyshev lowpass filter has an amplitude response function given by

$$|H_a(f)| = \frac{G}{\sqrt{1 + \varepsilon^2 C_N^2\left(\dfrac{f}{f_c}\right)}} \tag{4–6}$$

where G is the passband gain, ε is the *ripple factor*, and f_c is the cutoff frequency.
(Strictly speaking, this is a Chebyshev "type I" lowpass filter.) The function $C_N()$ is an
Nth order Chebyshev polynomial of the first kind. These polynomials can be constructed
as follows:

$$C_0(x) = 1$$
$$C_1(x) = x$$
$$C_2(x) = 2x^2 - 1$$
$$C_3(x) = 4x^3 - 3x \tag{4–7}$$
$$\vdots$$
$$C_{N+1}(x) = 2xC_N(x) - C_{N-1}(x)$$

Note that the $(N + 1)$th order polynomial can be constructed from the Nth and $(N - 1)$th order polynomials.

A specific example of Equation (4–6) may be helpful. The amplitude response function of a third order Chebyshev lowpass filter with cutoff frequency f_c is

$$|H_a(f)| = \frac{G}{\sqrt{1 + \varepsilon^2 C_3^2\left(\dfrac{f}{f_c}\right)}}$$

$$= \frac{G}{\sqrt{1 + \varepsilon^2\left[4\left(\dfrac{f}{f_c}\right)^3 - 3\left(\dfrac{f}{f_c}\right)\right]^2}} \tag{4–8}$$

This amplitude response function is shown in Figure 4–3 for the specific case $G = 1$, $\varepsilon = 0.764831$, and $f_c = 1000$ Hz. This plot clearly shows the characteristic passband ripple in a Chebyshev lowpass filter. Within the passband, that is, for frequencies less than or equal to f_c, $|H_a(f)|$ varies within a range bounded by G and $G/\sqrt{1 + \varepsilon^2}$. The larger ε is, the larger this ripple is. For frequencies greater than f_c, $|H_a(f)|$ decreases monotonically. Note that for a Chebyshev lowpass filter, the cutoff frequency f_c is *not* the 3 dB frequency f_{3dB}, that is, f_c is *not* the frequency at which $|H_a(f)| = G/\sqrt{2}$, unless $\varepsilon = 1$. The usual choice for ε is a number less than or equal to 1. When ε is less than 1, f_c is less than f_{3dB}. It can be shown that for a Chebyshev lowpass filter with $\varepsilon \le 1$, f_c and f_{3dB} are related by

$$\frac{f_{3dB}}{f_c} = \cosh\left\{\frac{1}{N}\cosh^{-1}\left(\frac{1}{\varepsilon}\right)\right\}, \varepsilon \le 1 \tag{4–9}$$

FIGURE 4–3

Amplitude response function of a third order Chebyshev lowpass filter with 2 dB of ripple.

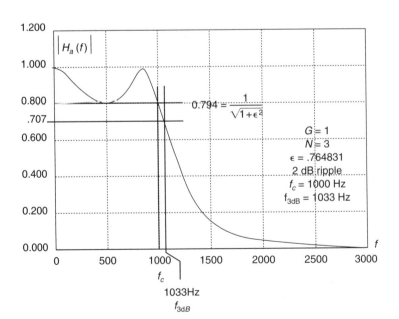

For the filter shown in Figure 4–3, $f_{3dB} = 1033$ Hz.

In terms of decibels, within the passband the gain varies within a range bounded by $20\log G$ dB and $(20\log G - 20\log\sqrt{1+\varepsilon^2})$dB. Therefore, the passband ripple in dB, denoted r_{dB}, is

$$r_{dB} = 20\log\sqrt{1+\varepsilon^2} \qquad (4\text{--}10)$$

Chebyshev lowpass filter passband ripple is usually specified in dB. If r_{dB} is given, the ripple factor ε can be solved for

$$\varepsilon = \sqrt{10^{\left(\frac{rdB}{10}\right)} - 1} \qquad (4\text{--}11)$$

The third order filter shown in Figure 4–3 has 2 dB of passband ripple. The reader can confirm this using either Equation (4–10) or Equation (4–11).

Figure 4–4(a) shows $|H_a(f)|$ for a seventh order Chebyshev lowpass filter with $G = 1$, $r_{dB} = 0.5$ dB, and $f_c = 1000$ Hz. ($\varepsilon = 0.1526204$.) When compared with Figure 4–3 (third order example), it is apparent that the number of ripple peaks plus the number of ripple valleys in the filter passband is equal to the filter order (N). Figures 4–4(b) and 4–4(c) show the dB amplitude response plot ($20\log|H_a(f)|$) for the seventh order filter described previously. In Figure 4–4(c), the vertical scale has been expanded to illustrate the 0.5 dB passband ripple.

Figures 4–5(a) and 4–5(b) illustrate the fact that for a Chebyshev lowpass filter with a fixed amount of passband ripple, increasing the order (N) results in a sharper transition from passband to stopband. For the two filters shown ($N = 3$ and $N = 4$), the parameters are $G = 1$, $f_c = 1000$ Hz, and $r_{dB} = 1.0$ dB. Note that increasing the filter order causes f_{3dB} to move closer to f_c. Figure 4–5(a) also shows that when N is *odd*, the gain at $f = 0$ is equal to G, but when N is *even*, the gain at $f = 0$ is equal to $G/\sqrt{1+\varepsilon^2}$.

FIGURE 4–4

A seventh order Chebyshev lowpass filter with 0.5 dB of ripple. (a) Amplitude response function. (b) Decibel amplitude response function. (c) Passband detail of decibel amplitude response function.

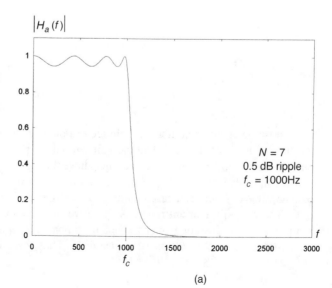

(a)

FIGURE 4–4
(continued)

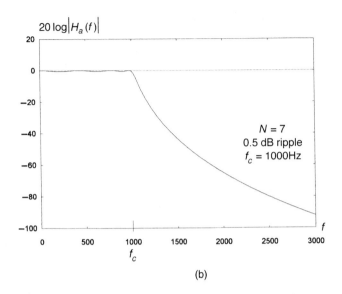

(b)

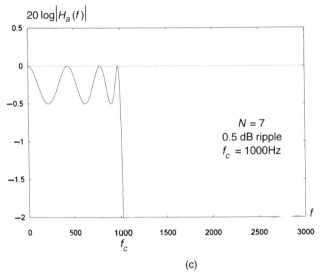

(c)

It is interesting to compare amplitude response functions for Chebyshev lowpass filters having different orders (N) but the same amount of passband ripple with the cutoff frequency f_c adjusted such that all of the filters have the same f_{3dB}. This is done in Figure 4–6, which compares filters with $r_{dB} = 1.0$ dB and $f_{3dB} = 1000$ Hz, with orders N = 2, 3, 4, and 5. (The value of f_c for each case is shown on the figure.)

There is an important relationship between the amount of passband ripple and the sharpness of the transition from passband to stopband: *for a given order (N), increasing the passband ripple will result in a sharper filter*. Thus there is a design tradeoff: To get a

FIGURE 4–5
Comparing third and fourth order Chebyshev lowpass filters having the same cutoff frequency and ripple. (a) Amplitude response function. (b) Decibel amplitude response function.

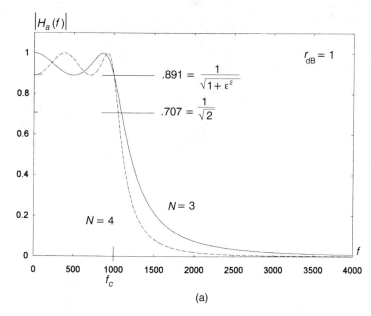

(a)

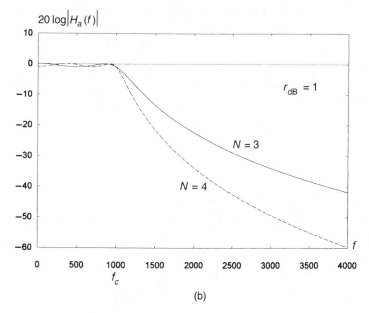

(b)

sharper filter without increasing the filter order, we can increase the amount of passband ripple. Increasing passband ripple is usually not desirable but it may be more attractive than the other way to increase sharpness, which is to increase N. It is easy to use Equation (4–6) to generate theoretical amplitude response functions for Chebyshev filters of large

FIGURE 4–6
Comparing Chebyshev
lowpass filters of different
order, with all filters having
the same 3 dB frequency and
ripple.

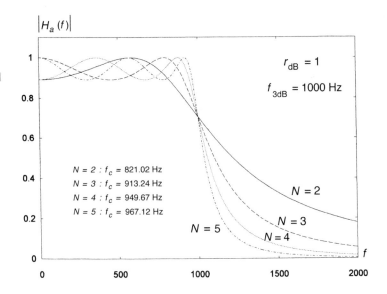

order (say, $N = 8$ or larger), but designing actual circuits to realize these high-order filters can be tricky, because these circuits are very sensitive to small variations in component values.

The relationship between passband ripple and filter sharpness for a fixed N is illustrated in Figure 4–7, which compares fifth order Chebyshev lowpass filters having r_{dB} of 0.1, 1.0, and 3.0 dB. The cutoff frequencies (f_c) have been adjusted so that f_{3dB} is the same for all filters (1000 Hz). (The actual f_c values are indicated on the figure.) Also shown on

FIGURE 4–7
Comparing Butterworth
and Chebyshev lowpass
filters of the same order
($N = 5$) and 3 dB frequency.
The more ripple allowed in
the passband, the sharper the
filter.

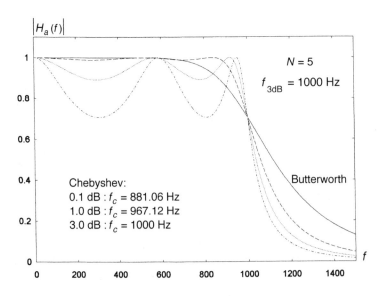

FIGURE 4–8
Roll off characteristics for fifth order Butterworth and Chebyshev lowpass filters having the same cutoff frequency. (The Chebyshev filter has 1 dB of ripple.)

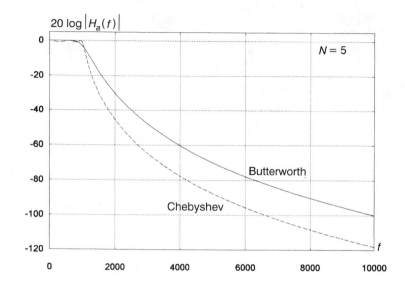

the same plot is a fifth order *Butterworth* lowpass filter having $f_c = f_{3dB} = 1000$ Hz. Note that a Chebyshev filter having even a small amount of passband ripple (0.1 dB) is a definite improvement over a Butterworth filter of the same order, at least in terms of the sharpness of $|H_a(f)|$.

For frequencies much higher than f_c, the dB plot for a Chebyshev lowpass filter rolls off at a rate of approximately $20N$ dB per decade, which is the same as a Butterworth filter. However, since the Chebyshev filter exhibits a sharper rolloff characteristic near the cutoff frequency, it turns out that for $f \gg f_c$ (for filters having the same N and f_c), $|H_a(f)|$ is smaller for the Chebyshev filter. This is illustrated in Figure 4–8, which shows the dB plot for fifth order Butterworth and Chebyshev lowpass filters, both having $f_c = 1000$ Hz. The Chebyshev filter has 1 dB of passband ripple (and $f_{3dB} = 1034$ Hz). Observe that for frequencies above approximately 4000 Hz, the two curves are separated by a gap of approximately 18 dB. The two curves are rolling off at the same rate, but the Chebyshev curve is 18 dB below the Butterworth curve. It can be shown that the gap (in dB) between the two curves, for $f \gg f_c$, for an Nth order Butterworth lowpass filter and an Nth order Chebyshev filter with specified ε, if both filters have the same f_c, is

$$gap\,(\text{dB}) \cong 6(N-1) + 20\log(\varepsilon) \qquad (4\text{–}12)$$

It should be emphasized that Equation (4–12) is for filters having the same f_c, *not* the same f_{3dB}. If the filters have the same f_{3dB}, where $f_c = f_{3dB}$ for the Chebyshev filter, the gap will be slightly larger than Equation (4–12) suggests. This is illustrated in Figure 4–9, which compares a fifth order Butterworth lowpass filter ($f_c = f_{3dB} = 1000$ Hz) and a fifth order Chebyshev lowpass filter ($f_c = 967.12$ Hz, $f_{3dB} = 1000$ Hz, and $r_{dB} = 1$ dB). In this case, the gap is approximately 20 dB instead of 18 dB.

FIGURE 4–9
Roll off characteristics for fifth order Butterworth and Chebyshev lowpass filters having the same 3 dB frequency. (The Chebyshev filter has 1 dB of ripple.)

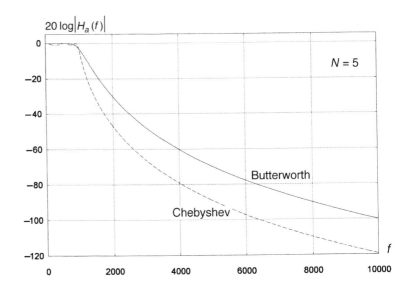

4–3   TRANSFER FUNCTIONS FOR NORMALIZED LOWPASS FILTERS

Nth order Butterworth and Chebyshev lowpass filters both have all-pole transfer functions of the form

$$H(s) = \frac{C}{s^N + a_{N-1}s^{N-1} + a_{N-2}s^{N-2} + \ldots + a_1 s + a_0} \tag{4–13}$$

where C is some constant, and the a_is are all real coefficients. These transfer functions are usually tabulated in *normalized* form. A normalized Butterworth or Chebyshev lowpass filter has a passband gain of $G = 1$ and a cutoff frequency of $f_c = 1/(2\pi)$ Hz. If the cutoff frequency is expressed in radians ($\Omega = 2\pi f$), the cutoff frequency of the normalized lowpass filter is $\Omega_c = 1$. A normalized transfer function can be *unnormalized* (i.e., the cutoff frequency can be changed to the desired value) by a lowpass-to-lowpass frequency transformation that will be described in Section 4–4.

The poles of the normalized transfer function are the roots of the denominator polynomial of $H(s)$ (Equation [4–13]). By factoring this polynomial, the transfer function can be expressed in cascade form. For even order filters ($N = 2, 4, 6, \ldots$), the normalized transfer function in cascade form for both Butterworth and Chebyshev lowpass filters is

$$H(s) = \frac{1}{\sqrt{1 + \varepsilon^2}} \prod_{k=1}^{N/2} \frac{|p_k|^2}{(s - p_k)(s - p_k^*)}$$

$$= \frac{1}{\sqrt{1 + \varepsilon^2}} \prod_{k=1}^{N/2} \frac{|p_k|^2}{s^2 - 2\,\mathrm{Re}(p_k)s + |p_k|^2} \tag{4–14}$$

For odd order filters ($N = 1, 3, 5, \ldots$), the normalized transfer function in cascade form for both Butterworth and Chebyshev lowpass filters is

$$H(s) = \frac{|p_0|}{s - p_0} \prod_{k=1}^{(N-1)/2} \frac{|p_k|^2}{(s - p_k)(s - p_k^*)}$$

$$= \frac{|p_0|}{s - p_0} \prod_{k=1}^{(N-1)/2} \frac{|p_k|^2}{s^2 - 2\,\mathrm{Re}(p_k)s + |p_k|^2}$$

(4–15)

For the even order filter, the N poles are p_k and p_k^*, $k = 1, 2, \ldots N/2$. p_k is a complex number; p_k^* is its complex conjugate.

For the odd order filter, the N poles are p_0 and p_k and p_k^*, $k = 1, 2, \ldots (N-1)/2$. p_0 is a real number; the remaining $N - 1$ poles are all complex, as in the even order case.

Conventional notation for complex numbers is used in these transfer functions. $|p_k|$ is the magnitude of complex number p_k. $\mathrm{Re}\{p_k\}$ is the real part of p_k.

For the even order $H(s)$, there is a constant multiplier term $1/\sqrt{1 + \varepsilon^2}$. For a Chebyshev filter, ε is the passband ripple factor discussed in Section 4–2. For a Butterworth filter, set $\varepsilon = 0$ in Equation (4–14).

The poles for normalized Butterworth lowpass filters of orders $N = 1, 2, \ldots, 8$ are tabulated in Table 4–1. For each N, complex poles are listed first in the form (a, b), which denotes the complex conjugate pair $(a \pm jb)$. If N is odd, the last pole listed is the real pole p_0. Note that $p_0 = -1$ for all odd order normalized Butterworth lowpass filters. Note also that for normalized Butterworth lowpass filters, $|p| = 1$ for all of the poles.

The poles for normalized Chebyshev lowpass filters of orders $N = 2, 3, \ldots, 8$ have been tabulated for several different passband ripples (0.1, 0.5, 1.0, 1.5, 2.0, and 3.0 dB) in Tables 4–2(a) through 4–2(f). Also shown in these tables is the corresponding value of ripple factor ε. (If N is fixed but the passband ripple changes, the poles also change.) The $f_{3\mathrm{dB}}/f_c$ ratio (discussed in Section 4–3) is listed for each case. For each N, complex poles are listed in the form (a, b), which denotes the complex conjugate pair $(a \pm jb)$. If N is odd, the last pole listed is the real pole p_0. Pole magnitude $|p|$ is listed for each pole. (For the Chebyshev case, $|p|$ is *not* the same for all cases.)

Note that in both the Butterworth and Chebyshev tables, the real part of every pole is negative. This, of course, is a necessary condition for stability (the poles must all be located in the left half of the s plane).

TABLE 4.1
Butterworth Lowpass Filter Data

$N = 1$		
pole p	pole magnitude $\lvert p \rvert$	ψ
$(-1.000000, 0.000000E+00)$	1.000000	
$N = 2$		
pole p	pole magnitude $\lvert p \rvert$	ψ
$(-7.071068E{-}01, 7.071068E{-}01)$	1.000000	1.414214
$N = 3$		
pole p	pole magnitude $\lvert p \rvert$	ψ
$(-5.000000E{-}01, 8.660254E{-}01)$	1.000000	1.000000
$(-1.000000, 0.000000E+00)$	1.000000	
$N = 4$		
pole p	pole magnitude $\lvert p \rvert$	ψ
$(-3.826835E{-}01, 9.238795E{-}01)$	1.000000	7.653669E{-}01
$(-9.238796E{-}01, 3.826834E{-}01)$	1.000000	1.847759
$N = 5$		
pole p	pole magnitude $\lvert p \rvert$	ψ
$(-3.090170E{-}01, 9.510565E{-}01)$	1.000000	6.180340E{-}01
$(-8.090170E{-}01, 5.877852E{-}01)$	1.000000	1.618034
$(-1.000000, 0.000000E+00)$	1.000000	
$N = 6$		
pole p	pole magnitude $\lvert p \rvert$	ψ
$(-2.588190E{-}01, 9.659258E{-}01)$	1.000000	5.176381E{-}01
$(-7.071068E{-}01, 7.071068E{-}01)$	1.000000	1.414214
$(-9.659258E{-}01, 2.588190E{-}01)$	1.000000	1.931852
$N = 7$		
pole p	pole magnitude $\lvert p \rvert$	ψ
$(-2.225209E{-}01, 9.749279E{-}01)$	1.000000	4.450419E{-}01
$(-6.234898E{-}01, 7.818314E{-}01)$	1.000000	1.246980
$(-9.009689E{-}01, 4.338837E{-}01)$	1.000000	1.801938
$(-1.000000, 0.000000E+00)$	1.000000	
$N = 8$		
pole p	pole magnitude $\lvert p \rvert$	ψ
$(-1.950903E{-}01, 9.807853E{-}01)$	1.000000	3.901806E{-}01
$(-5.555702E{-}01, 8.314696E{-}01)$	1.000000	1.111140
$(-8.314697E{-}01, 5.555702E{-}01)$	1.000000	1.662939
$(-9.807853E{-}01, 1.950903E{-}01)$	1.000000	1.961571

For each N, complex poles are listed first as (a, b), which is the complex conjugate pair $a \pm jb$. If N is odd, the last pole listed is real. Pole magnitude $\lvert p \rvert$ is listed for each pole. $\psi = -2\text{Re}\{p\}/\lvert p \rvert$ is listed for each complex pole.

TABLE 4–2(a)
Chebyshev Lowpass Filter Data

Passband RIPPLE = 1.000000E–01 dB
epsilon = 1.526204E–01

$N = 2$ $f_{3db}/f_c = 1.943$

pole p	pole magnitude $|p|$	ψ
(–1.186178,1.380948)	1.820450	1.303171

$N = 3$ $f_{3db}/f_c = 1.389$

pole p	pole magnitude $|p|$	ψ
(–4.847029E–01,1.206155)	1.299903	7.457524E–01
(–9.694057E–01,0.000000E+00)	9.694057E–01	

$N = 4$ $f_{3db}/f_c = 1.213$

pole p	pole magnitude $|p|$	ψ
(–2.641564E–01,1.122610)	1.153270	4.580998E–01
(–6.377299E–01,4.650002E–01)	7.892557E–01	1.616029

$N = 5$ $f_{3db}/f_c = 1.135$

pole p	pole magnitude $|p|$	ψ
(–1.665337E–01,1.080372)	1.093132	3.046910E–01
(–4.359908E–01,6.677066E–01)	7.974460E–01	1.093468
(–5.389143E–01,0.000000E+00)	5.389143E–01	

$N = 6$ $f_{3db}/f_c = 1.093$

pole p	pole magnitude $|p|$	ψ
(–1.146934E–01,1.056519)	1.062726	2.158474E–01
(–3.133481E–01,7.734255E–01)	8.344903E–01	7.509928E–01
(–4.280415E–01,2.830934E–01)	5.131875E–01	1.668168

$N = 7$ $f_{3db}/f_c = 1.068$

pole p	pole magnitude $|p|$	ψ
(–8.384097E–02,1.041833)	1.045202	1.604302E–01
(–2.349172E–01,8.354855E–01)	8.678836E–01	5.413564E–01
(–3.394651E–01,4.636595E–01)	5.746449E–01	1.181478
(–3.767779E–01,0.000000E+00)	3.767779E–01	

$N = 8$ $f_{3db}/f_c = 1.052$

pole p	pole magnitude $|p|$	ψ
(–6.398012E–02,1.032181)	1.034162	1.237332E–01
(–1.822000E–01,8.750411E–01)	8.938085E–01	4.076935E–01
(–2.726815E–01,5.846837E–01)	6.451436E–01	8.453359E–01
(–3.216498E–01,2.053135E–01)	3.815917E–01	1.685832

For each N, complex poles are listed first as (a,b), which is the complex conjugate pair $a \pm jb$. If N is odd, the last pole listed is real. Pole magnitude $|p|$ is listed for each pole. $\psi = -2\text{Re}\{p\}/|p|$ is listed for each complex pole.

TABLE 4.2(b)
Chebyshev Lowpass Filter Data

Passband RIPPLE = 5.000000E–01 dB
epsilon = 3.493114E–01

$N = 2$ $f_{3db}/f_c = 1.390$

pole p	pole magnitude $|p|$	ψ
(–7.128122E–01,1.004042)	1.231342	1.157781

$N = 3$ $f_{3db}/f_c = 1.167$

pole p	pole magnitude $|p|$	ψ
(–3.132282E–01,1.021927)	1.068853	5.861014E–01
(–6.264565E–01,0.000000E+00)	6.264565E–01	

$N = 4$ $f_{3db}/f_c = 1.093$

pole p	pole magnitude $|p|$	ψ
(–1.753531E–01,1.016253)	1.031270	3.400719E–01
(–4.233398E–01,4.209457E–01)	5.970024E–01	1.418218

$N = 5$ $f_{3db}/f_c = 1.059$

pole p	pole magnitude $|p|$	ψ
(–1.119629E–01,1.011557)	1.017735	2.200238E–01
(–2.931227E–01,6.251768E–01)	6.904832E–01	8.490365E–01
(–3.623196E–01,0.000000E+00)	3.623196E–01	

$N = 6$ $f_{3db}/f_c = 1.041$

pole p	pole magnitude $|p|$	ψ
(–7.765008E–02,1.008461)	1.011446	1.535427E–01
(–2.121440E–01,7.382445E–01)	7.681211E–01	5.523711E–01
(–2.897940E–01,2.702163E–01)	3.962290E–01	1.462760

$N = 7$ $f_{3db}/f_c = 1.030$

pole p	pole magnitude $|p|$	ψ
(–5.700319E–02,1.006409)	1.008022	1.130991E–01
(–1.597194E–01,8.070770E–01)	8.227293E–01	3.882672E–01
(–2.308012E–01,4.478939E–01)	5.038633E–01	9.161264E–01
(–2.561700E–01,0.000000E+00)	2.561700E–01	

$N = 8$ $f_{3db}/f_c = 1.023$

pole p	pole magnitude $|p|$	ψ
(–4.362008E–02,1.005002)	1.005948	8.672430E–02
(–1.242195E–01,8.519996E–01)	8.610075E–01	2.885445E–01
(–1.859076E–01,5.692879E–01)	5.988742E–01	6.208568E–01
(–2.192929E–01,1.999073E–01)	2.967361E–01	1.478033

For each *N*, complex poles are listed first as (*a*,*b*), which is the complex
conjugate pair $a \pm jb$. If *N* is odd, the last pole listed is real. Pole magnitude
$|p|$ is listed for each pole. $\psi = -2\mathrm{Re}\{p\}/|p|$ is listed for each complex pole.

TABLE 4.2(c)
Chebyshev Lowpass Filter Data

Passband RIPPLE = 1.000000 dB
epsilon = 5.088471E–01

$N = 2$ $f_{3db}/f_c = 1.218$

pole p	pole magnitude $|p|$	Ψ
(–5.488672E–01,8.951285E–01)	1.050005	1.045456

$N = 3$ $f_{3db}/f_c = 1.095$

pole p	pole magnitude $|p|$	Ψ
(–2.470853E–01,9.659986E–01)	9.970980E–01	4.956088E–01
(–4.941706E–01,0.000000E+00)	4.941706E–01	

$N = 4$ $f_{3db}/f_c = 1.053$

pole p	pole magnitude $|p|$	Ψ
(–1.395360E–01,9.833792E–01)	9.932295E–01	2.809743E–01
(–3.368697E–01,4.073290E–01)	5.285812E–01	1.274619

$N = 5$ $f_{3db}/f_c = 1.034$

pole p	pole magnitude $|p|$	Ψ
(–8.945837E–02,9.901071E–01)	9.941403E–01	1.799713E–01
(–2.342051E–01,6.119198E–01)	6.552083E–01	7.149026E–01
(–2.894934E–01,0.000000E+00)	2.894934E–01	

$N = 6$ $f_{3db}/f_c = 1.023$

pole p	pole magnitude $|p|$	Ψ
(–6.218103E–02,9.934111E–01)	9.953553E–01	1.249424E–01
(–1.698817E–01,7.272274E–01)	7.468062E–01	4.549553E–01
(–2.320628E–01,2.661837E–01)	3.531387E–01	1.314287

$N = 7$ $f_{3db}/f_c = 1.017$

pole p	pole magnitude $|p|$	Ψ
(–4.570898E–02,9.952839E–01)	9.963329E–01	9.175444E–02
(–1.280737E–01,7.981557E–01)	8.083659E–01	3.168707E–01
(–1.850719E–01,4.429430E–01)	4.800522E–01	7.710490E–01
(–2.054143E–01,0.000000E+00)	2.054143E–01	

$N = 8$ $f_{3db}/f_c = 1.013$

pole p	pole magnitude $|p|$	Ψ
(–3.500824E–02,9.964513E–01)	9.970661E–01	7.022250E–02
(–9.969502E–02,8.447506E–01)	8.506132E–01	2.344074E–01
(–1.492041E–01,5.644443E–01)	5.838315E–01	5.111206E–01
(–1.759983E–01,1.982064E–01)	2.650683E–01	1.327947

For each N, complex poles are listed first as (a,b), which is the complex conjugate pair $a \pm jb$. If N is odd, the last pole listed is real. Pole magnitude $|p|$ is listed for each pole. $\Psi = -2\text{Re}\{p\}/|p|$ is listed for each complex pole.

TABLE 4.2(d)
Chebyshev Lowpass Filter Data

Passband RIPPLE = 1.500000 dB
epsilon = 6.422908E–01

$N = 2$ $f_{3db}/f_c = 1.131$

pole p	pole magnitude $|p|$	ψ
(–4.610888E–01,8.441581E–01)	9.618762E–01	9.587280E–01

$N = 3$ $f_{3db}/f_c = 1.057$

pole p	pole magnitude $|p|$	ψ
(–2.100562E–01,9.393460E–01)	9.625458E–01	4.364596E–01
(–4.201124E–01,0.000000E+00)	4.201124E–01	

$N = 4$ $f_{3db}/f_c = 1.032$

pole p	pole magnitude $|p|$	ψ
(–1.191307E–01,9.676111E–01)	9.749171E–01	2.443915E–01
(–2.876070E–01,4.007976E–01)	4.933118E–01	1.166025

$N = 5$ $f_{3db}/f_c = 1.021$

pole p	pole magnitude $|p|$	ψ
(–7.652815E–02,9.797871E–01)	9.827712E–01	1.557395E–01
(–2.003533E–01,6.055416E–01)	6.378261E–01	6.282380E–01
(–2.476503E–01,0.000000E+00)	2.476503E–01	

$N = 6$ $f_{3db}/f_c = 1.014$

pole p	pole magnitude $|p|$	ψ
(–5.325112E–02,9.861585E–01)	9.875952E–01	1.078400E–01
(–1.454848E–01,7.219181E–01)	7.364317E–01	3.951073E–01
(–1.987359E–01,2.642404E–01)	3.306341E–01	1.202150

$N = 7$ $f_{3db}/f_c = 1.010$

pole p	pole magnitude $|p|$	ψ
(–3.917030E–02,9.899175E–01)	9.906921E–01	7.907663E–02
(–1.097527E–01,7.938522E–01)	8.014031E–01	2.739014E–01
(–1.585973E–01,4.405547E–01)	4.682324E–01	6.774298E–01
(–1.760297E–01,0.000000E+00)	1.760297E–01	

$N = 8$ $f_{3db}/f_c = 1.008$

pole p	pole magnitude $|p|$	ψ
(–3.001307E–02,9.923236E–01)	9.927774E–01	6.046283E–02
(–8.546998E–02,8.412514E–01)	8.455821E–01	2.021566E–01
(–1.279149E–01,5.621061E–01)	5.764768E–01	4.437815E–01
(–1.508859E–01,1.973854E–01)	2.484502E–01	1.214616

For each N, complex poles are listed first as (a,b), which is the complex
conjugate pair $a \pm jb$. If N is odd, the last pole listed is real. Pole magnitude
$|p|$ is listed for each pole. $\psi = -2\text{Re}\{p\}/|p|$ is listed for each complex pole.

TABLE 4.2(e)
Chebyshev Lowpass Filter Data

Passband RIPPLE = 2.000000 dB
epsilon = 7.647831E–01

$N = 2$ $f_{3db}/f_c = 1.074$

pole p	pole magnitude $|p|$	Ψ
(–4.019082E–01,8.133451E–01)	9.072268E–01	8.860149E–01

$N = 3$ $f_{3db}/f_c = 1.033$

pole p	pole magnitude $|p|$	Ψ
(–1.844554E–01,9.230772E–01)	9.413263E–01	3.919053E–01
(–3.689108E–01,0.000000E+00)	3.689108E–01	

$N = 4$ $f_{3db}/f_c = 1.018$

pole p	pole magnitude $|p|$	Ψ
(–1.048873E–01,9.579530E–01)	9.636780E–01	2.176811E–01
(–2.532202E–01,3.967971E–01)	4.707106E–01	1.075906

$N = 5$ $f_{3db}/f_c = 1.012$

pole p	pole magnitude $|p|$	Ψ
(–6.746098E–02,9.734557E–01)	9.757904E–01	1.382694E–01
(–1.766151E–01,6.016287E–01)	6.270168E–01	5.633506E–01
(–2.183083E–01,0.000000E+00)	2.183083E–01	

$N = 6$ $f_{3db}/f_c = 1.008$

pole p	pole magnitude $|p|$	Ψ
(–4.697322E–02,9.817051E–01)	9.828283E–01	9.558784E–02
(–1.283332E–01,7.186580E–01)	7.300265E–01	3.515851E–01
(–1.753064E–01,2.630471E–01)	3.161109E–01	1.109145

$N = 7$ $f_{3db}/f_c = 1.006$

pole p	pole magnitude $|p|$	Ψ
(–3.456636E–02,9.866205E–01)	9.872258E–01	7.002726E–02
(–9.685279E–02,7.912082E–01)	7.971141E–01	2.430086E–01
(–1.399563E–01,4.390875E–01)	4.608531E–01	6.073794E–01
(–1.553398E–01,0.000000E+00)	1.553398E–01	

$N = 8$ $f_{3db}/f_c = 1.005$

pole p	pole magnitude $|p|$	Ψ
(–2.649238E–02,9.897870E–01)	9.901415E–01	5.351232E–02
(–7.544392E–02,8.391009E–01)	8.424857E–01	1.790984E–01
(–1.129098E–01,5.606692E–01)	5.719253E–01	3.948410E–01
(–1.331862E–01,1.968808E–01)	2.376986E–01	1.120631

For each N, complex poles are listed first as (a,b), which is the complex conjugate pair $a \pm jb$. If N is odd, the last pole listed is real. Pole magnitude $|p|$ is listed for each pole. $\Psi = -2\text{Re}\{p\}/|p|$ is listed for each complex pole.

TABLE 4–2(f)
Chebyshev Lowpass Filter Data

Passband RIPPLE = 3.000000 dB
epsilon = 9.976283E–01

$N = 2$ $f_{3db}/f_c = 1.001$

pole p	pole magnitude $|p|$	ψ
(–3.224498E–01,7.771575E–01)	8.413963E–01	7.664636E–01

$N = 3$ $f_{3db}/f_c = 1.000$

pole p	pole magnitude $|p|$	ψ
(–1.493101E–01,9.038144E–01)	9.160644E–01	3.259817E–01
(–2.986202E–01,0.000000E+00)	2.986202E–01	

$N = 4$ $f_{3db}/f_c = 1.000$

pole p	pole magnitude $|p|$	ψ
(–8.517040E–02,9.464844E–01)	9.503087E–01	1.792479E–01
(–2.056195E–01,3.920467E–01)	4.426963E–01	9.289418E–01

$N = 5$ $f_{3db}/f_c = 1.000$

pole p	pole magnitude $|p|$	ψ
(–5.485988E–02,9.659274E–01)	9.674841E–01	1.134073E–01
(–1.436250E–01,5.969760E–01)	6.140102E–01	4.678262E–01
(–1.775303E–01,0.000000E+00)	1.775303E–01	

$N = 6$ $f_{3db}/f_c = 1.000$

pole p	pole magnitude $|p|$	ψ
(–3.822952E–02,9.764060E–01)	9.771541E–01	7.824665E–02
(–1.044450E–01,7.147788E–01)	7.223693E–01	2.891734E–01
(–1.426745E–01,2.616272E–01)	2.980013E–01	9.575426E–01

$N = 7$ $f_{3db}/f_c = 1.000$

pole p	pole magnitude $|p|$	ψ
(–2.814564E–02,9.826958E–01)	9.830987E–01	5.725904E–02
(–7.886235E–02,7.880608E–01)	7.919969E–01	1.991481E–01
(–1.139594E–01,4.373408E–01)	4.519444E–01	5.043072E–01
(–1.264854E–01,0.000000E+00)	1.264854E–01	

$N = 8$ $f_{3db}/f_c = 1.000$

pole p	pole magnitude $|p|$	ψ
(–2.157816E–02,9.867664E–01)	9.870023E–01	4.372464E–02
(–6.144940E–02,8.365402E–01)	8.387941E–01	1.465184E–01
(–9.196553E–02,5.589582E–01)	5.664732E–01	3.246951E–01
(–1.084807E–01,1.962800E–01)	2.242630E–01	9.674421E–01

For each *N*, complex poles are listed first as (*a,b*), which is the complex
conjugate pair $a \pm jb$. If *N* is odd, the last pole listed is real. Pole magnitude
$|p|$ is listed for each pole. $\psi = -2\mathrm{Re}\{p\}/|p|$ is listed for each complex pole.

4–4   LOWPASS-TO-LOWPASS TRANSFORMATION

A normalized ($\Omega_c = 1$) lowpass filter transfer function can be *unnormalized*, that is, converted to the desired cutoff frequency $\Omega_c = 2\pi f_c$ by replacing the variable s with s/Ω_c. In other words,

$$
\underset{\substack{\text{unnormalized} \\ \Omega_c \text{ as specified}}}{H(s)} = \underset{\text{normalized}}{H(s)}\Bigg|_{s \to \frac{s}{\Omega_c}} = \underset{\text{normalized}}{H\!\left(\frac{s}{\Omega_c}\right)}
\tag{4-16}
$$

This is called the *lowpass-to-lowpass transformation*. This transformation does not change the filter order or type. For example, an $N = 3$, 1 dB passband ripple Chebyshev lowpass filter is still an $N = 3$, 1 dB passband ripple Chebyshev lowpass filter after the transformation. In effect, the only thing that has changed is the scaling of the frequency axis for the frequency response function $H_a(f)$.

Applying the lowpass-to-lowpass transformation to normalized filters in cascade form is straightforward because we can carry out the transformation on each section individually. From Equations (4–14) and (4–15), we see that the normalized transfer functions consist of a product of second order sections, all having the same form:

$$
\underset{\substack{\text{2nd order section} \\ \text{normalized}}}{H(s)} = \frac{|p_k|^2}{s^2 - 2\,\mathrm{Re}(p_k)s + |p_k|^2}
\tag{4-17}
$$

If N is odd, there is also one first order section of the form:

$$
\underset{\substack{\text{1st order section} \\ \text{normalized}}}{H(s)} = \frac{|p_0|}{s - p_0}
\tag{4-18}
$$

Let us first consider unnormalizing the first order section. In addition to the frequency transformation, let us also introduce a scaling constant A_0:

$$
\underset{\substack{\text{1st order section} \\ \Omega_c \text{ as specified}}}{H(s)} = \frac{A_0|p_0|}{\dfrac{s}{\Omega_c} - p_0} = \frac{A_0|p_0|\Omega_c}{s + |p_0|\Omega_c}
\tag{4-19}
$$

This can be expressed as

$$
\underset{\substack{\text{1st order section} \\ \Omega_c \text{ as specified}}}{H(s)} = \frac{A_0\Omega_0}{s + \Omega_0}
\tag{4-20}
$$

where

$$
\Omega_0 = |p_0|\Omega_c = |p_0|(2\pi f_c)
\tag{4-21}
$$

NOTE: For a Butterworth filter, $\Omega_0 = \Omega_c$, since $|p_0| = 1$.

Next we apply the lowpass-to-lowpass transformation to the kth second order section, and also introduce a scaling constant A_k:

$$\underset{\substack{\text{kth 2nd order section} \\ \Omega_c \text{ as specified}}}{H(s)} = \frac{A_k |p_k|^2}{\left(\dfrac{s}{\Omega_c}\right)^2 - 2\,\mathrm{Re}(p_k)\left(\dfrac{s}{\Omega_c}\right) + |p_k|^2} \tag{4-22}$$

$$= \frac{A_k |p_k|^2 \Omega_c^2}{s^2 - 2\,\mathrm{Re}(p_k)\Omega_c s + |p_k|^2 \Omega_c^2}$$

It is convenient to express this as follows:

$$\underset{\substack{\text{kth 2nd order section} \\ \Omega_c \text{ as specified}}}{H(s)} = \frac{A_k |p_k|^2 \Omega_c^2}{s^2 - \left(\dfrac{2\,\mathrm{Re}(p_k)}{|p_k|}\right)|p_k|\Omega_c s + |p_k|^2 \Omega_c^2} \tag{4-23}$$

which in turn can be put into the following convenient form:

$$\underset{\substack{\text{kth 2nd order section} \\ \Omega_c \text{ as specified}}}{H(s)} = \frac{A_k \Omega_k^2}{s^2 + \Psi_k \Omega_k s + \Omega_k^2} \tag{4-24}$$

where

$$\Omega_k = |p_k|\Omega_c \tag{4-25}$$

$$\Psi_k = \frac{-2\,\mathrm{Re}(p_k)}{|p_k|} \tag{4-26}$$

NOTE: For a Butterworth filter, $\Omega_k = \Omega_c$ and $\Psi_k = -2\mathrm{Re}(p_k)$.

Let us summarize these results. The transfer function for an $N =$ even order Butterworth or Chebyshev lowpass filter with cutoff frequency f_c Hz is

$$H(s) = \frac{1}{\sqrt{1 + \varepsilon^2}} \prod_{k=1}^{N/2} \frac{A_k \Omega_k^2}{s^2 + \Psi_k \Omega_k s + \Omega_k^2} \tag{4-27}$$

The transfer function for an $N =$ odd order Butterworth or Chebyshev lowpass filter with cutoff frequency f_c Hz is

$$H(s) = \frac{A_0 \Omega_0}{s + \Omega_0} \prod_{k=1}^{(N-1)/2} \frac{A_k \Omega_k^2}{s^2 + \Psi_k \Omega_k s + \Omega_k^2} \tag{4-28}$$

The supporting formulas for both $N =$ even and $N =$ odd cases are

$$\Omega_c = 2\pi f_c \tag{4-29}$$

$$\Omega_k = |p_k|\Omega_c, k = 0, 1, 2, \ldots \tag{4-30}$$

$$\Psi_k = \frac{-2\operatorname{Re}(p_k)}{|p_k|}, k = 1, 2, \ldots \qquad (4\text{--}31)$$

The Butterworth and Chebyshev tables introduced in Section 4–3 list the pole magnitude $|p_k|$ for each pole. Also listed in the tables are the values of coefficients Ψ_k, under the heading "Ψ." In Equation (4–27), use $\varepsilon = 0$ for the Butterworth case.

For the N = even case, the filter passband gain is

$$\underset{N\,\text{even}}{G} = \frac{1}{\sqrt{1+\varepsilon^2}}\prod_{k=1}^{N/2} A_k \qquad (4\text{--}32)$$

(For the Butterworth case, set $\varepsilon = 0$ in this formula.) For the N = odd case, the filter passband gain is

$$\underset{N\,\text{odd}}{G} = A_0 \prod_{k=1}^{(N-1)/2} A_k \qquad (4\text{--}33)$$

Now let's look at three examples that illustrate the use of the tables and formulas.

EXAMPLE 4–4.1

Consider a fifth order Butterworth lowpass filter with $f_c = 1000$ Hz and passband gain $G = 5$. The transfer function is

$$H(s) = \left(\frac{A_0\Omega_0}{s+\Omega_0}\right)\left(\frac{A_1\Omega_1^2}{s^2+\Psi_1\Omega_1 s+\Omega_1^2}\right)\left(\frac{A_2\Omega_2^2}{s^2+\Psi_2\Omega_2 s+\Omega_2^2}\right) \qquad (4\text{--}34)$$

From Table 4–1, the following information is obtained:

$$|p_0| = 1$$
$$|p_1| = 1;\ \Psi_1 = 1.618034 \qquad (4\text{--}35)$$
$$|p_2| = 1;\ \Psi_2 = 0.618034$$

Equations (4–29) and (4–30) are used to obtain values for Ω_0, Ω_1, and Ω_2:

$$\Omega_c = 2\pi f_c = 2\pi \times 1000$$
$$\Omega_0 = \Omega_1 = \Omega_2 = \Omega_c \qquad (4\text{--}36)$$

The gain constants A_0, A_1, and A_2 must be selected such that $A_1 A_2 A_3 = 5$.

EXAMPLE 4–4.2

Consider a third order Chebyshev lowpass filter with $r_{dB} = 1.0$ dB, $f_c = 2000$ Hz, and passband gain $G = 10$. The transfer function is

$$H(s) = \left(\frac{A_0\Omega_0}{s+\Omega_0}\right)\left(\frac{A_1\Omega_1^2}{s^2+\Psi_1\Omega_1 s+\Omega_1^2}\right) \qquad (4\text{--}37)$$

The following information can be obtained from Table 4–2(c):

$$|p_0| = 0.4941706$$
$$|p_1| = 0.997098; \ \Psi_1 = 0.4956088 \tag{4–38}$$

Equations (4–29) and (4–30) are now used to find values for Ω_0 and Ω_1:

$$\Omega_c = 2\pi f_c = 2\pi \times 2000$$
$$\Omega_0 = |p_0|\Omega_c = (0.4941706)(2\pi)(2000) \tag{4–39}$$
$$\Omega_1 = |p_1|\Omega_c = (0.997098)(2\pi)(2000)$$

The gain constants A_0 and A_1 must be selected such that $A_0 A_1 = 10$.

Table 4–2(c) also shows that for this filter, $(f_{3dB}/f_c) = 1.095$. Therefore, the 3 dB frequency for this filter is $f_{3dB} = 2190$ Hz.

EXAMPLE 4–4.3

Consider a sixth order Chebyshev lowpass filter with $r_{dB} = 0.5dB$, $f_c = 500$ Hz, and passband gain $G = 10$. The transfer function is

$$H(s) = \left(\frac{1}{\sqrt{1+\varepsilon^2}} \right) \left(\frac{A_1 \Omega_1^2}{s^2 + \Psi_1 \Omega_1 s + \Omega_1^2} \right) \left(\frac{A_2 \Omega_2^2}{s^2 + \Psi_2 \Omega_2 s + \Omega_2^2} \right) \left(\frac{A_3 \Omega_3^2}{s^2 + \Psi_3 \Omega_3 s + \Omega_3^2} \right) \tag{4–40}$$

From Table 4–2(b), the following information can be obtained:

$$\varepsilon = 0.3493114$$
$$|p_1| = 0.396229; \quad \Psi_1 = 1.46276$$
$$|p_2| = 0.7681211; \Psi_2 = 0.5523711 \tag{4–41}$$
$$|p_3| = 1.011446; \quad \Psi_3 = 0.1535427$$

Equations (4–29) and (4–30) are now used to obtain values for Ω_1, Ω_2, and Ω_3:

$$\Omega_c = 2\pi f_c = (2\pi)(500)$$
$$\Omega_1 = |p_1|\Omega_c = (0.396229)(2\pi)(500)$$
$$\Omega_2 = |p_2|\Omega_c = (0.7681211)(2\pi)(500) \tag{4–42}$$
$$\Omega_3 = |p_3|\Omega_c = (1.011446)(2\pi)(500)$$

The gain constants A_1, A_2, and A_3 must be selected such that $A_1 A_2 A_3 / \sqrt{1+\varepsilon^2} = G = 10$. Since $\varepsilon = 0.3493114$, we must have $A_1 A_2 A_3 = 10.5925$.

Table 4–2(b) also shows that for this filter, $(f_{3dB}/f_c) = 1.041$. Therefore, for this filter, $f_{3dB} = 520.5$ Hz.

4–5 ACTIVE LOWPASS FILTER CIRCUITS

The transfer function for Butterworth and Chebyshev lowpass filters is given in Equation (4–27) (for N = even) and Equation (4–28) (for N = odd). For both even and odd order filters, all of the second order sections have the following form:

$$H(s) = \frac{A_k \Omega_k^2}{s^2 + \Psi_k \Omega_k s + \Omega_k^2} \qquad (4\text{–}43)$$

If N is odd, there is one remaining first order section:

$$H(s) = \frac{A_0 \Omega_0}{s + \Omega_0} \qquad (4\text{–}44)$$

One way to design an active (i.e., opamp-based) circuit that realizes the filter transfer function (Equation [4–27] or [4–28]) is to design separate circuits for each second order section (and for the first order section if N is odd), and then *cascade* the sections. If this approach is taken, there are only two circuit configurations that need to be designed: one that realizes the second order transfer function of Equation (4–43), and one that realizes the first order transfer function of Equation (4–44). (Of course, since each section has different values for coefficients A_k, Ψ_k, and Ω_k, the resistor and capacitor values in each circuit will be different in general.)

Let's start with the first order section. Assuming the opamp is ideal, it can be shown using standard opamp circuit analysis techniques that the circuit of Figure 4–10 has the following transfer function:

$$H(s) = \frac{\left(\dfrac{R_f}{R_a} + 1\right)\left(\dfrac{1}{RC}\right)}{s + \left(\dfrac{1}{RC}\right)} \qquad (4\text{–}45)$$

FIGURE 4–10
Lowpass filter first order section.

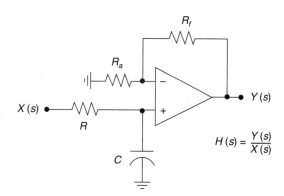

Equation (4–45) is the same as Equation (4–44) if we choose components R_a, R_f, R, and C to satisfy the following conditions:

$$\left(\frac{R_f}{R_a} + 1\right) = A_0 \tag{4-46}$$

$$\frac{1}{RC} = \Omega_0 \tag{4-47}$$

The second order section of Equation (4–43) can be realized using the equal component VCVS (voltage-controlled-voltage-source) circuit of Figure 4–11. It can be shown (again, assuming an ideal opamp) that the transfer function of this circuit is

$$H(s) = \frac{\left(\frac{R_f}{R_a} + 1\right)\left(\frac{1}{RC}\right)^2}{s^2 + \left[3 - \left(\frac{R_f}{R_a} + 1\right)\right]\left(\frac{1}{RC}\right)s + \left(\frac{1}{RC}\right)^2} \tag{4-48}$$

Equation (4–48) is the same as Equation (4–43) if we choose components R_a, R_f, R, and C to satisfy the following conditions:

$$\left(\frac{R_f}{R_a} + 1\right) = A_k \tag{4-49}$$

$$\frac{1}{RC} = \Omega_k \tag{4-50}$$

$$\left[3 - \left(\frac{R_f}{R_a} + 1\right)\right] = 3 - A_k = \Psi_k \tag{4-51}$$

FIGURE 4–11
Lowpass filter second order section.

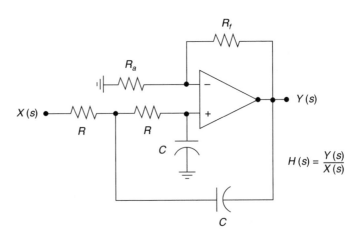

$$H(s) = \frac{Y(s)}{X(s)}$$

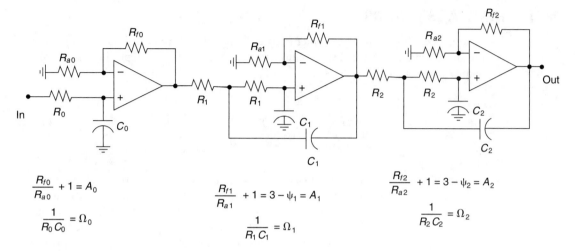

$$\frac{R_{f0}}{R_{a0}} + 1 = A_0$$

$$\frac{1}{R_0 C_0} = \Omega_0$$

$$\frac{R_{f1}}{R_{a1}} + 1 = 3 - \psi_1 = A_1$$

$$\frac{1}{R_1 C_1} = \Omega_1$$

$$\frac{R_{f2}}{R_{a2}} + 1 = 3 - \psi_2 = A_2$$

$$\frac{1}{R_2 C_2} = \Omega_2$$

FIGURE 4–12
Fifth order lowpass filter.

It is important to note that with the *equal component* version of the VCVS circuit being used here, we are *not* free to select gain factor A_k arbitrarily. Instead, A_k *must* be selected to satisfy Equation (4–51); we *must* use the correct value of Ψ_k.

There are other opamp circuits that can be used to realize a second order transfer function like Equation (4–43). Refer to the active filter design literature for more information on alternative circuits. It is not the intention of this author to claim that the equal component VCVS circuit is the best choice. However, it is a reasonably good choice, and the design equations are relatively straightforward.

The opamp circuits described here are appropriate for audio frequencies (say, 20 KHz or less) if the opamp is a type 741 or similar. The transfer function expressions of Equations (4–45) and (4–48) were derived based on an ideal opamp assumption; this assumption becomes less valid at higher frequencies.

As an example, consider the circuit shown in Figure 4–12, which is a fifth order lowpass filter. There is one first order section, followed by two second order sections. The component value design equations for each section are shown on the figure. As shown in Sections 4–3 and 4–4, the values for Ψ_k and Ω_k are determined by the filter type (Butterworth versus Chebyshev), cutoff frequency, etc. Since the second order sections are of the equal component VCVS type, the gain constant A_k for these sections is dictated by the value of coefficient Ψ_k. However, the gain constant A_0 for the first order section can be selected to obtain the desired overall passband gain. (For an even order filter realized using these equal component VCVS sections, the passband gain is in effect dictated by the Ψ_k values. However, if for some reason this is a problem, an additional amplifier stage can be added before or after the filter itself.)

4–6 HIGHPASS FILTERS

A normalized ($\Omega_c = 1$) *lowpass* filter transfer function can be converted to a *highpass* filter transfer function with the desired cutoff frequency $\Omega_c = 2\pi f_c$ by replacing the variable s with Ω_c/s. In other words;

$$\underset{\substack{\text{Highpass} \\ \Omega_C \text{ as specified}}}{H(s)} = \underset{\substack{\text{normalized} \\ \text{lowpass}}}{H(s)} \Bigg|_{s \to \frac{\Omega_C}{s}} = \underset{\substack{\text{normalized} \\ \text{lowpass}}}{H\left(\frac{\Omega_c}{s}\right)} \tag{4–52}$$

This is called the *lowpass-to-highpass transformation*. Filter order, ripple characteristics, and stopband roll off characteristics are preserved by this transformation. For example, a normalized fifth order Chebyshev lowpass filter with 1 dB of passband ripple becomes a fifth order Chebyshev highpass filter with 1 dB of passband ripple after the transformation.

The transfer function for both Butterworth and Chebyshev normalized lowpass filters is given in Equation (4–14) (for N = even) or Equation (4–15) (for N = odd). If the lowpass-to-highpass transformation is applied to these transfer functions, the following results are obtained. For an N = even order Butterworth or Chebyshev highpass filter with cutoff frequency f_c Hz, the transfer function is

$$H(s) = \frac{1}{\sqrt{1+\varepsilon^2}} \prod_{k=1}^{N/2} \frac{A_k s^2}{s^2 + \Psi_k \Omega_k s + \Omega_k^2} \tag{4–53}$$

The transfer function for an N = odd order Butterworth or Chebyshev highpass filter with cutoff frequency f_c Hz is

$$H(s) = \frac{A_0 s}{s + \Omega_0} \prod_{k=1}^{(N-1)/2} \frac{A_k s^2}{s^2 + \Psi_k \Omega_k s + \Omega_k^2} \tag{4–54}$$

In both cases, the A_ks are gain constants. (The passband gain is given by Equations [4–32] and [4–33]). The supporting formulas for both N = even and N = odd cases are

$$\Omega_c = 2\pi f_c \tag{4–55}$$

$$\Omega_k = \frac{\Omega_c}{|p_k|}, \ k = 0, 1, 2, \ldots \tag{4–56}$$

$$\Psi_k = \frac{-2 \, \text{Re}(p_k)}{|p_k|}, \ k = 1, 2, \ldots \tag{4–57}$$

The p_ks are the poles listed in the tables for normalized Butterworth and Chebyshev lowpass filters (Tables 4–1 and 4–2). Values for $|p_k|$ and Ψ_k (*psi*) are also listed in these tables. (In Equation [4–53], use $\varepsilon = 1$ for the Butterworth case.)

NOTE: While the formulas for Ψ_k are the same for both lowpass and highpass filters, the formulas for Ω_k are different.

It can be shown that the amplitude response function of an Nth order Butterworth highpass filter with passband gain G and cutoff frequency f_c Hz is

$$|H_a(f)| = \frac{G}{\sqrt{1 + \left(\dfrac{f_c}{f}\right)^{2N}}}$$

(4–58)

It can also be shown that the amplitude response function of an Nth order Chebyshev highpass filter with passband gain G, cutoff frequency f_c Hz, and ripple factor ε is:

$$|H_a(f)| = \frac{G}{\sqrt{1 + \varepsilon^2 C_N^2\left(\dfrac{f_c}{f}\right)}}$$

(4–59)

When Equations (4–58) and (4–59) are compared with their lowpass versions (Equations [4–1] and [4–6], respectively), one can see that the only difference is that the factor (f/f_c) in the lowpass version becomes (f_c/f) in the highpass version.

Figure 4–13 shows the amplitude response functions of fifth order Butterworth lowpass and highpass filters plotted together. Both filters have the same cutoff frequency ($f_c = 1000$ Hz). Figure 4–14 shows the dB amplitude response plots for the same filters (plotted on a logarithmic frequency scale). Observe that both filters have the same rolloff rate in their respective stopbands.

FIGURE 4–13
Amplitude response functions for fifth order Butterworth lowpass and highpass filters.

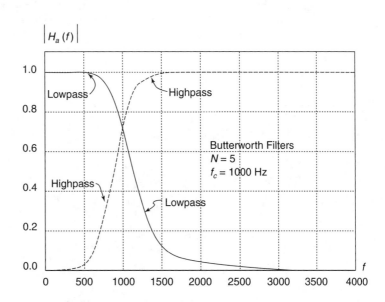

FIGURE 4–14
Decibel amplitude response functions for fifth order Butterworth lowpass and highpass filters.

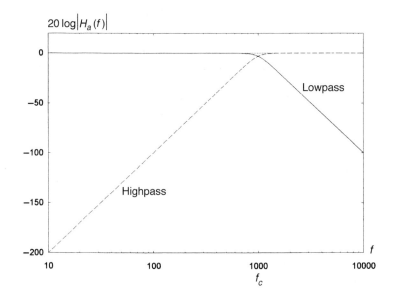

Figure 4–15 shows the amplitude response functions of fifth order Chebyshev lowpass and highpass filters plotted together. Both filters have the same cutoff frequency (f_c = 1000 Hz) and the same passband ripple (1.0 dB). Figure 4–16 shows the dB amplitude response plots for the same filters (plotted on a logarithmic frequency scale). Observe that both filters have the same rolloff rate in their respective stopbands. Figure 4–17 shows the amplitude response function for the Chebyshev highpass filter plotted on a different frequency scale, so that the reader can clearly see the behavior for $f \gg f_c$; it is still true that the number of peaks plus the number of valleys in the passband is equal to the filter order (N).

FIGURE 4–15
Amplitude response functions for fifth order Chebyshev lowpass and highpass filters.

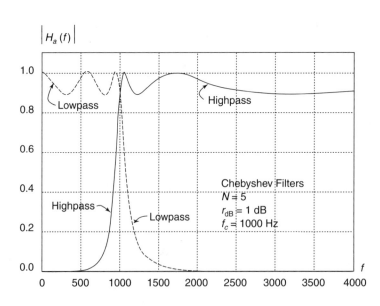

FIGURE 4–16
Decibel amplitude response functions for fifth order Chebyshev lowpass and highpass filters.

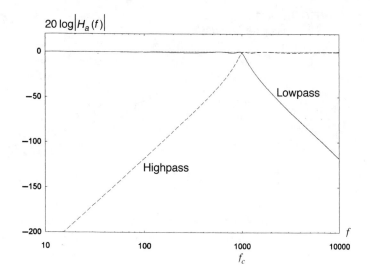

FIGURE 4–17
Amplitude response function of a fifth order Chebyshev highpass filter plotted on a wider frequency scale, showing the behavior of the passband ripple at frequencies much higher than the cutoff frequency.

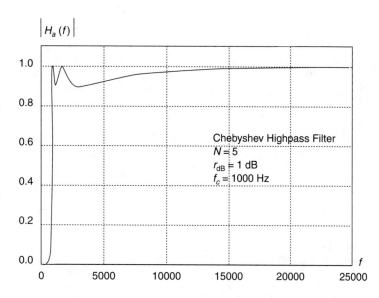

Active (opamp-based) highpass filter circuits can be realized by designing separate circuits for each second order section of the transfer function and then cascading the sections. There will also be one first order section to design if N is odd. The opamp circuit in Figure 4–18 can be used to realize a first order section. The transfer function for this circuit (assuming an ideal opamp) is

$$H(s) = \frac{\left(\dfrac{R_f}{R_a} + 1\right)s}{s + \dfrac{1}{RC}} \qquad (4\text{–}60)$$

FIGURE 4–18
Highpass filter first order
section.

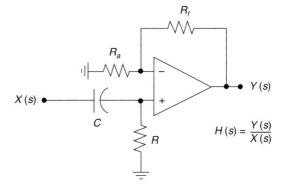

Equation (4–60) has the same form as the first order section in Equation (4–54), and is identical to it provided we choose R_a, R_a, R, and C to satisfy the following conditions:

$$\frac{R_f}{R_a} + 1 = A_0 \qquad \text{(4–61)}$$

$$\frac{1}{RC} = \Omega_0 \qquad \text{(4–62)}$$

The equal component VCVS circuit in Figure 4–19 has the following transfer function (again, assuming an ideal opamp):

$$H(s) = \frac{\left(\dfrac{R_f}{R_a} + 1\right)s^2}{s^2 + \left[3 - \left(\dfrac{R_f}{R_a} + 1\right)\right]\left(\dfrac{1}{RC}\right)s + \left(\dfrac{1}{RC}\right)^2} \qquad \text{(4–63)}$$

FIGURE 4–19
Highpass filter second order
section.

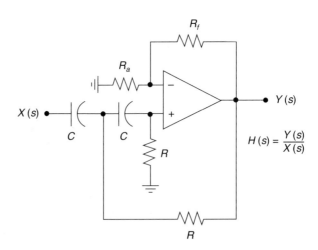

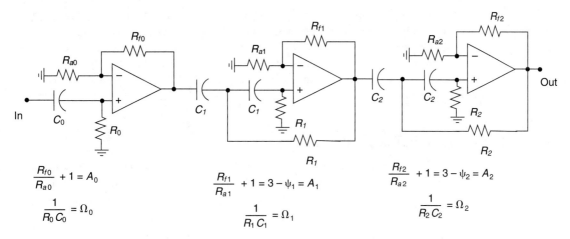

$$\frac{R_{f0}}{R_{a0}} + 1 = A_0$$

$$\frac{1}{R_0 C_0} = \Omega_0$$

$$\frac{R_{f1}}{R_{a1}} + 1 = 3 - \psi_1 = A_1$$

$$\frac{1}{R_1 C_1} = \Omega_1$$

$$\frac{R_{f2}}{R_{a2}} + 1 = 3 - \psi_2 = A_2$$

$$\frac{1}{R_2 C_2} = \Omega_2$$

FIGURE 4–20
Fifth order highpass filter.

Equation (4–63) has the same form as the second order sections of Equations (4–53) and (4–54) and can be made identical to any particular second order section by choosing the resistors and capacitors to satisfy the following conditions:

$$\frac{R_f}{R_a} + 1 = A_k \qquad (4\text{--}64)$$

$$\frac{1}{RC} = \Omega_k \qquad (4\text{--}65)$$

$$\left[3 - \left(\frac{R_f}{R_a} + 1 \right) \right] = 3 - A_k = \Psi_k \qquad (4\text{--}66)$$

Note that with the *equal component* VCVS circuit being used here, we are not free to select A_k arbitrarily. Instead, A_k *must* be selected to satisfy Equation (4–66), so that the correct value of Ψ_k is obtained. (As mentioned in Section 4–5, the author does not claim that the equal component VCVS circuit is the best choice. Nevertheless, it is often a reasonably good choice.)

The circuit shown in Figure 4–20 is presented as an example; it is a fifth order highpass filter. The component values are selected based on Equations (4–55), (4–56), (4–57), (4–61), (4–62), (4–64), (4–65), and (4–66).

The reader is again reminded that the transfer functions for the opamp circuits shown in this chapter were derived based on an ideal opamp assumption, which is reasonably accurate in the audio frequency range. When considering highpass filters, one must remember that the opamps have a finite *gain bandwidth product* that will ultimately impact the high-frequency response of the circuit. For example, suppose the highpass filter circuit of Figure 4–20 was designed for a cutoff frequency (f_c) of 1000 Hz. This means that this circuit will *pass* signals at frequencies greater than 1000 Hz, up to some upper limit, which might be on the order of 100 kHz to 1 MHz or more. (The actual location of this upper frequency limit depends on the nominal passband gain of the filter and the gain bandwidth product of the particular opamp being used in the circuit.) For frequencies above this upper limit, the filter gain will roll off. Thus the circuit actually behaves as a highpass filter with respect to signals bandlimited to less than this ultimate upper limit frequency.

4–7   REQUIRED FILTER ORDER

Formulas for determining the required order (N) for Butterworth and Chebyshev lowpass and highpass filters are presented here without derivation. These formulas are based on information in *Fundamentals of Digital Signal Processing* by Ludeman (see Bibliography).

Butterworth Lowpass Filter

Figure 4–21 shows the amplitude response function (in dB) for a Butterworth lowpass filter with a cutoff frequency of f_c Hz. Also indicated is a higher frequency, denoted f_u, at which the desired attenuation is A dB. By specifying parameters f_u and A, a sharpness requirement is being imposed. To determine the required order, use the following equations:

$$a = 10^{\left(\frac{-A}{20}\right)} \tag{4–67}$$

$$x = \frac{f_u}{f_c} \tag{4–68}$$

$$N = \frac{\frac{1}{2}\log\left[\left(\frac{1}{a}\right)^2 - 1\right]}{\log[x]} \tag{4–69}$$

Of course, Equation (4–69) will not result in an integer value for N; therefore, the result must be rounded to the nearest integer.

FIGURE 4–21
Determining the required order for a Butterworth lowpass filter. (See text for equations.)

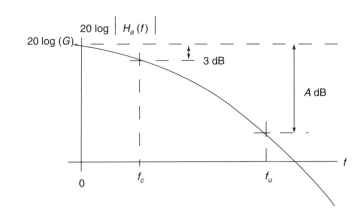

The following examples illustrate the use of these formulas.

EXAMPLE 4–7.1
The specifications for a Butterworth lowpass filter are

$$f_c = 3000 \text{ Hz}$$
$$f_u = 30000 \text{ Hz} \tag{4–70}$$
$$A = 60 \text{ dB}$$

Therefore,

$$a = 10^{\left(\frac{-60}{20}\right)} = .001$$
$$x = \frac{30000}{3000} = 10 \tag{4–71}$$
$$N = \frac{\frac{1}{2}\log\left[\left(\frac{1}{.001}\right)^2 - 1\right]}{\log[10]}$$
$$= 2.9999 \cong 3$$

The required filter order is $N = 3$. This should be no surprise, since the specifications amount to a 60 dB per decade rolloff in the stopband.

EXAMPLE 4–7.2
The specifications for a Butterworth lowpass filter are

$$f_c = 3000 \text{ Hz}$$
$$f_u = 3500 \text{ Hz} \tag{4–72}$$
$$A = 20 \text{ dB}$$

Therefore,

$$a = 10^{\left(\frac{-20}{20}\right)} = 0.1$$
$$x = \frac{3500}{3000} = 1.166667 \tag{4–73}$$
$$N = \frac{\frac{1}{2}\log\left[\left(\frac{1}{.1}\right)^2 - 1\right]}{\log[1.166667]}$$
$$= 14.9 \cong 15$$

The required filter order is $N = 15$. (Building such a filter is nontrivial, to say the least. The answer "$N = 15$" should serve as a motive to investigate other possibilities, such as a Chebyshev lowpass filter.)

Butterworth Highpass Filter

Figure 4–22 shows the amplitude response function (in dB) for a Butterworth highpass filter with a cutoff frequency of f_c Hz. Also indicated is a lower frequency, denoted f_r, at which the desired attenuation is A dB. To determine the required filter order, use the following equations:

$$a = 10^{\left(\frac{-A}{20}\right)} \qquad \text{(4–74)}$$

$$y = \frac{f_c}{f_r} \qquad \text{(4–75)}$$

$$N = \frac{\frac{1}{2}\log\left[\left(\frac{1}{a}\right)^2 - 1\right]}{\log[y]} \qquad \text{(4–76)}$$

FIGURE 4–22

Determining the required order for a Butterworth highpass filter. (See text for equations.)

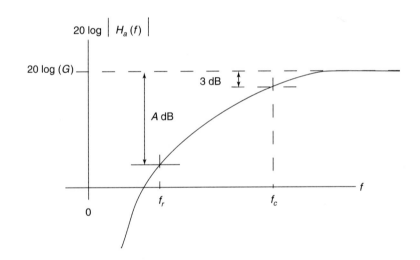

(The result must be rounded to the nearest integer.)

EXAMPLE 4–7.3

The specifications for a Butterworth highpass filter are

$$f_c = 3000 \text{ Hz}$$
$$f_r = 300 \text{ Hz} \qquad \text{(4–77)}$$
$$A = 80 \text{ dB}$$

Therefore,

$$a = 10^{\left(\frac{-80}{20}\right)} = .0001 \qquad \text{(4–78)}$$

$$y = \frac{3000}{300} = 10$$

$$N = \frac{\frac{1}{2}\log\left[\left(\frac{1}{.0001}\right)^2 - 1\right]}{\log[10]}$$

$$= 3.9999 \cong 4$$

The required order is $N = 4$.

Chebyshev Lowpass Filter

Figure 4–23 shows the amplitude response function (in dB) for a Chebyshev lowpass filter with a cutoff frequency of f_c Hz. The passband ripple, in dB, is denoted r_{dB}. Also indicated is a higher frequency, denoted f_u, at which the desired attenuation is A dB. By specifying parameters f_u and A, a sharpness requirement is being imposed. To determine the required filter order, use the following equations:

$$a = 10^{\left(\frac{-A}{20}\right)} \tag{4–79}$$

$$\varepsilon^2 = 10^{\frac{r_{dB}}{10}} - 1 \tag{4–80}$$

$$g = \sqrt{\frac{\left(\frac{1}{a}\right)^2 - 1}{\varepsilon^2}} \tag{4–81}$$

$$x = \frac{f_u}{f_c} \tag{4–82}$$

FIGURE 4–23
Determining the required order for a Chebyshev lowpass filter. (See text for equations.)

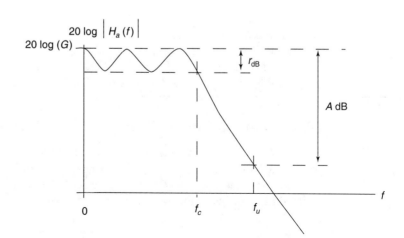

$$N = \frac{\log\left[g + \sqrt{g^2 - 1}\right]}{\log\left[x + \sqrt{x^2 - 1}\right]} \tag{4–83}$$

(The result must be rounded to the nearest integer.)

EXAMPLE 4–7.4

The specifications for a Chebyshev lowpass filter are

$$\begin{aligned} f_c &= 3000 \text{ Hz} \\ f_u &= 3500 \text{ Hz} \\ A &= 20 \text{ dB} \\ r_{\text{dB}} &= 1 \text{ dB} \end{aligned} \tag{4–84}$$

Therefore,

$$a = 10^{\left(\frac{-20}{20}\right)} = 0.1$$

$$\varepsilon^2 = 10^{\left(\frac{1}{10}\right)} - 1 = 0.258925$$

$$g = \sqrt{\frac{\left(\frac{1}{.1}\right)^2 - 1}{0.258925}} = 19.55377 \tag{4–85}$$

$$x = \frac{3500}{3000} = 1.166667$$

$$N = \frac{\log\left[g + \sqrt{g^2 - 1}\right]}{\log\left[x + \sqrt{x^2 - 1}\right]} = 6.43$$

If we choose $N = 7$, the specifications with respect to sharpness will be exceeded; on the other hand, if we choose $N = 6$, the specifications will not quite be satisfied. The reader should compare this example to Example 4–7.2, in which we determined that a Butterworth lowpass filter with analogous sharpness specifications would have to be fifteenth order.

Chebyshev Highpass Filter

Figure 4–24 shows the amplitude response function (in dB) for a Chebyshev highpass filter with a cutoff frequency of f_c Hz. The passband ripple, in dB, is denoted r_{dB}. Also indicated is a lower frequency, denoted f_r, at which the desired attenuation is A dB. To determine the required filter order, use the following equations:

$$a = 10^{\left(\frac{-A}{20}\right)} \tag{4–86}$$

FIGURE 4–24
Determining the required
order for a Chebyshev
highpass filter. (See text for
equations.

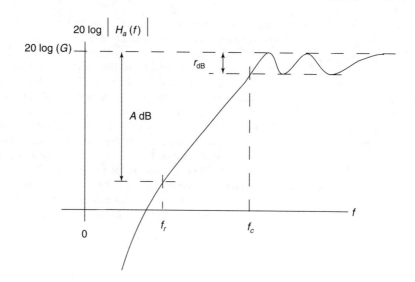

$$\varepsilon^2 = 10^{\left(\frac{r_{dB}}{10}\right)} - 1 \qquad (4\text{--}87)$$

$$g = \sqrt{\frac{\left(\frac{1}{a}\right)^2 - 1}{\varepsilon^2}} \qquad (4\text{--}88)$$

$$y = \frac{f_c}{f_r} \qquad (4\text{--}89)$$

$$N = \frac{\log\left[g + \sqrt{g^2 - 1}\right]}{\log\left[y + \sqrt{y^2 - 1}\right]} \qquad (4\text{--}90)$$

(The result must be rounded to the nearest integer.)

PROBLEMS

4–1. Find the fifth order Chebyshev polynomial. Then find the amplitude response function (i.e., find $|H_a(f)|$) for a fifth order Chebyshev lowpass filter.

4–2. Find the transfer function $H(s)$ for a seventh order Butterworth lowpass filter with a cutoff frequency of $f_c = 3000$ Hz and a passband gain of $G = 10$.

4–3. Find the transfer function $H(s)$ for a fifth order Chebyshev lowpass filter with 0.5 dB of passband ripple, a cutoff frequency of $f_c = 3000$ Hz, and a passband gain of $G = 10$. Also find the 3 dB frequency of this filter.

4–4. Find the transfer function $H(s)$ for an eighth order Chebyshev lowpass filter with 1.5 dB of passband ripple, a cutoff frequency of $f_c = 4000$ Hz, and a passband gain of $G = 20$. Also find the 3 dB frequency of this filter.

4–5. On a piece of graph paper, carefully plot the location of the poles on the s plane for the following filters (all normalized for $\Omega_c = 1$):
  a. fifth order Butterworth lowpass filter
  b. fifth order Chebyshev lowpass filter, with 0.1 dB passband ripple
  c. fifth order Chebyshev lowpass filter, with 1.5 dB passband ripple
  You should observe that the poles of the normalized Butterworth lowpass filter lie on a circle on the s plane with a radius of one, centered at the origin. On the other hand, the poles of a Chebyshev lowpass filter lie on an ellipse, the shape of which depends on the amount of passband ripple. (Of course, these poles are in the left half of the s plane.)

4–6. Find the transfer function $H(s)$ of a seventh order Butterworth highpass filter having a cutoff frequency of 1000 Hz, and a passband gain of $G = 25$.

4–7. Find the transfer function $H(s)$ of a fifth order Chebyshev highpass filter having a cutoff frequency of 1000 Hz, 1 dB of passband ripple, and a passband gain of $G = 25$.

4–8. Suppose a lowpass filter must be designed to satisfy the following specifications:

$$f_c = 4000 \text{ Hz}$$
$$f_u = 4500 \text{ Hz}$$
$$A = 25 \text{ dB}$$

(see Figures 4–21 and 4–23.) Find the required filter order (N) for:
  a. Butterworth lowpass filter
  b. Chebyshev lowpass filter, 0.1 dB passband ripple
  c. Chebyshev lowpass filter, 1.5 dB passband ripple

4–9. Suppose a lowpass filter must be designed having a cutoff frequency of 4000 Hz, with 15 dB of attenuation at 4500 Hz. Find the required filter order (N) for:
  a. Butterworth lowpass filter
  b. Chebyshev lowpass filter, 0.1 dB passband ripple
  c. Chebyshev lowpass filter, 1.5 dB passband ripple

4–10. Suppose a highpass filter must be designed having a cutoff frequency of 1000 Hz, with 20 dB of attenuation at 800 Hz. Find the required filter order (N) for:
  a. Butterworth highpass filter
  b. Chebyshev highpass filter, 0.1 dB ripple
  c. Chebyshev highpass filter, 1.5 dB ripple

5

LINEAR, SHIFT-INVARIANT DISCRETE TIME SYSTEMS

5–1 TIME DOMAIN DESCRIPTION

We begin by considering discrete time signals, or *sequences*. Let $x(n)$ be a discrete time signal. The time index (n) is an integer, which in general can take on both negative and nonnegative values. The sequence $x(n)$ is obtained by sampling a continuous time signal $x_a(t)$. The relationship between $x(n)$ and $x_a(t)$ is

$$x(n) = x_a(nT) \tag{5–1}$$

where T is the sampling period. Figure 5–1 shows how $x_a(t)$ is sampled every T seconds. Figure 5–2 shows how the resulting sequence $x(n)$ is often represented graphically. Actually, $x(n)$ is a sequence of numbers stored in computer memory.

NOTE: The lines with the dots on top in Figure 5–2 are *not* the impulse functions (Dirac delta functions) described in Chapter 1.

The sampling frequency, denoted f_s, is related to the sampling period as follows:

$$f_s = \frac{1}{T} \tag{5–2}$$

FIGURE 5–1
The sampling process.
Continuous time signal $x_a(t)$
is sampled every T seconds.
The sampling frequency is
$f_s = 1/T$.

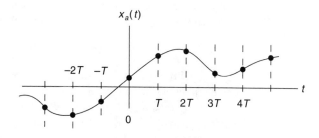

FIGURE 5–2

Discrete time signal
(sequence) $x(n)$ resulting from
sampling $x_a(t)$ in Figure 5–1.
$x(n) = x_a(nT)$.

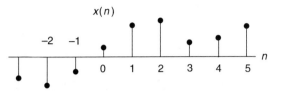

The sampling frequency is an extremely important parameter in discrete time signal processing, or *digital signal processing.*

A specific example may be helpful. Suppose the continuous time signal is given by

$$x_a(t) = \begin{cases} \cos(2\pi f_o t), t \geq 0 \\ 0, \text{ otherwise} \end{cases} \tag{5-3}$$

Suppose $f_o = 10$ Hz. Suppose further that this signal is sampled at $f_s = 100$ Hz (thus the sampling period is $T = .01$ seconds). The resulting sequence is obtained by replacing t with nT and evaluating the expression for $n = 0, 1, \ldots$:

$$x(n) = \begin{cases} \cos(2\pi f_o nT), n = 0, 1, \ldots \\ 0, \qquad\qquad n < 0 \end{cases} \tag{5-4}$$

That is,

$$\begin{aligned} x(0) &= \cos(0) = 1 \\ x(1) &= \cos(2\pi \times 10 \times 1 \times .01) = .809 \\ x(2) &= \cos(2\pi \times 10 \times 2 \times .01) = .309 \\ x(3) &= \cos(2\pi \times 10 \times 3 \times .01) = -.309 \end{aligned} \tag{5-5}$$

etc.

In practice, the sampling operation is accomplished by an analog-to-digital converter (A/D converter). Equation (5–1) and Figure 5–1 both imply that the sequence values that appear at the A/D converter output are precisely equal to $x_a(nT)$. In reality, an A/D converter has a finite number of bits available to represent each sample value; therefore, the *infinite precision* implied by Equation (5–1) is not literally possible. For example, the output from a 10-bit A/D converter will have to be one of $2^{10} = 1024$ possible discrete values. An A/D converter must therefore *quantize* each sample value. Quantize means, in effect, that the A/D converter assigns to $x(n)$ the possible discrete value nearest to $x_a(nT)$. This means that there will virtually always be a *quantization error*, which is the difference between the assigned value and the actual value. The operation can be represented as follows:

$$x(n) = Q[x_a(nT)] = x_a(nT) + e(n) \tag{5-6}$$

where $Q[\ ]$ represents the quantization process, and $e(n)$ is the quantization error, which can be thought of as added *noise.*

A *digital signal* is usually defined as a discrete time signal that has been quantized. Strictly speaking, if we completely ignore the effects of quantization and other finite precision effects (or finite word-length effects) introduced by the processor in our theoretical discussion of the processing of sequence $x(n)$, we are discussing *discrete time signal processing*. In the real world these finite precision effects cannot be completely ignored, and if they are at least acknowledged we are then discussing *digital signal processing*.

In much of this book there is a tacit assumption that in many cases these finite precision effects do not play a major role and can be safely pushed into the background for practical purposes or for the purpose of making the analysis mathematically tractable. For example, it will often be assumed that Equation (5–1) describes the sampling process for practical purposes. Philosophically this is similar to arguing that small nonlinear effects can in many cases be safely ignored in circuit analysis, or that noise generated by transistors need not be considered every time an amplifier circuit is analyzed.

Quantization noise is considered in more depth in Chapter 12. Certain other finite precision effects are covered in Sections 9–7 and 9–8 of Chapter 9. However, a complete and rigorous discussion of these effects is beyond the intended scope of this book.

A/D converters can of course create problems more profound than the mere addition of noise. If an A/D converter is overloaded by the input signal, the effect can be extremely nonlinear. For example, suppose an A/D converter can handle input signals varying between plus and minus 2.5 volts. If the actual signal has excursions beyond these bounds, then $x(n)$ will not be a sampled version of $x_a(t)$. If $x_a(t)$ is a sinewave with a peak value of 5 volts, the resulting $x(n)$ will not represent a sampled sinusoid; instead, it will represent a sampled, clipped sinusoid. If we ignore quantization errors and assume that Equation (5–1) is literally correct, then of course we are also assuming that the more pathological problems (like the overload problem described above) are not present either.

Before turning attention to discrete time signal processing, we need to examine the concept of *time shifting* of a sequence. Let k be a positive integer. The sequence $x(n - k)$, as a function of time index n, is the sequence $x(n)$ shifted to the right (i.e., delayed) by k time units. For example, if $x(n)$ appears as in Figure 5–3(a), then $x(n - 3)$ appears as in Figure 5–3(b). The sequence $x(n + k)$ is $x(n)$ shifted to the left by k time units; Figure 5–3(c) shows the sequence $x(n + 3)$.

A discrete time system (a digital filter is essentially a discrete time system) is really nothing more than a mathematical algorithm that operates on an input sequence $x(n)$ to create an output sequence $y(n)$. For example,

$$y(n) = 5x(n) + 2x(n - 1) + \tfrac{1}{2} y(n - 1) \tag{5–7}$$

In this example, the current value of $y(n)$ is linear combination of the current input $x(n)$, previous input $x(n - 1)$, and previous output $y(n - 1)$. The previous input and output values must be stored in memory. The output sequence $y(n)$ can be generated recursively with Equation (5–7), assuming that the input $x(n)$ is initially applied at time $n = 0$, and that the algorithm memories, that is, $x(n - 1)$ and $y(n - 1)$, are initially set to zero. (In other words, the discrete time system is initially at rest.) The recursion is as follows:

$$y(0) = 5x(0)$$
$$y(1) = 5x(1) + 2x(0) + \tfrac{1}{2} y(0)$$
$$y(2) = 5x(2) + 2x(1) + \tfrac{1}{2} y(1) \tag{5–8}$$
$$y(3) = 5x(3) + 2x(2) + \tfrac{1}{2} y(2)$$

etc.

FIGURE 5–3
Shifting in the time domain.

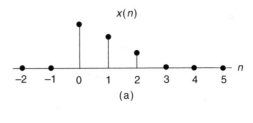

$x(n)$

(a)

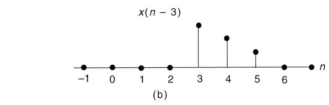

$x(n - 3)$

(b)

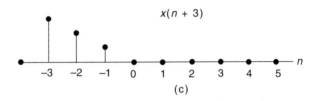

$x(n + 3)$

(c)

An algorithm such as this one can be programmed to run on a computer. The program can access the file containing the sequence $x(n)$ and store the resulting sequence $y(n)$ in another file. If the system is operating in real time, then instead of accessing a file to obtain the input, the program will access an A/D converter. If the output sequence $y(n)$ is to be converted back to a continuous time signal $y_a(t)$, then $y(n)$ is sent to a digital-to-analog converter (D/A converter). The block diagram of a real-time discrete time system is shown in Figure 5–4. In this book we are mainly concerned with the purely discrete time part of the system: $x(n)$ is the input; $y(n)$ is the output. It should also be noted that the system in Figure 5–4 suggests *ideal* A/D and D/A conversion processes; in the real world, neither process is ideal. Some of the nonideal characteristics of A/D conversion were discussed previously in this chapter. The nonideal characteristics of the D/A conversion process are considered in detail in Chapter 7.

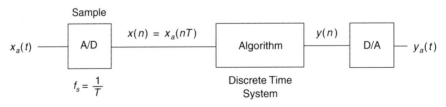

FIGURE 5–4
Real-time digital signal processing.

Let us now define a *linear, shift invariant* (LSI) discrete time system. Let $x(n)$ be the input to a discrete time system. There is some algorithm, denoted $T[\ ]$, that creates an output signal $y(n)$:

$$y(n) = T[x(n)] \tag{5-9}$$

$T[\ ]$ represents a *linear* discrete time system if superposition holds; that is, if

$$T[a_1x_1(n) + a_2x_2(n)] = a_1T[x_1(n)] + a_2T[x_2(n)] \tag{5-10}$$

A discrete time system is *shift invariant* if $T[\ ]$ does not depend on time. That is, for a shift invariant system, if $T[x(n)] = y(n)$, then $T[x(n-k)] = y(n-k)$. A shift invariant system is also called a *time invariant* system.

The LSI systems we are generally interested in can be represented by a linear difference equation with constant coefficients, as shown in Equation (5-11), in which $x(n)$ and $y(n)$ are the input and output signals, respectively:

$$y(n) = \sum_{k=0}^{M} b_k x(n-k) - \sum_{k=1}^{N} a_k y(n-k) \tag{5-11}$$

Equation (5-11) says that the current output $y(n)$ is formed as a weighted sum (or linear combination) of present and past input values and past output values. Equation (5-11) represents an *infinite impulse response* (IIR) system. An IIR system is characterized by the inclusion of feedback terms, those involving past values $y(n-k)$. (Note that the example presented earlier (Equation [5-7]) is an example of an IIR system.) If the feedback terms are not included (or in other words, if all the coefficients a_k are set to zero), we have a *finite impulse response* (FIR) system, in which the output is formed as a weighted sum of present and past inputs only:

$$y(n) = \sum_{k=0}^{M} b_k x(n-k) \tag{5-12}$$

A block diagram of an IIR system with $M=2$ and $N=2$ is shown in Figure 5-5. (Note that z^{-1} represents a *unit delay operator*. The reason for this notation will become clear when Z transforms are considered in Section 5-3.) The difference equation for this system is:

$$y(n) = b_0x(n) + b_1x(n-1) + b_2x(n-2) - a_1y(n-1) - a_2y(n-2) \tag{5-13}$$

A string of unit delay (z^{-1}) operators, as appears on the left and right of Figure 5-5, is basically a shift register, or memory, that contains previous values.

It must be emphasized that a system such as that depicted in Figure 5-5 and described by Equation (5-13) is actually realized by a computer program. Let us write a pseudocode program to realize this system. We assume that the input $x(n)$ is available from an existing file or directly from an A/D converter, and that the input begins at $n=0$, that is, $x(n)=0$ for $n<0$. *In this book it is assumed that any system under consideration is initially at rest, meaning that the shift register values are initially zero.* Specifically, for the example being

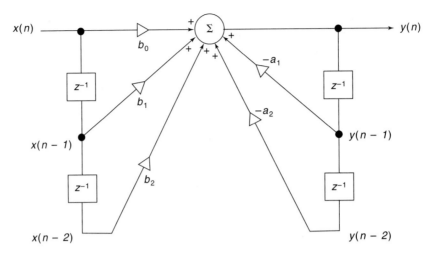

FIGURE 5–5

Infinite impulse response (IIR) discrete time system for the specific case of $N = M = 2$ in Equation (5–11).

considered here, initial values $x(-1)$, $x(-2)$, $y(-1)$, and $y(-2)$ are all set to zero. The pseudocode program would be as follows:

```
DEFINITION OF VARIABLES:
    X = current value of input x(n)
    Y = current value of output y(n)
    XNM1  contains x(n-1)
    XNM2  contains x(n-2)
    YNM1  contains y(n-1)
    YNM2  contains y(n-2)
INITIALIZE SHIFT REGISTER CONTENTS (ALL ZERO):
    XNM2 = 0       ! x(-2)
    XNM1 = 0       ! x(-1)
    YNM2 = 0       ! y(-2)
    YNM1 = 0       ! y(-1)
DEFINE SYSTEM COEFFICIENTS:
    B0 =
    B1 =
    B2 =
    A1 =
    A2 =
BEGIN ITERATIONS:  N = 0, 1, 2, 3, . . .
N = 0
READ X     (i.e., get x(0) )
Y = B0*X + B1*XNM1 + B2*XNM2 - A1*YNM1 - A2*YNM2
WRITE OUT Y    (this is y(0) )
UPDATE SHIFT REGISTERS:
    XNM2 = XNM1
    XNM1 = X
    YNM2 = YNM1
    YNM1 = Y
```

```
N = 1
READ X    (i.e., get x(1) )
Y = B0*X + B1*XNM1 + B2*XNM2 - A1*YNM1 - A2*YNM2
WRITE OUT Y      (this is y(1) )
UPDATE SHIFT REGISTERS:
    XNM2 = XNM1
    XNM1 = X
    YNM2 = YNM1
    YNM1 = Y
N = 2
READ X    (i.e., get x(2) )
Y = B0*X + B1*XNM1 + B2*XNM2 - A1*YNM1 - A2*YNM2
WRITE OUT Y      (this is y(2) )
UPDATE SHIFT REGISTERS:
    XNM2 = XNM1
    XNM1 = X
    YNM2 = YNM1
    YNM1 = Y
N = 3
    .
    .
    .
etc.
```

A simple example involving actual numbers may be helpful. Suppose the system difference equation is

$$y(n) = x(n) + \tfrac{1}{2} y(n - 1) \tag{5–14}$$

where input signal $x(n)$ is given by:

$$x(n) = \begin{cases} 1, n \ge 0 \\ 0, n < 0 \end{cases} \tag{5–15}$$

The output $y(n)$ can be generated recursively as follows (assume the initial conditions are zero):

$$
\begin{aligned}
y(0) &= x(0) + \tfrac{1}{2} y(-1) = 1 + \tfrac{1}{2}(0) = 1 \\
y(1) &= x(1) + \tfrac{1}{2} y(0) = 1 + \tfrac{1}{2}(1) = \tfrac{3}{2} \\
y(2) &= x(2) + \tfrac{1}{2} y(1) = 1 + \tfrac{1}{2}(\tfrac{3}{2}) = \tfrac{7}{4} \\
y(3) &= x(3) + \tfrac{1}{2} y(2) = 1 + \tfrac{1}{2}(\tfrac{7}{4}) = \tfrac{15}{8} \\
y(4) &= x(4) + \tfrac{1}{2} y(3) = 1 + \tfrac{1}{2}(\tfrac{15}{8}) = \tfrac{31}{16}
\end{aligned}
\tag{5–16}
$$

$$\vdots$$

In fact, $y(n)$ turns out to be

$$y(n) = 2 - \left(\tfrac{1}{2}\right)^n, \quad n = 0, 1, 2, \ldots \tag{5–17}$$

Figure 5–6 shows the block diagram for this system along with plots of the discrete time signals $x(n)$ and $y(n)$.

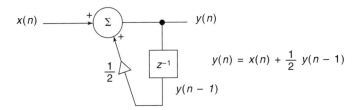

$$y(n) = x(n) + \frac{1}{2} y(n - 1)$$

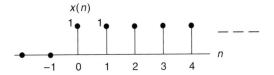

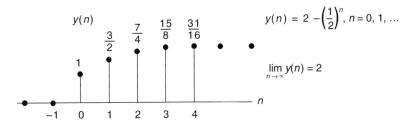

$$y(n) = 2 - \left(\frac{1}{2}\right)^n, \; n = 0, 1, \ldots$$

$$\lim_{n \to \infty} y(n) = 2$$

FIGURE 5–6
Simple example of a discrete time system with a specific input $x(n)$ and a resulting output $y(n)$. The output signal can be calculated recursively, given the input signal and the system difference equation.

In Chapter 1 the concept of causality was introduced in the context of continuous time systems. The concept needs to be examined here as well. Consider a discrete time system initially at rest. Suppose a signal is first applied to the system input at time $n = 0$. If the system is *causal*, there will be no output signal until $n = 0$ at the earliest. In other words, a causal discrete time system cannot see into the future and anticipate future inputs (or future outputs, for that matter). The system described in the above example (Equations [5–14] to [5–17]) is an example of a causal system. It is possible to discuss *noncausal* systems in theory, but noncausal systems cannot be realized. An example of a theoretical non-causal system is an ideal lowpass filter (which is considered in Chapter 6).

5–2 IMPULSE RESPONSE AND CONVOLUTION

Of special interest is the discrete time signal called the unit *impulse*, denoted $\delta(n)$. By definition

$$\delta(n) = \begin{cases} 1, & n = 0 \\ 0, & \text{otherwise} \end{cases} \tag{5–18}$$

The shifted impulse $\delta(n - k)$ is given by

$$\delta(n - k) = \begin{cases} 1, & n = k \\ 0, & \text{otherwise} \end{cases} \tag{5-19}$$

$\delta(n)$ and $\delta(n - k)$ (for positive k) are shown in Figure 5–7. Note that this discrete time impulse, which is sometimes called a *Kronecker delta*, is conceptually much simpler than the continuous time impulse $\delta(t)$ introduced in Chapter 1. It should also be noted that the discrete time impulse $\delta(n)$ is *not* a sampled version of the continuous time impulse $\delta(t)$; such a sample would have to have a value of infinity, not one.

If the input to a discrete time system is $x(n) = \delta(n)$, applied when the system is at rest (shift register memories zeroed out), then the system output is by definition called the *impulse response* of the system, denoted $h(n)$. That is,

$$T[\delta(n)] = h(n) \tag{5-20}$$

If the system is shift invariant, then

$$T[\delta(n - k)] = h(n - k) \tag{5-21}$$

Let us consider a simple example. Suppose the system difference equation is

$$y(n) = x(n) + \tfrac{1}{2} y(n - 1) \tag{5-22}$$

If $x(n) = \delta(n)$, then $y(n) = h(n)$ (by definition), and we have

$$h(n) = \delta(n) + \tfrac{1}{2} h(n - 1) \tag{5-23}$$

The impulse response can be generated recursively as follows; remember that the system is initially at rest, that is, $h(-1) = 0$:

$$h(0) = \delta(0) + \tfrac{1}{2} h(-1) = 1 + \tfrac{1}{2}(0) = 1$$
$$h(1) = \delta(1) + \tfrac{1}{2} h(0) = 0 + \tfrac{1}{2}(1) = \tfrac{1}{2}$$
$$h(2) = \delta(2) + \tfrac{1}{2} h(1) = 0 + \tfrac{1}{2}(\tfrac{1}{2}) = \tfrac{1}{4} \tag{5-24}$$
$$h(3) = \delta(3) + \tfrac{1}{2} h(2) = 0 + \tfrac{1}{2}(\tfrac{1}{4}) = \tfrac{1}{8}$$

$$\vdots$$

FIGURE 5–7
Discrete time impulse
(Kronecker delta).

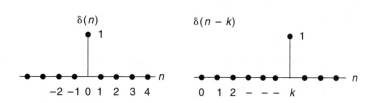

It is easy to see the pattern in this example:

$$h(n) = \begin{cases} \left(\frac{1}{2}\right)^n, n \geq 0 \\ 0, n < 0 \end{cases} \qquad \text{(5–25)}$$

If a system is *causal*, its impulse response doesn't start until the impulse is applied to the input. Therefore, for a causal system, $h(n) = 0$ for $n < 0$. Any theoretical system having an impulse response that starts before $n = 0$ is noncausal, and therefore not realizable.

For a linear shift invariant discrete time system, the relationship between the input and output sequences can be expressed in terms of *discrete time convolution*. To begin with, observe that any sequence $x(n)$ can be expressed as a sum of weighted and shifted impulses:

$$x(n) = \sum_{k=-\infty}^{\infty} x(k)\delta(n - k) \qquad \text{(5–26)}$$

That is,

$$x(n) = \cdots x(-1)\delta(n + 1) + x(0)\delta(n) + x(1)\delta(n - 1) + x(2)\delta(n - 2) + \cdots \qquad \text{(5–27)}$$

This concept is illustrated in Figure 5–8, in which $x(n)$ is a three-point sequence.

Now consider a discrete time system with input signal $x(n)$ and output signal $y(n)$, where

$$y(n) = T\big[x(n)\big] \qquad \text{(5–28)}$$

Substituting Equation (5–26) into Equation (5–28) results in

$$y(n) = T\left[\sum_{k=-\infty}^{\infty} x(k)\delta(n - k)\right] \qquad \text{(5–29)}$$

If the system $T[\ ]$ is *linear*, then superposition is valid, and we can write

$$y(n) = \sum_{k=-\infty}^{\infty} x(k)T\big[\delta(n - k)\big] \qquad \text{(5–30)}$$

Recall that $T[\delta(n)] = h(n)$, by definition. If the system is *shift invariant*, then

$$T\big[\delta(n - k)\big] = h(n - k) \qquad \text{(5–31)}$$

Thus if the system is shift invariant, Equation (5–30) becomes

$$y(n) = \sum_{k=-\infty}^{\infty} x(k)h(n - k) \qquad \text{(5–32)}$$

Equation (5–32) represents the *discrete time convolution* of $x(n)$ and $h(n)$:

$$y(n) = x(n) * h(n) = \sum_{k=-\infty}^{\infty} x(k)h(n - k) \qquad \text{(5–33)}$$

FIGURE 5–8
Any sequence can be thought
of as a sum of weighted and
shifted discrete time impulses.

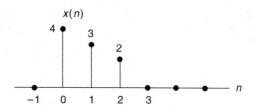

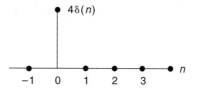

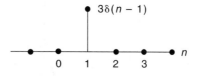

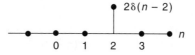

The operator ∗ denotes convolution. Thus, for a linear shift invariant discrete time system, the output can be found by convolving the input signal with the system impulse response. Equation (5–33) can be interpreted in the following way: The system output can be found by adding together weighted and shifted copies of the system impulse response. That is, $y(n) = \ldots$ $x(-1)h(n + 1) + x(0)h(n) + x(1)h(n - 1) + x(2)h(n - 2) + \ldots$. If we think of the input $x(n)$ as consisting of a sum of weighted and shifted impulses (as in Equation [5–27]), then convolution becomes superposition (we add up all of the individual system responses to the individual weighted and shifted impulse inputs). This concept is illustrated in Figure 5–9, in which $h(n)$ and $x(n)$ are both three-point sequences.

Observe that in the example depicted in Figure 5–9, the result of convolving two three-point sequences is a five-point sequence. This serves to illustrate the following fact: When two *finite duration* sequences with lengths N points and M points, respectively, are convolved, the resulting sequence has a length of $N + M - 1$ points.

It should be noted here that discrete time convolution commutes, that is,

$$y(n) = h(n) * x(n) = \sum_{k=-\infty}^{\infty} h(k)x(n - k) \qquad (5\text{–}34)$$

FIGURE 5–9

Discrete time convolution.

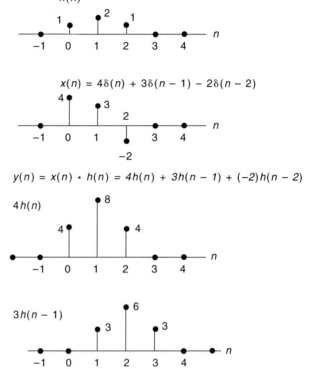

$$x(n) = 4\delta(n) + 3\delta(n - 1) - 2\delta(n - 2)$$

$$y(n) = x(n) * h(n) = 4h(n) + 3h(n - 1) + (-2)h(n - 2)$$

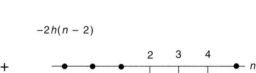

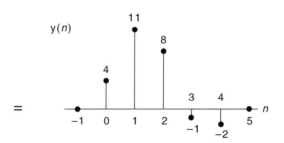

Equation (5–34) says that the output can be found by adding together weighted and shifted copies of the input signal. That is, $y(n) = \ldots h(-1)x(n + 1) + h(0)x(n) + h(1)x(n - 1) + \ldots$.

The summation limits $k = -\infty$ to ∞ are general. For specific convolution problems, the exact limits need to be determined. If the input $x(n)$ is applied to the system at time $n = 0$, and if the system is causal (meaning that the impulse response doesn't begin until the impulse is applied), then the summation limits in Equations (5–33) and (5–34) can be restricted to $k = 0$ to n, that is,

$$y(n) = \sum_{k=0}^{n} x(k)h(n - k) = \sum_{k=0}^{n} h(k)x(n - k) \qquad (5\text{–}35)$$

It should be noted that for an FIR system, the convolution expression is really the same thing as the system difference equation, which is given by Equation (5–12). By letting $x(n) = \delta(n)$ in Equation (5–12), the output becomes $h(n)$ by definition. Therefore, the impulse response of an FIR system is

$$h(n) = \sum_{k=0}^{M} b_k \delta(n - k) \qquad (5\text{–}36)$$

From Equation (5–36) it is apparent that $h(k) = b_k$, $k = 0,1, \ldots, M$; but $h(k) = 0$ if k is outside of the interval $[0, M]$. (The justification for the name "FIR" becomes clear: The impulse response is of finite time duration.) Therefore, the convolution expression for an FIR system can be written as

$$y(n) = h(n) * x(n) = \sum_{k=0}^{M} h(k)x(n - k) \qquad (5\text{–}37)$$

Equation (5–37) is the same as Equation (5–12) since $h(k) = b_k$. Figure 5–10 shows the system block diagram for the FIR system of Equation (5–37).

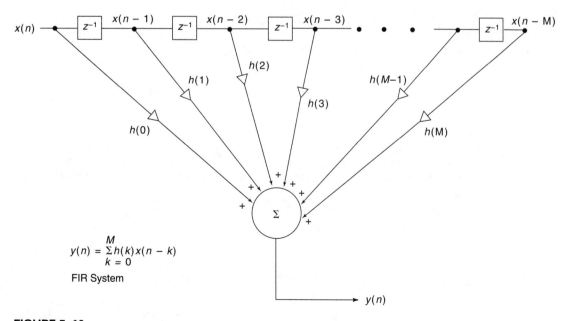

$$y(n) = \sum_{k=0}^{M} h(k)x(n - k)$$

FIR System

FIGURE 5–10
Finite impulse response (FIR) discrete time system.

5-3   THE Z TRANSFORM

In discrete time system theory, the Z transform plays a role analogous to that played by the Laplace transform in continous time system theory. The Z transform of a signal $x(n)$, denoted $X(z)$, is defined as

$$Z\{x(n)\} = X(z) = \sum_{n=-\infty}^{\infty} x(n)z^{-n} \tag{5-38}$$

where $Z\{x(n)\}$ is shorthand for "the Z transform of $x(n)$." The transform pair relationship is indicated by

$$x(n) \leftrightarrow X(z) \tag{5-39}$$

The usual convention with respect to lower- and uppercase letters holds here: lowercase for the *time domain,* and uppercase for the *transform domain.*

In order for the Z transform to exist, it must be possible to define a region in the z plane (i.e., a range of values for z, where z is in general complex) in which the Z transform summation (Equation [5-38]) converges. It is possible to find such a region of convergence for the Z transform of any signal $x(n)$ that is of practical interest to us as engineers. As is the case with the Laplace transform, the region of convergence issue rarely impacts the utility of the Z transform, so it will not be given heavy emphasis here.

There are two Z transform properties of immediate interest. The first is the linearity property. If $x_1(n) \leftrightarrow X_1(z)$ and $x_2(n) \leftrightarrow X_2(z)$, then

$$A_1 x_1(n) + A_2 x_2(n) \leftrightarrow A_1 X_1(z) + A_2 X_2(z) \tag{5-40}$$

The proof of the linearity property follows from the definition of the Z transform:

$$Z\{A_1 x_1(n) + A_2 x_2(n)\} = \sum_{n=-\infty}^{\infty} \{A_1 x_1(n) + A_2 x_2(n)\}z^{-n}$$

$$= A_1 \sum_{n=-\infty}^{\infty} x_1(n)z^{-n} + A_2 \sum_{n=-\infty}^{\infty} x_2(n)z^{-n} \tag{5-41}$$

$$= A_1 X_1(z) + A_2 X_2(z)$$

Equation (5-40) shows the linearity property with respect to two signals, but it can be extended to three or more signals. The linearity property essentially means that the Z transform of a weighted sum of several signals is equal to the weighted sum of the Z transforms of those signals.

The second property of immediate interest is the *time shifting property.* If $x(n) \leftrightarrow X(z)$, then

$$x(n - k) \leftrightarrow z^{-k} X(z) \tag{5-42}$$

This is relatively easy to show. Take the Z transform of $x(n - k)$:

$$Z\{x(n - k)\} = \sum_{n=-\infty}^{\infty} \{x(n - k)\}z^{-n} \tag{5-43}$$

Now change variables in the summation: Let $n - k = m$. This results in

$$\mathbf{Z}\{x(n-k)\} = \sum_{m=-\infty}^{\infty} x(m)z^{-(m+k)} = z^{-k}\left[\sum_{m=-\infty}^{\infty} x(m)z^{-m}\right] = z^{-k}X(z) \tag{5-44}$$

Before deriving some commonly used Z transform pairs (see Table 5–1), it is useful to introduce the *discrete time unit step function*, denoted $u(n)$, which is defined as follows:

$$u(n) = \begin{cases} 1, n \geq 0 \\ 0, n < 0 \end{cases} \tag{5-45}$$

If a discrete time signal is multiplied by $u(n)$, the effect is to set the signal equal to zero when $n < 0$. (A signal having this property is sometimes called a *causal signal*.) For example,

$$x(n) = \begin{cases} a^n, n \geq 0 \\ 0, n < 0 \end{cases} \tag{5-46}$$

can also be expressed conveniently as

$$x(n) = a^n u(n) \tag{5-47}$$

To generate most of the Z transform pairs shown in Table 5–1, it is necessary to appeal to certain well-known formulas relating to geometric series. The reader has probably seen these formulas before, but may have had little occasion to use them; therefore, a brief review is in order.

Consider the finite geometric series

$$S_{N-1} = \sum_{n=0}^{N-1} a^n = 1 + a + a^2 + \cdots + a^{N-1} \tag{5-48}$$

The notation S_{N-1} means "sum out to the $(N-1)$th term." A closed-form formula can be derived for this sum. The first step is to observe that

$$aS_{N-1} = a\sum_{n=0}^{N-1} a^n = \sum_{n=0}^{N-1} a^{n+1} \tag{5-49}$$

That is,

$$aS_{N-1} = a + a^2 + a^3 + \cdots + a^{N-1} + a^N \tag{5-50}$$

Therefore,

$$S_{N-1} - aS_{N-1} = 1 - a^N \tag{5-51}$$

which can be written as

$$S_{N-1}(1-a) = 1 - a^N \tag{5-52}$$

Therefore, we can solve for S_{N-1}

$$S_{N-1} = \frac{1 - a^N}{1 - a} \tag{5-53}$$

This gives us the well-known formula for a finite geometric series:

$$\sum_{n=0}^{N-1} a^n = \frac{1 - a^N}{1 - a} \tag{5-54}$$

Note that with respect to Equations (5–53) and (5–54), the only restriction on a is that it cannot be equal to 1. Other than that, a could be any real, imaginary, or complex number. (If $a = 1$, then it is easy to show that $S_{N-1} = N$).

Let us now consider the infinite geometric series

$$S = \sum_{n=0}^{\infty} a^n = \lim_{N \to \infty} S_{N-1} \tag{5-55}$$

The closed formula for S is

$$S = \lim_{N \to \infty} \left[\frac{1 - a^N}{1 - a} \right] \tag{5-56}$$

In order for the infinite geometric series to converge (that is, in order for S to be finite), a^N must become zero as $N \to \infty$. That is, the infinite geometric series converges if

$$\lim_{N \to \infty} a^N = 0 \tag{5-57}$$

Clearly, the infinite geometric series will converge if $|a| < 1$. The closed-form formula for the infinite geometric series is therefore

$$\sum_{n=0}^{\infty} a^n = \frac{1}{1 - a}, \text{provided } |a| < 1 \tag{5-58}$$

Recall that a can be real or complex. If a is complex, then it can be expressed in polar form as

$$a = re^{j\theta} \tag{5-59}$$

where r is the magnitude of a, and is therefore nonnegative. In this case, the requirement $|a| < 1$ means the same as $r < 1$.

We are now ready to generate a short table of Z transforms. We begin with the discrete time unit impulse $\delta(n)$

$$\mathbf{Z}\{\delta(n)\} = \sum_{n=-\infty}^{\infty} \delta(n)z^{-n} \tag{5-60}$$

Since $\delta(n) = 0$ when $n \neq 0$, only one term in the summation is nonzero (the $n = 0$ term). Since $\delta(0) = 1$ and $z^0 = 1$, we have

$$\delta(n) \leftrightarrow 1 \qquad (5\text{--}61)$$

The Z transform of a shifted impulse, $\delta(n - k)$, is

$$\mathbf{Z}\{\delta(n - k)\} = \sum_{n=-\infty}^{\infty} \delta(n - k)z^{-n} \qquad (5\text{--}62)$$

Since $\delta(n - k) = 0$ when $n \neq k$, only one term in the summation is nonzero (the $n = k$ term). Therefore,

$$\sum_{n=-\infty}^{\infty} \delta(n - k)z^{-n} = \delta(k - k)z^{-k} = \delta(0)z^{-k} = z^{-k} \qquad (5\text{--}63)$$

We therefore have the following Z transform pair:

$$\delta(n - k) \leftrightarrow z^{-k} \qquad (5\text{--}64)$$

Note that Z transforms (5–61) and (5–64) were obtained without having to impose any restrictions on z. Therefore, for these two Z transforms, the region of convergence is the entire z plane.

The next example is highly useful because it can be used to generate several Z transform pairs:

$$x(n) = c^n u(n) \qquad (5\text{--}65)$$

where c is allowed to take on either real or complex values. In this case, since $x(n)$ is a causal sequence, the Z transform summation can start at $n = 0$:

$$\mathbf{Z}\{c^n u(n)\} = \sum_{n=0}^{\infty} c^n z^{-n} = \sum_{n=0}^{\infty} \left(cz^{-1}\right)^n \qquad (5\text{--}66)$$

Equation (5–66) is an infinite geometric series; it converges if $|cz^{-1}| < 1$, that is, if $|z| > |c|$. This constraint describes the region of convergence for this Z transform. The closed-form formula for an infinite geometric series can be applied, resulting in

$$\sum_{n=0}^{\infty} \left(cz^{-1}\right)^n = \frac{1}{1 - cz^{-1}}, \text{ provided } |z| > |c| \qquad (5\text{--}67)$$

Therefore, we have derived the following Z transform pair:

$$c^n u(n) \leftrightarrow \frac{1}{1 - cz^{-1}} \qquad (5\text{--}68)$$

The region of convergence for this Z transform pair, as noted above, is $|z| > |c|$.

Observe that if $c = 1$, then $c^n u(n)$ is just $u(n)$. Therefore, the Z transform pair for the unit step sequence is

$$u(n) \leftrightarrow \frac{1}{1 - z^{-1}} \tag{5–69}$$

The region of convergence in this case is $|z| > 1$.

If c is a real constant (positive or negative), it is convenient to directly apply the Z transform relationship of Equation (5–68). For example,

$$\left(-\tfrac{1}{4}\right)^n u(n) \leftrightarrow \frac{1}{1 + \tfrac{1}{4} z^{-1}} \qquad \left(|z| > \tfrac{1}{4}\right) \tag{5–70}$$

(The region of convergence is shown at the right.) Another example is as follows:

$$2^n u(n) \leftrightarrow \frac{1}{1 - 2z^{-1}} \qquad \left(|z| > 2\right) \tag{5–71}$$

Let us now consider the following signal:

$$x(n) = \lambda^n \cos(\beta n) u(n) \tag{5–72}$$

where λ is a real constant (positive or negative). The Z transform relationship of Equation (5–68), along with the linearity property, can be used to find the Z transform in this case. We begin by observing that Equation (5–72) can be rewritten using Euler's theorem:

$$\lambda^n \cos(\beta n) u(n) = \lambda^n \left[\tfrac{1}{2} e^{j\beta n} + \tfrac{1}{2} e^{-j\beta n} \right] u(n) \tag{5–73}$$

Therefore,

$$\lambda^n \cos(\beta n) u(n) = \tfrac{1}{2} \left(\lambda e^{j\beta} \right)^n u(n) + \tfrac{1}{2} \left(\lambda e^{-j\beta} \right)^n u(n) \tag{5–74}$$

The functions $[\lambda \exp(\pm j\beta n)]^n$ are special cases of c^n where c is complex. Therefore,

$$Z\{\lambda^n \cos(\beta n) u(n)\} = \frac{\tfrac{1}{2}}{1 - \lambda e^{j\beta} z^{-1}} + \frac{\tfrac{1}{2}}{1 - \lambda e^{-j\beta} z^{-1}} \tag{5–75}$$

The right side of Equation (5–75) can be combined as follows:

$$Z\{\lambda^n \cos(\beta n) u(n)\} = \frac{\tfrac{1}{2} - \tfrac{1}{2} \lambda e^{-j\beta} z^{-1} + \tfrac{1}{2} - \tfrac{1}{2} \lambda e^{j\beta} z^{-1}}{\left(1 - \lambda e^{j\beta} z^{-1}\right)\left(1 - \lambda e^{-j\beta} z^{-1}\right)} \tag{5–76}$$

The right side of Equation (5–76) can be cleaned up as follows:

$$Z\{\lambda^n \cos(\beta n) u(n)\} = \frac{1 - \lambda\left(\tfrac{1}{2} e^{j\beta} + \tfrac{1}{2} e^{-j\beta}\right) z^{-1}}{1 - \lambda\left(e^{j\beta} + e^{-j\beta}\right) z^{-1} + \lambda^2 z^{-2}} \tag{5–77}$$

$$= \frac{1 - \lambda \cos(\beta)z^{-1}}{1 - 2\lambda \cos(\beta)z^{-1} + \lambda^2 z^{-2}} \qquad (5\text{--}78)$$

Therefore, we have the following Z transform pair:

$$\lambda^n \cos(\beta n)u(n) \leftrightarrow \frac{1 - \lambda \cos(\beta)z^{-1}}{1 - 2\lambda \cos(\beta)z^{-1} + \lambda^2 z^{-2}} \qquad (5\text{--}79)$$

The region of convergence for this Z transform pair is $|z| > |\lambda|$, since this is the region of convergence for both terms on the right side of Equation (5–75).

A special case is when $\lambda = 1$:

$$\cos(\beta n)u(n) \leftrightarrow \frac{1 - \cos(\beta)z^{-1}}{1 - 2\cos(\beta)z^{-1} + z^{-2}} \qquad (5\text{--}80)$$

In a similar manner, we can find the Z transform of the following signal:

$$x(n) = \lambda^n \sin(\beta n)u(n) = \lambda^n \left[\tfrac{1}{2} e^{j\beta n} - \tfrac{1}{2} e^{-j\beta n} \right] u(n) \qquad (5\text{--}81)$$

That is,

$$x(n) = \tfrac{1}{2}\left(\lambda e^{j\beta}\right)^n u(n) - \tfrac{1}{2}\left(\lambda e^{-j\beta}\right)^n u(n) \qquad (5\text{--}82)$$

The Z transform of this signal is

$$\mathbf{Z}\left\{\lambda^n \sin(\beta n)u(n)\right\} = \frac{\tfrac{1}{2}}{1 - \lambda e^{j\beta} z^{-1}} + \frac{-\tfrac{1}{2}}{1 - \lambda e^{-j\beta} z^{-1}} \qquad (5\text{--}83)$$

The region of convergence is $|z| > |\lambda|$. It is left as an exercise for the reader to show that this cleans up to

$$\lambda^n \sin(\beta n)u(n) \leftrightarrow \frac{\lambda \sin(\beta)z^{-1}}{1 - 2\lambda \cos(\beta)z^{-1} + \lambda^2 z^{-2}} \qquad (5\text{--}84)$$

A special case is when $\lambda = 1$:

$$\sin(\beta n)u(n) \leftrightarrow \frac{\sin(\beta)z^{-1}}{1 - 2\cos(\beta)z^{-1} + z^{-2}} \qquad (5\text{--}85)$$

A short table of Z transforms is presented in Table 5–1. Several Z transform properties are also listed in Table 5–1.

All of the Z transform pairs in Table 5–1 are for causal signals. However, the Z transform (as defined in Equation [5.38]) is two-sided, that is, not restricted to causal signals. We can find the Z transform of a signal $x(n)$ having nonzero values when $n < 0$. As is the case with the two-sided Laplace transform, there is a theoretical uniqueness problem with

TABLE 5–1
Transforms and Properties

$$x(n) \leftrightarrow X(z)$$

$$\delta(n) \leftrightarrow 1$$

$$\delta(n - k) \leftrightarrow z^{-k}$$

$$u(n) \leftrightarrow \frac{1}{1 - z^{-1}}$$

$$c^n u(n) \leftrightarrow \frac{1}{1 - cz^{-1}}$$

$$\cos(\beta n)u(n) \leftrightarrow \frac{1 - \cos(\beta)z^{-1}}{1 - 2\cos(\beta)z^{-1} + z^{-2}}$$

$$\sin(\beta n)u(n) \leftrightarrow \frac{\sin(\beta)z^{-1}}{1 - 2\cos(\beta)z^{-1} + z^{-2}}$$

$$\lambda^n \cos(\beta n)u(n) \leftrightarrow \frac{1 - \lambda\cos(\beta)z^{-1}}{1 - 2\lambda\cos(\beta)z^{-1} + \lambda^2 z^{-2}}$$

$$\lambda^n \sin(\beta n)u(n) \leftrightarrow \frac{\lambda\sin(\beta)z^{-1}}{1 - 2\lambda\cos(\beta)z^{-1} + \lambda^2 z^{-2}}$$

$$na^n u(n) \leftrightarrow \frac{az^{-1}}{\left(1 - az^{-1}\right)^2}$$

Properties

Linearity : $\displaystyle\sum_{k=1}^{M} A_k x_k(n) \leftrightarrow \sum_{k=1}^{M} A_k X_k(z)$

Time Shifting : $x(n - k) \leftrightarrow z^{-k} X(z)$

Convolution : $x(n) * g(n) \leftrightarrow X(z)G(z)$

the two-sided Z transform. Any causal signal has an anticausal twin that has the same Z transform; the two transforms differ only with respect to the region of convergence. As an example, consider the following anticausal signal:

$$x(n) = \begin{cases} -c^n, & n < 0 \\ 0, & n \geq 0 \end{cases} \tag{5–86}$$

Equation (5–86) can also be expressed as follows:

$$x(n) = -c^n u(-1 - n) \tag{5–87}$$

The Z transform in this case can be found as follows:

$$Z\{-c^n u(-1 - n)\} = -\sum_{n=-\infty}^{-1} c^n z^{-n} = -\sum_{n=-\infty}^{-1} \left(cz^{-1}\right)^n \tag{5–88}$$

which can be manipulated as follows:

$$-\sum_{n=-\infty}^{-1}\left(cz^{-1}\right)^{n} = -\sum_{n=1}^{\infty}\left(cz^{-1}\right)^{-n} = -\sum_{n=1}^{\infty}\left(c^{-1}z\right)^{n} = -\sum_{n=0}^{\infty}\left(c^{-1}z\right)^{n}+\left(c^{-1}z\right)^{0} \quad (5\text{--}89)$$

Therefore,

$$\mathbf{Z}\left\{-c^{n}u(-1-n)\right\} = 1 - \sum_{n=0}^{\infty}\left(c^{-1}z\right)^{n} \quad (5\text{--}90)$$

The formula for the infinite geometric series comes into play once again:

$$\mathbf{Z}\left\{-c^{n}u(-1-n)\right\} = 1 - \frac{1}{1-c^{-1}z} \quad (5\text{--}91)$$

with a region of convergence defined by $|c^{-1}z| < 1$, that is, $|z| < |c|$. The right side of Equation (5–91) can be further manipulated as follows:

$$1 - \frac{1}{1-c^{-1}z} = \frac{1-c^{-1}z-1}{1-c^{-1}z} = \frac{-c^{-1}z}{1-c^{-1}z} = \frac{1}{1-cz^{-1}} \quad (5\text{--}92)$$

Therefore,

$$-c^{n}u(-1-n) \leftrightarrow \frac{1}{1-cz^{-1}} \quad (5\text{--}93)$$

This Z transform pair has a region of convergence defined by $|z| < |c|$. The reader should compare this result with Equation (5–68). Clearly, $c^{n}u(n)$ and $-c^{n}u(-1-n)$ have the same Z transform, but different regions of convergence. (In fact, the two regions of convergence are mutually exclusive.)

From a mathematician's perspective, the inverse Z transform cannot be found unless a region of convergence is specified. However, in actual engineering applications there will be known constraints that resolve the uniqueness problem without directly considering the region of convergence. For example, if a signal $x(n)$ is first applied to a causal system at time $n = 0$, the output signal *must* be a causal signal; the anticausal twin can be ruled out. There are advanced applications in which anticausal sequences must be considered, but these cases are beyond the scope of this book.

Presented here are some relatively simple Z transform examples that illustrate the use of Table 5–1. With respect to the inverse Z transform examples, note that all sequences are assumed to be causal. More complicated examples requiring partial fraction expansion are presented in Section 5–6.

EXAMPLE 5–3.1
Find the Z transform of the following sequence:

$$x(n) = 2\delta(n) + 3\delta(n-1) + \tfrac{1}{2}\delta(n-10) \quad (5\text{--}94)$$

To find $X(z)$, we must use the linearity property and the Z transform pair $\delta(n-k) \leftrightarrow z^{-k}$:

$$X(z) = 2 + 3z^{-1} + \tfrac{1}{2}z^{-10} \quad (5\text{--}95)$$

EXAMPLE 5–3.2

Find the inverse Z transform of

$$X(z) = -\tfrac{1}{2} + 2z^{-2} + 2z^{-4} - \tfrac{1}{2}z^{-6} \tag{5–96}$$

To find $x(n)$, we must use the linearity property and the Z transform pair $\delta(n-k) \leftrightarrow z^{-k}$:

$$x(n) = -\tfrac{1}{2}\delta(n) + 2\delta(n-2) + 2\delta(n-4) - \tfrac{1}{2}\delta(n-6) \tag{5–97}$$

EXAMPLE 5–3.3

Find the inverse Z transform of

$$X(z) = \frac{20}{1-z^{-1}} - \frac{\tfrac{1}{2}z^{-2}}{1-\tfrac{1}{3}z^{-1}} \tag{5–98}$$

To find $x(n)$, we must use the linearity property and apply the Z transform pair $c^n u(n) \leftrightarrow 1/(1-cz^{-1})$ to each term on the right side of Equation (5–98). In addition, the time shifting property must be applied to the second term. The solution is

$$x(n) = 20u(n) - \tfrac{1}{2}\left(\tfrac{1}{3}\right)^{n-2}u(n-2) \tag{5–99}$$

Note that $u(n-2)$ is a shifted unit step function.

EXAMPLE 5–3.4

Find the Z transform of the following sequence:

$$x(n) = 10\left(-\tfrac{1}{2}\right)^n \cos\left(\tfrac{\pi}{3}n\right)u(n) \tag{5–100}$$

By a straightforward application of one of the entries in Table 5–1, we find the solution:

$$X(z) = \frac{10\left[1 + \tfrac{1}{2}\cos\left(\tfrac{\pi}{3}\right)z^{-1}\right]}{1 + 2\left(\tfrac{1}{2}\right)\cos\left(\tfrac{\pi}{3}\right)z^{-1} + \tfrac{1}{4}z^{-2}} \tag{5–101}$$

EXAMPLE 5–3.5

Find the inverse Z transform of

$$X(z) = \frac{5e^{j\frac{\pi}{3}}}{1 - \tfrac{1}{4}e^{j\frac{\pi}{10}}z^{-1}} + \frac{5e^{-j\frac{\pi}{3}}}{1 - \tfrac{1}{4}e^{-j\frac{\pi}{10}}z^{-1}} \tag{5–102}$$

To find $x(n)$, we must use the linearity property and apply the Z transform pair $c^n u(n) \leftrightarrow 1/(1 - cz^{-1})$ to each term on the right side of Equation (5–102):

$$x(n) = 5e^{j\frac{\pi}{3}} \left(\frac{1}{4} e^{j\frac{\pi}{10}} \right)^n u(n) + 5e^{-j\frac{\pi}{3}} \left(\frac{1}{4} e^{-j\frac{\pi}{10}} \right)^n u(n) \tag{5–103}$$

The right side of Equation (5–103) can be written as follows:

$$x(n) = 5\left(\frac{1}{4}\right)^n \left[e^{j\left(\frac{\pi}{10}n + \frac{\pi}{3}\right)} + e^{-j\left(\frac{\pi}{10}n + \frac{\pi}{3}\right)} \right] u(n) \tag{5–104}$$

The right side of Equation (5–104) can be simplified as follows:

$$x(n) = 10\left(\frac{1}{4}\right)^n \cos\left(\frac{\pi}{10}n + \frac{\pi}{3}\right) u(n) \tag{5–105}$$

EXAMPLE 5–3.6

Find the Z transform of the following sequence:

$$x(n) = \left(\frac{1}{2}\right)^n u(n) + \left(\frac{1}{4}\right)^n u(n) \tag{5–106}$$

The solution can be found using the linearity property and the Z transform pair $c^n u(n) \leftrightarrow 1/(1 - cz^{-1})$:

$$X(z) = \frac{1}{1 - \frac{1}{2} z^{-1}} + \frac{1}{1 - \frac{1}{4} z^{-1}} \tag{5–107}$$

NOTE: This example can be used to illustrate the following concept: The region of convergence for a sum of Z transforms (such as Equation [5–107]) is the intersection of the individual regions of convergence. In this case, the individual regions of convergence are $|z| > 1/2$ and $|z| > 1/4$; therefore, the overall region of convergence is $|z| > 1/2$.

5–4   TRANSFER FUNCTION OF A DISCRETE TIME SYSTEM

The *transfer function H(z)* of an LSI discrete time system relates the input and output signals in the Z transform domain in a manner analogous to the way in which $H(s)$ relates the input and output signals of an LSI continuous time system in the Laplace transform domain. Consider an LSI discrete time system with input and output signals $x(n)$ and $y(n)$, respectively, that have Z transforms $X(z)$ and $Y(z)$, respectively. The transfer function $H(z)$ is defined as

$$H(z) = \frac{Y(z)}{X(z)} \tag{5–108}$$

Rearranging Equation (5–108) shows that the Z transform of the output signal can be found by multiplying the Z transform of the input signal by the system transfer function:

$$Y(z) = H(z)X(z) \tag{5–109}$$

It is important to note that when defining a system transfer function, the system is assumed to be at rest before the input signal is applied.

There are two ways that $H(z)$ can be determined:

1. Take the Z transform of both sides of the system difference equation (Equation [5–11]), using the time shifting and linearity properties discussed in the previous section, and solve for $Y(z)/X(z)$.
2. $H(z)$ is the Z transform of the system impulse response $h(n)$.

Let us consider the first method. The system difference equation is repeated here for the convenience of the reader:

$$y(n) = \sum_{k=0}^{M} b_k x(n-k) - \sum_{k=1}^{N} a_k y(n-k) \tag{5–110}$$

Take the Z transform of both sides of Equation (5–110) term by term, using the time-shifting and linearity properties:

$$Y(z) = \sum_{k=0}^{M} b_k z^{-k} X(z) - \sum_{k=1}^{N} a_k z^{-k} Y(z) \tag{5–111}$$

Since $X(z)$ and $Y(z)$ are not functions of summation index k, they can be brought outside the summations:

$$Y(z) = X(z) \sum_{k=0}^{M} b_k z^{-k} - Y(z) \sum_{k=1}^{N} a_k z^{-k} \tag{5–112}$$

Therefore,

$$Y(z)\left[1 + \sum_{k=1}^{N} a_k z^{-k}\right] = X(z) \sum_{k=0}^{M} b_k z^{-k} \tag{5–113}$$

Finally we can solve for $H(z)$:

$$H(z) = \frac{Y(z)}{X(z)} = \frac{\displaystyle\sum_{k=0}^{M} b_k z^{-k}}{1 + \displaystyle\sum_{k=1}^{N} a_k z^{-k}} \tag{5–114}$$

The transfer function can also be expressed in factored form as

$$H(z) = \frac{G \displaystyle\prod_{k=1}^{M}\left(1 - \lambda_k z^{-1}\right)}{\displaystyle\prod_{k=1}^{N}\left(1 - p_k z^{-1}\right)} \tag{5–115}$$

where λ_k, $k = 1, 2, \ldots, M$ are the *zeros* of the transfer function, and p_k, $k = 1, 2, \ldots, N$ are the *poles* of the transfer function. Note that the zeros of the transfer function are those values of z which cause the numerator of $H(z)$ to equal zero; the poles of the transfer function are those values of z that cause the denominator of $H(z)$ to equal zero. G is a gain constant. Assuming the coefficients a_k and b_k in Equation (5–110) are real, complex poles must exist in complex conjugate pairs. That is, if p_k is a pole, then p_k^* is also a pole. Complex zeros must also exist in complex conjugate pairs. Note that except for the gain constant G, a discrete time LSI system is completely represented by its poles and zeros. Note also that the transfer function of an FIR system has no poles, since in this case the denominator of Equations (5–114) and (5–115) is just equal to 1; for this reason, an FIR system is sometimes called an *all-zero system*. If a transfer function has no zeros (the numerator of Equations [5–114] and [5–115] is just a constant), but does have poles, then the system is called an *all-pole system*.

Poles and zeros can be plotted as points on the complex z plane. Consider the following example.

EXAMPLE 5–4.1

Suppose we have the following transfer function:

$$H(z) = \frac{\left(1 - z^{-1} - 2z^{-2}\right)}{\left(1 + \frac{1}{4}z^{-1} - \frac{3}{8}z^{-2}\right)\left(1 - \frac{1}{2}z^{-1} + \frac{1}{4}z^{-2}\right)}$$

$$= \frac{\left(1 + z^{-1}\right)\left(1 - 2z^{-1}\right)}{\left(1 + \frac{3}{4}z^{-1}\right)\left(1 - \frac{1}{2}z^{-1}\right)\left(1 - \frac{1}{2}e^{j\frac{\pi}{3}}z^{-1}\right)\left(1 - \frac{1}{2}e^{-j\frac{\pi}{3}}z^{-1}\right)}$$

(5–116)

This transfer function has two real zeros at $z = -1$ and $z = 2$; two real poles at $z = -3/4$ and $z = 1/2$; and a pair of complex conjugate poles at

$$z = \frac{1}{2}\exp\left(\pm j\frac{\pi}{3}\right)$$

that is,

$$z = \frac{1}{4} \pm j\frac{\sqrt{3}}{4}$$

Using the notation suggested by Equation (5–115), the poles and zeros can be listed as follows:

$$\lambda_1 = -1$$
$$\lambda_2 = 2$$
$$p_1 = -\frac{3}{4}$$
$$p_2 = \frac{1}{2}$$

(5–117)

$$p_3 = \frac{1}{2}e^{j\frac{\pi}{3}} = \frac{1}{4} + j\frac{\sqrt{3}}{4}$$
$$p_4 = \frac{1}{2}e^{-j\frac{\pi}{3}} = \frac{1}{4} - j\frac{\sqrt{3}}{4} = p_3^*$$

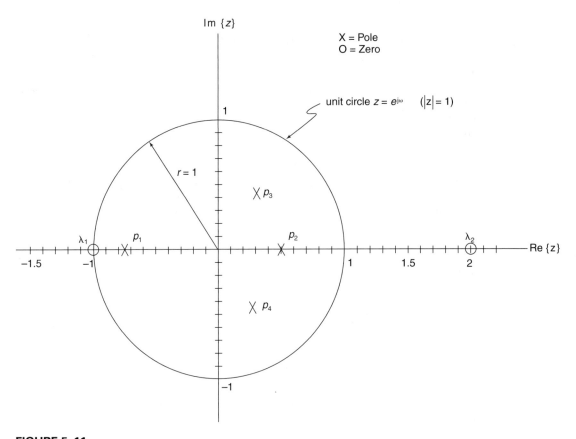

FIGURE 5–11
Poles and zeros on the z plane. Note the location of the unit circle.

These poles and zeros are shown plotted on the z plane in Figure 5–11. The axis labeled $\text{Re}\{z\}$ is the real axis. The axis labeled $\text{Im}\{z\}$ is the imaginary axis. The circle centered on the origin with a radius equal to 1 is called the *unit circle*. The unit circle is defined by

$$|z| = 1 \qquad\qquad (5\text{–}118)$$

The unit circle can also be defined by

$$z = e^{j\omega} \qquad\qquad (5\text{–}119)$$

where ω is an angle in radians. Note that $|e^{j\omega}| = 1$ for any value of ω. For a particular value of ω, $e^{j\omega}$ is a particular point on the unit circle. Any point z *inside* the unit circle has a magnitude of less than one. Any point z *outside* of the unit circle has a magnitude of greater than one. In Section 5–5 we will discuss the significance of the unit circle with respect to the stability of a discrete time system. In Chapter 6, you will see that the *discrete time Fourier transform* of a sequence $x(n)$ is the Z transform of the sequence evaluated on the unit circle in the z plane ($z = e^{j\omega}$).

It is important to note that if $H(z)$ is known (and has the form of Equation [5–114]), the system difference equation (Equation [5–110]) can be obtained. Starting with Equation (5–114), cross multiply to obtain Equation (5–113); rearrange to obtain Equation (5–111); then take the inverse Z transform term by term (applying the shifting property) to obtain the system difference equation. This is important because with some digital filter design methods, the transfer function $H(z)$ is obtained first, and the designer must then derive the system difference equation. Remember that this difference equation is the actual algorithm that must be programmed.

Let us now consider some additional example problems.

EXAMPLE 5–4.2

Given the following system difference equation, find the system transfer function. Also find the poles and zeros:

$$y(n) = x(n) + \tfrac{1}{2} x(n-1) - \tfrac{1}{2} y(n-1) - \tfrac{1}{4} y(n-2) \qquad \textbf{(5–120)}$$

To find the transfer function, take the Z transform of both sides of Equation (5–120):

$$Y(z) = X(z) + \tfrac{1}{2} z^{-1} X(z) - \tfrac{1}{2} z^{-1} Y(z) - \tfrac{1}{4} z^{-2} Y(z) \qquad \textbf{(5–121)}$$

Now solve for the ratio $Y(z)/X(z)$, which by definition is $H(z)$:

$$Y(z)\left[1 + \tfrac{1}{2} z^{-1} + \tfrac{1}{4} z^{-2}\right] = X(z)\left[1 + \tfrac{1}{2} z^{-1}\right]$$

$$\therefore \frac{Y(z)}{X(z)} = H(z) = \frac{1 + \tfrac{1}{2} z^{-1}}{1 + \tfrac{1}{2} z^{-1} + \tfrac{1}{4} z^{-2}} \qquad \textbf{(5–122)}$$

The zeros of the transfer function are those values of z that cause the numerator of $H(z)$ to equal zero. In this example, there is one zero at $z = -1/2$. The poles of the transfer function are those values of z that cause the denominator to equal zero. To find the poles, we must solve the equation

$$1 + \tfrac{1}{2} z^{-1} + \tfrac{1}{4} z^{-2} = 0 \qquad \textbf{(5–123)}$$

In other words, we must find the *roots* of the denominator. We can multiply both sides of Equation (5–123) by z^2 to clarify the problem:

$$z^2 + \tfrac{1}{2} z + \tfrac{1}{4} = 0 \qquad \textbf{(5–124)}$$

The well-known quadratic formula can be used to solve this equation:

$$z = \frac{-\tfrac{1}{2} \pm \sqrt{\left(\tfrac{1}{2}\right)^2 - (4)\left(\tfrac{1}{4}\right)}}{2} \qquad \textbf{(5–125)}$$

By evaluating the right side of Equation (5–125), we find that there is a complex conjugate pair of poles at

$$z = -\tfrac{1}{4} \pm j\,\tfrac{\sqrt{3}}{4} \qquad\qquad (5\text{--}126)$$

EXAMPLE 5–4.3

Given the following transfer function, find the system difference equation:

$$H(z) = \frac{2 + 3z^{-1} - \tfrac{1}{2}z^{-2}}{1 + \tfrac{5}{6}z^{-1} + \tfrac{1}{6}z^{-2}} \qquad\qquad (5\text{--}127)$$

To find the system difference equation, set $H(z)$ equal to $Y(z)/X(z)$, cross multiply, and take the inverse Z transform term by term:

$$\frac{Y(z)}{X(z)} = \frac{2 + 3z^{-1} - \tfrac{1}{2}z^{-2}}{1 + \tfrac{5}{6}z^{-1} + \tfrac{1}{6}z^{-2}}$$

$$\therefore Y(z) + \tfrac{5}{6}z^{-1}Y(z) + \tfrac{1}{6}z^{-2}Y(z) = 2X(z) + 3z^{-1}X(z) - \tfrac{1}{2}z^{-2}X(z) \qquad (5\text{--}128)$$

$$\therefore y(n) + \tfrac{5}{6}y(n-1) + \tfrac{1}{6}y(n-2) = 2x(n) + 3x(n-1) - \tfrac{1}{2}x(n-2)$$

The system difference equation is therefore

$$y(n) = 2x(n) + 3x(n-1) - \tfrac{1}{2}x(n-2) - \tfrac{5}{6}y(n-1) - \tfrac{1}{6}y(n-2) \qquad (5\text{--}129)$$

EXAMPLE 5–4.4

Consider an FIR system with the following transfer function:

$$H(z) = 1 + 2z^{-1} + z^{-2} \qquad\qquad (5\text{--}130)$$

Suppose the input to this system is a finite duration sequence $x(n)$:

$$x(n) = 4\delta(n) + 2\delta(n-1) \qquad\qquad (5\text{--}131)$$

Use the Z transform to find the output signal $y(n)$.

To solve this problem we must first find $\mathbf{Z}\{x(n)\} = X(z)$; we then must use the transfer function relationship $Y(z) = X(z)H(z)$; finally, we must take the inverse Z transform of $Y(z)$:

$$X(z) = \mathbf{Z}\{4\delta(n) + 2\delta(n-1)\} = 4 + 2z^{-1}$$

$$Y(z) = X(z)H(z) = \left(4 + 2z^{-1}\right)\left(1 + 2z^{-1} + z^{-2}\right)$$

$$Y(z) = 4 + 8z^{-1} + 4z^{-2} + 2z^{-1} + 4z^{-2} + 2z^{-3} \qquad (5\text{--}132)$$

$$Y(z) = 4 + 10z^{-1} + 8z^{-2} + 2z^{-3}$$

$$\therefore y(n) = 4\delta(n) + 10\delta(n-1) + 8\delta(n-2) + 2\delta(n-3)$$

Let us now consider the relationship between the system impulse response $h(n)$ and system transfer function $H(z)$. Consider a discrete time LSI system with input $x(n) = \delta(n)$. This means that the system output $y(n)$ is the same as the impulse response $h(n)$. In the Z transform domain, the output $Y(z)$ is given as

$$Y(z) = X(z)H(z) \tag{5–133}$$

But since $x(n) = \delta(n)$, and thus $X(z) = 1$, Equation (5–133) reduces to

$$Y(z) = (1)H(z) = H(z) \tag{5–134}$$

Equation (5–134) implies that when $x(n)$ is the unit impulse, the Z transform of the output $y(n)$ is the same as $H(z)$. But when $x(n)$ is the unit impulse, $y(n)$ is by definition the impulse response $h(n)$. We therefore conclude that the Z transform of the system impulse response is the system transfer function:

$$h(n) \leftrightarrow H(z) \tag{5–135}$$

Note also that since $y(n) = h(n) * x(n)$, and $Y(z) = H(z)X(z)$, we have the following Z transform relationship:

$$h(n) * x(n) \leftrightarrow H(z)X(z) \tag{5–136}$$

In other words, the convolution of two sequences can be evaluated by finding the Z transform of each sequence, multiplying the two Z transforms, and finding the inverse Z transform of the result. The reader is invited to revisit Example 5–4.4, find the impulse response of the FIR system, use convolution instead of the Z transform to find $y(n)$, and compare the result with the solution presented in Equation (5–132). (Of course, the result should be the same.)

5–5 BIBO STABILITY OF LSI DISCRETE TIME SYSTEMS

A discrete time linear shift invariant (LSI) system is said to be *bounded input, bounded output* stable (BIBO stable) if, for any amplitude bounded input $x(n)$, the output $y(n)$ is also amplitude bounded. Signals $x(n)$ and $y(n)$ are amplitude bounded if there exist finite real positive numbers α and β such that

$$\begin{aligned} |x(n)| &< \alpha \quad \text{for all } n \\ |y(n)| &< \beta \quad \text{for all } n \end{aligned} \tag{5–137}$$

Essentially, this means that the output of a BIBO stable system does not blow up when the input is excited by an input signal that is bounded in ampltude. (As noted in Chapter 1, in this book the term *stable* means BIBO stable.)

It can be shown that a necessary and sufficient condition for the BIBO stability of a causal LSI system is that the impulse response is absolutely summable, that is,

$$\sum_{n=0}^{\infty} |h(n)| < \infty \tag{5–138}$$

For the causal LSI systems of interest to us here (causal systems represented by Equation [5–11] or [5–12]), it can be shown that BIBO stability is ensured if the system impulse response dies out as n gets large. That is, such a system is BIBO stable if and only if

$$\lim_{n \to \infty} h(n) = 0 \tag{5–139}$$

An FIR system automatically satisfies this requirement because its impulse response vanishes after a finite amount of time. The difference equation for an FIR system is

$$y(n) = \sum_{k=0}^{M} b_k x(n - k) \tag{5–140}$$

If $x(n) = \delta(n)$, then $y(n) = h(n)$

$$h(n) = \sum_{k=0}^{M} b_k \delta(n - k) \tag{5–141}$$

That is, for an FIR system,

$$h(n) = b_0 \delta(n) + b_1 \delta(n - 1) + \cdots + b_M \delta(n - M) \tag{5–142}$$

Since $h(n) = 0$ for $n > M$, the stability requirement is automatically satisfied.

On the other hand, an IIR system can be either stable or unstable. A simple example will serve to illustrate the point. Consider a causal IIR system represented by the following difference equation:

$$y(n) = x(n) + ay(n - 1) \tag{5–143}$$

where a is a real constant. Taking the Z transform on both sides of the equation results in

$$Y(z) = X(z) + az^{-1}Y(z) \tag{5–144}$$

The next step is to solve for the transfer function:

$$Y(z)\left[1 - az^{-1}\right] = X(z)$$

$$\therefore \frac{Y(z)}{X(z)} = H(z) = \frac{1}{1 - az^{-1}} \tag{5–145}$$

The system impulse response is therefore

$$h(n) = a^n u(n) \tag{5–146}$$

(See Table 5–1). Observe that the stability of this system depends on the constant a. If $|a| < 1$, the system is BIBO stable because $h(n)$ dies out as $n \to \infty$. On the other hand, if $|a| \geq 1$, the system is not BIBO stable.

This simple example also illustrates the fact that whether or not a causal IIR system (Equation [5–11]) is BIBO stable depends on the location of its poles on the z plane. In this example, the system has one pole at $z = a$. If $|a| < 1$, this pole is a point *inside* the unit circle on the z plane (See Figure 5–11). If $|a| = 1$, this pole is a point *on* the unit circle. If $|a| > 1$, this pole is a point *outside* the unit circle. This example therefore serves to illustrate the following rule:

> A causal IIR discrete time system represented by Equation (5–11) is BIBO stable if and only if *all* of its poles have a magnitude of less than 1. In other words, *all* of its poles must be *inside* the unit circle on the z plane.

This rule will be further illustrated by the examples in Section 5–6.

If we have $|a| = 1$ in the previous example, the impulse response of the system is $h(n) = u(n)$, which does not die out as $n \to \infty$, but does not blow up either. In some of the literature this is considered as an example of a *marginally stable* system, but a marginally stable system is *not* BIBO stable, because a bounded input signal can be found that will cause the output of a marginally stable system to blow up. For the specific example at hand, suppose $x(n) = u(n)$, which is a bounded input signal. The output of the system is $y(n) = x(n) * h(n) = u(n) * u(n)$. It is left to the reader to show that the result is $y(n) = (n + 1)u(n)$, which is rising without bound as $n \to \infty$.

Before moving on to the next section, let us summarize the analogies between continuous time system theory and discrete time system theory. (See Table 5–2.) There are

TABLE 5–2
Comparison of LSI systems

Continuous Time System	Discrete Time System
Circuit	Algorithm
System differential equation	System difference equation
Laplace transform	Z transform
$h(t) = T[\delta(t)]$	$h(n) = T[\delta(n)]$
$h(t) * x(t) = \int_{-\infty}^{\infty} h(\tau)x(t - \tau)d\tau$	$h(n) * x(n) = \sum_{k=-\infty}^{\infty} h(k)x(n - k)$
$H(s) = \int_{-\infty}^{\infty} h(t)e^{-st}dt$	$H(z) = \sum_{n=-\infty}^{\infty} h(n)z^{-n}$
$H(s) = \dfrac{\sum_{k=0}^{M} b_k s^k}{1 + \sum_{k=1}^{N} a_k s^k}$	$H(z) = \dfrac{\sum_{k=0}^{M} b_k z^{-k}}{1 + \sum_{k=1}^{N} a_k z^{-k}}$
$h(t) * x(t) \leftrightarrow H(s)X(s)$	$h(n) * x(n) \leftrightarrow H(z)X(z)$
Fourier transform	Discrete time Fourier transform
$H_a(f) = \int_{-\infty}^{\infty} h(t)e^{-j2\pi ft}dt$	$H(e^{j\omega}) = \sum_{n=-\infty}^{\infty} h(n)e^{-j\omega n}$
$h(t) * x(t) \leftrightarrow H_a(f)X_a(f)$	$h(n) * x(n) \leftrightarrow H(e^{j\omega})X(e^{j\omega})$
BIBO stable if all poles are in the left half of the s-plane	BIBO stable if all poles are inside the unit circle on the z-plane

obviously several similar concepts. With respect to both types of systems, we have the concepts of system impulse response, convolution, transfer function, and frequency response function. (The frequency response function of a discrete time system is considered in Chapter 6.) Of course, the mathematical details differ: we have differential equations versus difference equations, Laplace transforms versus Z transforms, etc. The biggest difference between the two types of systems is their actual implementation: circuits versus algorithms. In the design of continuous time systems, one must choose values for circuit components; in the design of discrete time systems, one must choose the values of the coefficients in the algorithm.

5–6 THE Z TRANSFORM AND SYSTEM ANALYSIS

Z transform analysis can be used to determine the output, $y(n)$, from an LSI discrete time system, given the system transfer function and an input signal, $x(n)$, for which the Z transform $X(z)$ can be found. The steps in the problem are as follows:

1. Find the transfer function $H(z)$ of the system.
2. Find the Z transform of the input signal $X(z)$
3. $Y(z) = X(z)H(z)$
4. $y(n) = \mathbf{Z}^{-1}\{Y(z)\}$

$\mathbf{Z}^{-1}\{\ \}$ denotes the inverse Z transform. We will restrict our attention to causal systems initially at rest, where the input signal is first applied at time $n = 0$. There are three types of input signals we will consider here:

1. $x(n) = \delta(n)$, in which case $y(n)$ is the impulse response $h(n)$
2. $x(n) = u(n)$, in which case $y(n)$ is the system *step response*
3. $x(n) = \cos(\beta n)u(n)$

The Z transforms of these signals were derived in Section 5–3, and are shown in Table 5–1. The system transfer function can be obtained from the system difference equation as shown in Section 5–4. Therefore, the main problem at hand is to find the inverse Z transform of $Y(z)$, which in general will have the form

$$Y(z) = \frac{N(z)}{D(z)} = \frac{c_0 + c_1 z^{-1} + c_2 z^{-2} + \cdots + c_M z^{-M}}{1 + d_1 z^{-1} + d_2 z^{-2} + \ldots + d_N z^{-N}} \qquad (5\text{–}147)$$

For the special case $Y(z) = N(z)$, that is, $D(z) = 1$, the procedure for finding $y(n)$ is straightforward, and has already been explained by example (see Example 5–4.4). Therefore, we will assume here that at least some of the coefficients d_k in Equation (5–147) are nonzero.

Things can always be arranged such that $D(z)$ leads off with a 1, as shown in Equation (5–147). For example, if $Y(z)$ is

$$Y(z) = \frac{5 + 3z^{-1}}{2 + 4z^{-1} + 9z^{-2}} \qquad (5\text{–}148)$$

we can multiply the numerator and denominator by 1/2 to obtain the desired form:

$$Y(z) = \frac{\frac{5}{2} + \frac{3}{2} z^{-1}}{1 + 2z^{-1} + \frac{9}{2} z^{-2}} \tag{5-149}$$

In general, the denominator $D(z)$ can be factored as

$$D(z) = \left(1 - p_1 z^{-1}\right)\left(1 - p_2 z^{-1}\right)...\left(1 - p_N z^{-1}\right) \tag{5-150}$$

where $p_1, p_2, \ldots p_N$ are the roots of $D(z)$, that is, the values of z that satisfy the equation:

$$1 + d_1 z^{-1} + d_2 z^{-2} + ... + d_N z^{-N} = 0 \tag{5-151}$$

(One can multiply both sides of Equation [5–151] by z^N to convert to a problem involving positive powers of z.)

If $N > M$, that is, if $D(z)$ is of higher order than $N(z)$, and if the roots $p_1, p_2, \ldots p_N$ are *distinct* (meaning they are all different numbers), then the *partial fraction expansion* of $Y(z)$ is

$$\frac{N(z)}{\left(1 - p_1 z^{-1}\right)\left(1 - p_2 z^{-1}\right)...\left(1 - p_N z^{-1}\right)} = \frac{A_1}{1 - p_1 z^{-1}} + \frac{A_2}{1 - p_2 z^{-1}} + ... + \frac{A_N}{1 - p_N z^{-1}} \tag{5-152}$$

The case of *nondistinct* (i.e., *repeated*) roots will *not* be considered in this book. However, the case $M \geq N$ will be considered later.

To solve for the ith constant A_i in the partial fraction expansion, multiply both sides of Equation (5–152) by $(1 - p_i z^{-1})$, cancel out the common terms, and then let $z = p_i$. The result is

$$A_i = \left(1 - p_i z^{-1}\right)Y(z)\Big|_{z=p_i} = \frac{N(z)}{\prod_{\substack{k=1 \\ k \neq i}}^{N}\left(1 - p_k z^{-1}\right)}\Bigg|_{z=p_i} = \frac{N(p_i)}{\prod_{\substack{k=1 \\ k \neq i}}^{N}\left(1 - \frac{p_k}{p_i}\right)} \tag{5-153}$$

Note that the product in the denominator of Equation (5–153) does *not* include the $k = i$ term, because that term is canceled out by $(1 - p_i z^{-1})$.

Having found $A_1, A_2, \ldots A_N$, the inverse Z transform can be found term by term:

$$y(n) = \left[A_1\left(p_1\right)^n + A_2\left(p_2\right)^n + ... + A_N\left(p_N\right)^n\right]u(n) \tag{5-154}$$

Since the roots $p_1, p_2, \ldots p_N$ can be complex, the values $A_1, A_2, \ldots A_N$ can also be complex. Complex roots will come in complex conjugate pairs. The same thing will be true of the values A_i: if there are complex values of A_i, they will exist in complex conjugate pairs. For example, suppose $p_3 = p_2^*$; it will then turn out that $A_3 = A_2^*$. (This will be shown by example.) The corresponding time domain terms can then be combined in a way that will eliminate the complex numbers from that part of the solution. To see how this is done, start by expressing all of the associated complex numbers in polar form:

$$A_2 = |A_2|e^{j\alpha_2}$$
$$A_3 = A_2^* = |A_2|e^{-j\alpha_2}$$
$$p_2 = |p_2|e^{j\beta_2}$$
$$p_3 = p_2^* = |p_2|e^{-j\beta_2}$$

(5–155)

Then

$$A_2(p_2)^n + A_3(p_3)^n = A_2(p_2)^n + A_2^*(p_2^*)^n$$
$$= |A_2|e^{j\alpha_2}|p_2|^n e^{j\beta_2 n} + |A_2|e^{-j\alpha_2}|p_2|^n e^{-j\beta_2 n}$$
$$= |A_2||p_2|^n \left[e^{j(\beta_2 n + \alpha_2)} + e^{-j(\beta_2 n + \alpha_2)}\right]$$
$$= 2|A_2||p_2|^n \cos(\beta_2 n + \alpha_2)$$

(5–156)

(Again, this will be shown by example.)

If the order of $N(z)$ is equal to or greater than the order of $D(z)$, that is, if $M \geq N$ in Equation (5–147), then before attempting partial fraction expansion we must first divide $N(z)$ by $D(z)$, carrying out the operation until obtaining a remainder of the form

$$R(z) = q_0 + q_1 z^{-1} + \ldots q_L z^{-L}$$

(5–157)

where $L < N$. This is probably best shown by an example. This example will also be used to illustrate partial fraction expansion, etc.

EXAMPLE 5–6.1
Suppose

$$Y(z) = \frac{1 + 4z^{-1} + 4z^{-2} + z^{-3}}{1 + 2z^{-1} + 2z^{-2}}$$

(5–158)

The problem is to find $y(n)$. In Equation (5–158), $M = 3$ and $N = 2$, so we must perform the division before attempting partial fraction expansion. Note that the polynomials are written in reverse order to carry out this operation. The first step is

$$\frac{\frac{1}{2}z^{-1}}{2z^{-2} + 2z^{-1} + 1 \overline{\smash{\big)} z^{-3} + 4z^{-2} + 4z^{-1} + 1}}$$
$$\underline{z^{-3} + z^{-2} + \tfrac{1}{2}z^{-1}}$$
$$3z^{-2} + \tfrac{7}{2}z^{-1} + 1$$

(5–159)

The remainder has order $L = 2$, which is not less than N, so we must continue:

$$
\begin{array}{r}
\frac{1}{2}z^{-1} + \frac{3}{2} \\[2pt]
2z^{-2} + 2z^{-1} + 1 \overline{\smash{\big)}\, z^{-3} + 4z^{-2} + 4z^{-1} + 1} \\[2pt]
\underline{z^{-3} + \; z^{-2} + \frac{1}{2}z^{-1}} \\[2pt]
3z^{-2} + \frac{7}{2}z^{-1} + 1 \\[2pt]
\underline{3z^{-2} + 3z^{-1} + \frac{3}{2}} \\[2pt]
\frac{1}{2}z^{-1} - \frac{1}{2}
\end{array}
$$

(5–160)

The remainder now has order $L = 1 < N$; therefore, we are finished dividing and can express $Y(z)$ as

$$ Y(z) = \frac{3}{2} + \frac{1}{2}z^{-1} + \frac{-\frac{1}{2} + \frac{1}{2}z^{-1}}{1 + 2z^{-1} + 2z^{-2}} = \frac{3}{2} + \frac{1}{2}z^{-1} + \frac{R(z)}{D(z)} $$

(5–161)

Partial fraction expansion can now be carried out on $R(z)/D(z)$. It is left as an exercise for the reader to show that $D(z)$ has two distinct roots, p_1 and p_2, that are a complex conjugate pair:

$$ p_1 = -1 + j = 1.4142e^{j2.3562} = |p_1|e^{j\beta_1} $$
$$ p_2 = -1 - j = 1.4142e^{-j2.3562} = p_1^* = |p_1|e^{-j\beta_1} $$

(5–162)

The partial fraction expansion of $R(z)/D(z)$ is

$$ \frac{-\frac{1}{2} + \frac{1}{2}z^{-1}}{\left(1 - p_1 z^{-1}\right)\left(1 - p_2 z^{-1}\right)} = \frac{A_1}{\left(1 - p_1 z^{-1}\right)} + \frac{A_2}{\left(1 - p_2 z^{-1}\right)} $$

(5–163)

where

$$ A_1 = \frac{-\frac{1}{2} + \frac{\frac{1}{2}}{p_1}}{1 - \frac{p_2}{p_1}} = -0.25 - j0.5 = 0.559017e^{-j2.0344} = |A_1|e^{j\alpha_1} $$

(5–164)

$$ A_2 = \frac{-\frac{1}{2} + \frac{\frac{1}{2}}{p_2}}{1 - \frac{p_1}{p_2}} = -0.25 + j0.5 = 0.559017e^{j2.0344} = |A_1|e^{-j\alpha_1} $$

(5–165)

The reader might verify the results of Equations (5–164) and (5–165) as an exercise. We are now ready to write down the solution for $y(n)$:

$$ y(n) = \frac{3}{2}\delta(n) + \frac{1}{2}\delta(n-1) + A_1(p_1)^n u(n) + A_2(p_2)^n u(n) $$

(5–166)

Using the results shown in Equation (1–156), the two terms involving A_1 and A_2 can be combined as follows:

$$y(n) = \tfrac{3}{2}\delta(n) + \tfrac{1}{2}\delta(n-1) + 2|A_1||p_1|^n \cos(\beta_1 n + \alpha_1)u(n) \qquad \text{(5–167)}$$

That is,

$$y(n) = \tfrac{3}{2}\delta(n) + \tfrac{1}{2}\delta(n-1) + 1.118(1.4142)^n \cos(2.3562n - 2.0344)u(n) \qquad \text{(5–168)}$$

Now that the mathematical tools are in place, we can consider a few more example problems.

EXAMPLE 5–6.2

Given the following transfer function:

$$H(z) = \frac{3 + z^{-1}}{1 + 0.2z^{-1} - 0.48z^{-2}} = \frac{3 + z^{-1}}{\left(1 + 0.8z^{-1}\right)\left(1 - 0.6z^{-1}\right)} \qquad \text{(5–169)}$$

1. Find the impulse response of this system.
2. Find the step response of this system.

To find the impulse response, that is, $h(n)$, we must find the inverse Z transform of $H(z)$. We have the form

$$H(z) = \frac{N(z)}{D(z)} = \frac{3 + z^{-1}}{\left(1 - p_1 z^{-1}\right)\left(1 - p_2 z^{-1}\right)} \qquad \text{(5–170)}$$

where the roots of $D(z)$ are $p_1 = -0.8$ and $p_2 = 0.6$. In this case, these roots are also the poles of $H(z)$. Since both poles are inside the unit circle (i.e., have magnitudes less than 1), we anticipate that $h(n)$ will die out as $n \to \infty$.

Since $D(z)$ is of greater order than $N(z)$, the first step in the solution is partial fraction expansion:

$$H(z) = \frac{3 + z^{-1}}{\left(1 - p_1 z^{-1}\right)\left(1 - p_2 z^{-1}\right)} = \frac{A_1}{1 - p_1 z^{-1}} + \frac{A_2}{1 - p_2 z^{-1}} \qquad \text{(5–171)}$$

where

$$A_1 = \frac{3 + \frac{1}{p_1}}{1 - \frac{p_2}{p_1}} = \frac{3 - \frac{1}{.8}}{1 + \frac{.6}{.8}} = 1$$

$$A_2 = \frac{3 + \frac{1}{p_2}}{1 - \frac{p_1}{p_2}} = \frac{3 + \frac{1}{.6}}{1 + \frac{.8}{.6}} = 2 \qquad \text{(5–172)}$$

Therefore, the impulse response is

$$h(n) = \left[A_1(p_1)^n + A_2(p_2)^n\right]u(n) = \left[(-0.8)^n + 2(0.6)^n\right]u(n) \qquad (5\text{–}173)$$

Observe that $h(n)$ dies out as $n \to \infty$, as expected.

To find the step response of this system, we start by noting that since $x(n) = u(n)$,

$$X(z) = \frac{1}{1 - z^{-1}} \qquad (5\text{–}174)$$

Therefore,

$$Y(z) = H(z)X(z) = \frac{3 + z^{-1}}{\left(1 + 0.8z^{-1}\right)\left(1 - 0.6z^{-1}\right)\left(1 - z^{-1}\right)} = \frac{N(z)}{D(z)} \qquad (5\text{–}175)$$

This has the form

$$\frac{N(z)}{D(z)} = \frac{3 + z^{-1}}{\left(1 - p_1 z^{-1}\right)\left(1 - p_2 z^{-1}\right)\left(1 - p_3 z^{-1}\right)} \qquad (5\text{–}176)$$

where the roots of $D(z)$ are $p_1 = -0.8$, $p_2 = 0.6$, and $p_3 = 1$. Since $D(z)$ is of greater order than $N(z)$, we proceed at once to partial fraction expansion:

$$Y(z) = \frac{3 - z^{-1}}{\left(1 - p_1 z^{-1}\right)\left(1 - p_2 z^{-1}\right)\left(1 - p_3 z^{-1}\right)} = \frac{A_1}{1 - p_1 z^{-1}} + \frac{A_2}{1 - p_2 z^{-1}} + \frac{A_3}{1 - p_3 z^{-1}} \qquad (5\text{–}177)$$

where

$$A_1 = \frac{3 + \frac{1}{p_1}}{\left(1 - \frac{p_2}{p_1}\right)\left(1 - \frac{p_3}{p_1}\right)} = \frac{3 - \frac{1}{.8}}{\left(1 + \frac{.6}{.8}\right)\left(1 + \frac{1}{.8}\right)} = 0.44444$$

$$A_2 = \frac{3 + \frac{1}{p_2}}{\left(1 - \frac{p_1}{p_2}\right)\left(1 - \frac{p_3}{p_2}\right)} = \frac{3 + \frac{1}{.6}}{\left(1 + \frac{.8}{.6}\right)\left(1 - \frac{1}{.6}\right)} = -3.0 \qquad (5\text{–}178)$$

$$A_3 = \frac{3 + \frac{1}{p3}}{\left(1 - \frac{p_1}{p_3}\right)\left(1 - \frac{p_2}{p_3}\right)} = \frac{3 + \frac{1}{1}}{\left(1 + \frac{.8}{1}\right)\left(1 - \frac{.6}{1}\right)} = 5.5555$$

Therefore, the step response of this system is

$$y(n) = \left[A_1(p_1)^n + A_2(p_2)^n + A_3(p_3)^n\right]u(n)$$
$$= \left[0.4444(-0.8)^n - 30(0.6)^n + 5.5555\right]u(n) \qquad (5\text{–}179)$$

In this case, $y(n) \to 5.5555$ as $n \to \infty$.

In Section 5–5 it was noted that a causal discrete time system is BIBO stable if and only if all of its poles are inside the unit circle. This example illustrates that point, since the input is amplitude bounded, $x(n) = u(n)$, the system poles are inside the unit circle, and the output is amplitude bounded.

EXAMPLE 5–6.3

Given the following transfer function:

$$H(z) = \frac{\frac{1}{2} + z^{-2}}{\left(1 - p_1 z^{-1}\right)\left(1 - p_2 z^{-1}\right)} = \frac{\frac{1}{2} + z^{-2}}{1 - 1.2728 z^{-1} + 0.81 z^{-2}} \quad \text{(5–180)}$$

where the system poles, p_1 and p_2, are a complex conjugate pair:

$$p_1 = |p_1| e^{j\beta_1} = 0.9 e^{j\frac{\pi}{4}} = 0.636396 + j.636396$$

$$p_2 = |p_1| e^{-j\beta_1} = 0.9 e^{-j\frac{\pi}{4}} = 0.636396 - j.636396$$

$$\quad \text{(5–181)}$$

1. Find the impulse response of this system.
2. Find the step response of this system.

Since the poles of this system are inside the unit circle, we expect that the impulse response will die out as $n \to \infty$. To find the impulse response, we must find the inverse Z transform of $H(z) = N(z)/D(z)$. In this case, $D(z)$ has the same order as $N(z)$, so our first step must be to divide $N(z)$ by $D(z)$. The result is

$$H(z) = 1.23457 + \frac{-0.73457 + 1.57136 z^{-1}}{1 - 1.2728 z^{-1} + 0.81 z^{-2}} = 1.23457 + \frac{R(z)}{D(z)} \quad \text{(5–182)}$$

(The reader is invited to verify this result.) Since $D(z)$ has an order greater than $R(z)$, we can carry out the partial fraction expansion on $R(z)/D(z)$:

$$\frac{-0.73457 + 1.57136 z^{-1}}{\left(1 - p_1 z^{-1}\right)\left(1 - p_2 z^{-1}\right)} = \frac{A_1}{1 - p_1 z^{-1}} + \frac{A_2}{1 - p_2 z^{-1}} \quad \text{(5–183)}$$

where

$$A_1 = \frac{-0.73457 + \frac{1.57136}{p_1}}{1 - \frac{p_2}{p_1}} = 0.94186 e^{-j1.9714} = |A_1| e^{j\alpha_1}$$

$$A_2 = \frac{-0.73457 + \frac{1.57136}{p_2}}{1 - \frac{p_1}{p_2}} = 0.94186 e^{j1.9714} = A_1^* = |A_1| e^{-j\alpha_1}$$

$$\quad \text{(5–184)}$$

(Again, the reader is invited to verify these numbers.) Since $p_2 = p_1^*$ and $A_2 = A_1^*$, we can apply the results shown in Equation (1–156). The impulse response of this system is

$$h(n) = 1.23457\delta(n) + 2(0.94186)(0.9)^n \cos\left(\frac{\pi}{4} n - 1.9714\right) u(n) \quad \text{(5–185)}$$

Note that as expected, $h(n)$ dies out as $n \to \infty$. The impulse response is shown in Figure 5–12(a).

To find the step response of this system, we must find the inverse Z transform of

$$Y(z) = H(z)X(z) = \frac{\frac{1}{2} + z^{-2}}{\left(1 - p_1 z^{-1}\right)\left(1 - p_2 z^{-1}\right)\left(1 - p_3 z^{-1}\right)} = \frac{N(z)}{D(z)} \qquad (5\text{–}186)$$

where $p_3 = 1$.

NOTE: The Z transform of the unit step is $1/(1 - z^{-1})$. This time, the order of $D(z)$ is greater than the order of $N(z)$, so we can proceed directly to partial fraction expansion:

$$Y(z) = \frac{\frac{1}{2} + z^{-2}}{\left(1 - p_1 z^{-1}\right)\left(1 - p_2 z^{-1}\right)\left(1 - p_3 z^{-1}\right)} \qquad (5\text{–}187)$$

$$= \frac{A_1}{1 - p_1 z^{-1}} + \frac{A_2}{1 - p_2 z^{-1}} + \frac{A_3}{1 - p_3 z^{-1}}$$

where

$$A_1 = \frac{\frac{1}{2} + \frac{1}{p_1^2}}{\left(1 - \frac{p_2}{p_1}\right)\left(1 - \frac{p_3}{p_1}\right)} = 1.1565 e^{j3.0073}$$

$$A_2 = \frac{\frac{1}{2} + \frac{1}{p_2^2}}{\left(1 - \frac{p_1}{p_2}\right)\left(1 - \frac{p_3}{p_2}\right)} = 1.1565 e^{-j3.0073} = A_1^* \qquad (5\text{–}188)$$

$$A_3 = \frac{\frac{1}{2} + \frac{1}{p_3^2}}{\left(1 - \frac{p_1}{p_3}\right)\left(1 - \frac{p_2}{p_3}\right)} = 2.7923$$

The step response is therefore

$$y(n) = \left[A_1(p_1)^n + A_2(p_2)^n + A_3(p_3)^n\right] u(n) \qquad (5\text{–}189)$$

But since $p_2 = p_1^*$ and $A_2 = A_1^*$, this result can be expressed as

$$y(n) = 2|A_1||p_1|^n \cos(\beta_1 n + \alpha_1) u(n) + A_3(p_3)^n u(n) \qquad (5\text{–}190)$$

Plugging in the appropriate numbers, we get

$$y(n) = 2(1.1565)(0.9)^n \cos\left(\tfrac{\pi}{4} n + 3.0073\right) u(n) + 2.7923 u(n) \qquad (5\text{–}191)$$

In this case, note that $y(n) \to 2.7923$ as $n \to \infty$. This step response is shown in Figure 5–12 (b). This example illustrates a BIBO stable system: the input is bounded, $x(n) = u(n)$, the system poles are inside the unit circle, and the output is bounded.

FIGURE 5–12
(a) Impulse response for Example 5–6.3. (b) Step response for Example 5–6.3.

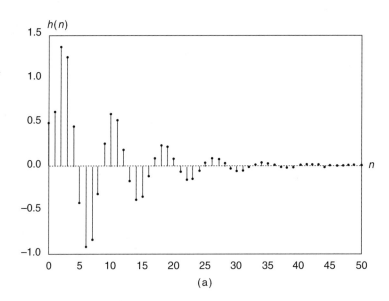

(a)

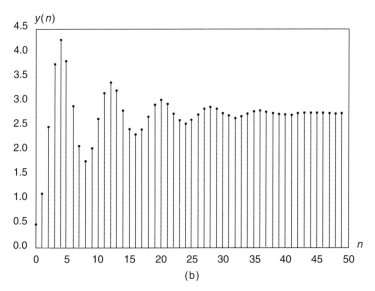

(b)

EXAMPLE 5–6.4

Given a causal, BIBO-stable system with the following transfer function:

$$H(z) = \frac{1}{1 - 0.8z^{-1}} \tag{5–192}$$

find the response of this system to the following input signal:

$$x(n) = \cos\left(\tfrac{\pi}{4} n\right)u(n) \tag{5–193}$$

To find the response, that is, output signal $y(n)$, we first note that the Z transform of the input signal is

$$X(z) = \frac{1 - \cos\left(\frac{\pi}{4}\right)z^{-1}}{1 - 2\cos\left(\frac{\pi}{4}\right)z^{-1} + z^{-2}} \qquad (5\text{--}194)$$

To find $y(n)$, we must find the inverse Z transform of $Y(z)$, where

$$Y(z) = H(z)X(z) = \frac{1 - \cos\left(\frac{\pi}{4}\right)z^{-1}}{\left(1 - 0.8z^{-1}\right)\left(1 - 2\cos\left(\frac{\pi}{4}\right)z^{-1} + z^{-2}\right)} \qquad (5\text{--}195)$$

$Y(z)$ can be expressed as

$$Y(z) = \frac{1 - \left(\frac{1}{\sqrt{2}}\right)z^{-1}}{\left(1 - p_1 z^{-1}\right)\left(1 - p_2 z^{-1}\right)\left(1 - p_3 z^{-1}\right)} = \frac{N(z)}{D(z)} \qquad (5\text{--}196)$$

The roots of $D(z)$ are

$$p_1 = 0.8$$
$$p_2 = e^{j\frac{\pi}{4}} \qquad (5\text{--}197)$$
$$p_3 = e^{-j\frac{\pi}{4}} = p_2^*$$

Since $D(z)$ has an order greater than that of $N(z)$, the next step is partial fraction expansion:

$$Y(z) = \frac{N(z)}{D(z)} = \frac{A_1}{1 - p_1 z^{-1}} + \frac{A_2}{1 - p_2 z^{-1}} + \frac{A_3}{1 - p_3 z^{-1}} \qquad (5\text{--}198)$$

The constants A_1, A_2, and A_3 are evaluated as follows:

$$A_1 = \frac{1 - \frac{1}{p_1 \sqrt{2}}}{\left(1 - \frac{p_2}{p_1}\right)\left(1 - \frac{p_3}{p_1}\right)} = 1.4611$$

$$A_2 = \frac{1 - \frac{1}{p_2 \sqrt{2}}}{\left(1 - \frac{p_1}{p_2}\right)\left(1 - \frac{p_3}{p_2}\right)} = 0.70107e^{-j0.91602} \qquad (5\text{--}199)$$

$$A_3 = \frac{1 - \frac{1}{p_3 \sqrt{2}}}{\left(1 - \frac{p_1}{p_3}\right)\left(1 - \frac{p_2}{p_3}\right)} = 0.70107e^{j0.91602} = A_2^*$$

The output is therefore

$$y(n) = \left[A_1(p_1)^n + A_2(p_2)^n + A_3(p_3)^n\right]u(n) \qquad (5\text{--}200)$$

But since $A_3 = A_2^*$, $p_3 = p_2^*$, and $|p_2| = |p_3| = 1$, $y(n)$ can be expressed as

$$y(n) = 1.4611(0.8)^n u(n) + 2(0.70107)\cos\left(\tfrac{\pi}{4}n - 0.91602\right)u(n) \tag{5-201}$$

In this case, observe that as $n \to \infty$, $y(n)$ takes on the same form as the input signal $x(n)$: a sampled sinusoid with frequency $\pi/4$ radians per sample. However, the amplitude and phase of the sinusoid have been altered by the system. This example also illustrates a BIBO stable system: the input

$$x(n) = \cos\left(\frac{\pi}{4}n\right)u(n)$$

is bounded, the system pole is inside the unit circle, and the output is bounded.

We are now ready to investigate the general form of the solution with respect to the impulse response, step response, and response to $x(n) = \cos(\omega_0 n)u(n)$ of a causal system having distinct poles. We assume that the transfer function of this system has the form

$$H(z) = \frac{N_H(z)}{D_H(z)} = \frac{b_0 + b_1 z^{-1} + b_2 z^{-2} + \ldots + b_M z^{-M}}{1 + a_1 z^{-1} + a_2 z^{-2} + \ldots + a_N z^{-N}} \tag{5-202}$$

where (with little loss of generality and in order to streamline the discussion) it is assumed that $N \geq M$. We assume that $H(z)$ has I real poles and L *pairs* of complex conjugate poles, for a total of $N = I + 2L$ distinct poles:

real poles : $\quad p_k, k = 1, 2, \ldots, I$

complex poles : $\quad p_k = |p_k|e^{j\beta_k}, k = I+1, \ldots, I+L$

$$p_k^* = |p_k|e^{-j\beta_k}, k = I+1, \ldots, I+L \tag{5-203}$$

In other words, $H(z)$ can be expressed as

$$H(z) = \frac{N_H(z)}{\prod\limits_{k=1}^{I}\left(1 - p_k z^{-1}\right)\prod\limits_{k=I+1}^{I+L}\left(1 - p_k z^{-1}\right)\left(1 - p_k^* z^{-1}\right)} \tag{5-204}$$

Furthermore, it is assumed that $N_H(z)$ and $D_H(z)$ have no roots in common.

The *impulse response* of this system is $h(n) = \mathbf{Z}^{-1}\{H(z)\}$. If the order of $D_H(z)$ is greater than the order of $N_H(z)$ (that is, if $N > M$), then $H(z)$ can be written as

$$H(z) = \sum_{k=1}^{I}\frac{A_k}{1 - p_k z^{-1}} + \sum_{k=I+1}^{I+L}\left(\frac{A_k}{1 - p_k z^{-1}} + \frac{A_k^*}{1 - p_k^* z^{-1}}\right) \tag{5-205}$$

where

$$A_i = \left(1 - p_i z^{-1}\right)H(z)\Big|_{z=p_i}, i = 1, 2, \ldots, I, I+1, \ldots, I+L \tag{5-206}$$

The values of A_k, $k = 1, 2, \ldots, I$ will be real. The values of A_k, $k = I + 1, \ldots I + L$ will be complex:

$$A_k \Big|_{k=I+1,\ldots I+L} = |A_k| e^{j\alpha_k} \Big|_{k=I+1,\ldots I+L} \tag{5–207}$$

The solution for $h(n)$ will have the form

$$h(n) = \left[\sum_{k=1}^{I} A_k (p_k)^n + \sum_{k=I+1}^{I+L} 2|A_k||p_k|^n \cos(\beta_k n + \alpha_k) \right] u(n) \tag{5–208}$$

Observe that if all of the system poles are inside the unit circle, that is, $|p_k| < 1$ for all k, then $h(n) \to 0$ as $n \to 0$. The form of the impulse response clearly depends on the location of the system poles.

If $D_H(z)$ and $N_H(z)$ have the same order, then $H(z)$ must be expressed as follows prior to partial fraction expansion:

$$H(z) = B + \frac{R(z)}{D_H(z)} \tag{5–209}$$

where $R(z)$ has an order less than that of $D_H(z)$. Partial fraction expansion is then carried out on $R(z)/D_H(z)$:

$$\frac{R(z)}{D_H(z)} = \sum_{k=1}^{I} \frac{A_k}{1 - p_k z^{-1}} + \sum_{k=I+1}^{I+L} \left(\frac{A_k}{1 - p_k z^{-1}} + \frac{A_k^*}{1 - p_k^* z^{-1}} \right) \tag{5–210a}$$

where

$$A_i = \left(1 - p_i z^{-1}\right) \frac{R(z)}{D_H(z)} \bigg|_{z=p_i} , i = 1, \ldots, I, I + 1, \ldots, I + L \tag{5–210b}$$

The impulse response has the form

$$h(n) = B\delta(n) + \sum_{k=1}^{I} A_k (p_k)^n u(n) + \sum_{k=I+1}^{I+L} 2|A_k||p_k|^n \cos(\beta_k n + \alpha_k) u(n) \tag{5–211}$$

which is the same form as Equation (5–208), except for the extra term $B\delta(n)$.

The *step response* of this system is the output $y(n)$ that results when $x(n) = u(n)$. The step response is $y(n) = \mathbf{Z}^{-1}\{X(z)H(z)\}$, where $X(z)$ is the Z transform of $u(n)$:

$$X(z) = \frac{1}{1 - z^{-1}} \tag{5–212}$$

To make this fit the form of our general expression for $Y(z)$, write $X(z)$ as

$$X(z) = \frac{1}{1 - p_0 z^{-1}} \tag{5–213}$$

where it is understood that $p_0 = 1$. We can now express $Y(z)$ as

$$Y(z) = X(z)H(z) = \frac{N_H(z)}{\left(1 - p_0 z^{-1}\right)D_H(z)} = \frac{N(z)}{D(z)} \qquad (5\text{–}214)$$

$D(z)$ will have greater order than $N(z)$, given the assumptions about $H(z)$. $D(z)$ has $(I + 1)$ real roots $p_0, p_1, \ldots, p_I$, and L pairs of complex conjugate roots p_k and p_k^*, $k = I + 1, \ldots, I + L$.

NOTE: The roots of $D(z)$ are the poles of $H(z)$, plus an additional root ($p_0 = 1$). We assume that these roots are distinct, so $Y(z)$ can be expressed as

$$Y(z) = \sum_{k=0}^{I} \frac{A_k}{1 - p_k z^{-1}} + \sum_{k=I+1}^{I+L} \left(\frac{A_k}{1 - p_k z^{-1}} + \frac{A_k^*}{1 - p_k^* z^{-1}} \right) \qquad (5\text{–}215)$$

where

$$A_i = \left(1 - p_i z^{-1}\right)Y(z)\Big|_{z=p_i} , i = 0, 1, \ldots, I, I + 1, \ldots, I + L \qquad (5\text{–}216)$$

The values A_k, $k = 0, 1, \ldots, I$ will be real. The values A_k, $k = I + 1, \ldots, I + L$ will be complex: $A_k = |A_k| \exp(j\alpha_k)$. The step response will have the form

$$y(n) = \sum_{k=0}^{I} A_k \left(p_k\right)^n u(n) + \sum_{k=I+1}^{I+L} 2|A_k| |p_k|^n \cos(\beta_k n + \alpha_k) u(n) \qquad (5\text{–}217)$$

Using the fact that $p_0 = 1$, we can write the step response as

$$y(n) = A_0 u(n) + \left\{ \sum_{k=1}^{I} A_k \left(p_k\right)^n u(n) + \sum_{k=I+1}^{I+L} 2|A_k| |p_k|^n \cos(\beta_k n + \alpha_k) u(n) \right\} \qquad (5\text{–}218)$$

Observe that $y(n)$ has two components. The first component, $A_0 u(n)$, has the same form as the input signal, $x(n) = u(n)$. This component is known as the *forced response* of the system. The second component is the sum of all the terms in the braces. This component is known as the *natural response* of the system. The form of the natural response depends on the location of the poles of the transfer function. If all of the system poles are inside the unit circle (that is, if the system is BIBO stable), the natural response is called a *transient response* because it dies out as n → ∞. If even one pole is outside of the unit circle, $y(n)$ will blow up. If none of the poles are outside the unit circle, but some are actually on the unit circle (a *marginally stable* system), then for this particular input signal the output signal does not blow up, but the components of the natural response associated with the poles on the unit circle will not die out. However, it is always possible to find a bounded input signal that will cause the output of a marginally stable system to blow up. As noted in Section 5–5, a marginally stable system is not BIBO stable.

The response of this system to a sinusoidal input of the form $x(n) = \cos(\omega_o n)u(n)$ is $y(n) = \mathbf{Z}^{-1}\{X(z)H(z)\}$, where

$$X(z) = \frac{1 - \cos(\omega_o)z^{-1}}{1 - 2\cos(\omega_o)z^{-1} + z^{-2}} = \frac{1 - \cos(\omega_o)z^{-1}}{\left(1 - p_0 z^{-1}\right)\left(1 - p_0^* z^{-1}\right)} \qquad (5\text{–}219)$$

where the roots of the denominator of $X(z)$ are

$$p_0 = e^{j\omega_o}$$
$$p_0^* = e^{-j\omega_o}$$

(5–220)

Therefore, $Y(z)$ can be written as

$$Y(z) = \frac{\left(1 - \cos(\omega_o)z^{-1}\right)N_H(z)}{\left(1 - p_0 z^{-1}\right)\left(1 - p_0^* z^{-1}\right)D_H(z)} = \frac{N(z)}{D(z)}$$

(5–221)

$D(z)$ will have an order greater than that of $N(z)$, given the assumptions about $H(z)$. $D(z)$ has I real roots $p_1, \ldots, p_I$, and $(L + 1)$ pairs of complex conjugate roots p_k and p_k^*, $k = 0$, and $k = I + 1, \ldots, I + L$. The roots of $D(z)$ are the poles of $H(z)$, plus an additional pair of roots (p_0 and p_0^*). We assume that the roots of $D(z)$ are distinct, so that $Y(z)$ can be written as

$$Y(z) = \left(\frac{A_0}{1 - p_0 z^{-1}} + \frac{A_0^*}{1 - p_0^* z^{-1}}\right) + \sum_{k=1}^{I} \frac{A_k}{1 - p_k z^{-1}} + \sum_{k=I+1}^{I+L}\left(\frac{A_k}{1 - p_k z^{-1}} + \frac{A_k^*}{1 - p_k^* z^{-1}}\right)$$

(5–222)

where

$$A_i = \left(1 - p_i z^{-1}\right)Y(z)\Big|_{z=p_i}, i = 0, 1, \ldots, I, I + 1, \ldots, I + L$$

(5–223)

The values A_k, $k = 1, \ldots, I$ will be real. The values A_k, $k = 0$, and $k = I + 1, \ldots, I + L$ will be complex: $A_k = |A_k| \exp(j\alpha_k)$. The solution for $y(n)$ will have the form

$$y(n) = 2|A_0| \cos(\omega_o n + \alpha_0)u(n)$$
$$+ \left\{\sum_{k=1}^{I} A_k (p_k)^n u(n) + \sum_{k=I+1}^{I+L} 2|A_k||p_k|^n \cos(\beta_k n + \alpha_k)u(n)\right\}$$

(5–224)

In writing down the first term on the right side, we have taken advantage of the fact that $|p_0|^n$ $= 1^n = 1$. Again we see that $y(n)$ has two components; the first, $2|A_0| \cos(\omega_o n + \alpha_0)u(n)$, has the same form as the input signal: a sampled sinusoid with the same frequency ω_o. The amplitude and phase of this sinusoid have been altered by the system. This is the *forced response* of the system. The second component is the sum of all the terms in the braces, the form of which depends on the location of the poles of the system. This component is the *natural response* of the system. If all of the system poles are inside the unit circle, the natural response is a *transient response* because it dies out as $n \to \infty$. If the natural response is actually a transient response, the forced response can be called the *steady-state response* of the system.

PROBLEMS

5–1. Given

$$x_a(t) = \begin{cases} e^{-4t}, & t \geq 0 \\ 0, & t < 0 \end{cases}$$

Suppose this signal is sampled at $f_s = 10$ Hz to create the sequence $x(n)$. Evaluate $x(n)$ for $n = 0, 1, \ldots, 10$, and graph $x(n)$.

5–2. Show that the discrete time system described below is *not* a linear system:

$$y(n) = T[x(n)] = x^2(n)$$

5–3. Given the following system difference equation:

$$y(n) = x(n) + 3x(n-1) + x(n-2)$$

where the input signal is

$$x(n) = \begin{cases} 1, & n = 0, 1, 2, 3, 4 \\ 0, & \text{otherwise} \end{cases}$$

a. Assume the system is initially at rest. Find $y(n)$ by direct evaluation of the difference equation, as shown in Section 5–1.
b. Draw the block diagram for this system.
c. Find and graph the impulse response of this system.

5–4. Given the following system difference equation:

$$y(n) = \tfrac{1}{2} x(n) + \tfrac{1}{4} x(n-1) - \tfrac{1}{4} y(n-1)$$

where the input signal is

$$x(n) = \begin{cases} 1, & n = 0, 1, 2, 3 \\ 0, & \text{otherwise} \end{cases}$$

a. Assume the system is initially at rest. Recursively calculate $y(0), y(1), \ldots, y(8)$, as shown in Section 5–1.
b. Find a compact expression for $y(n)$ when $n > 3$.
c. Draw the block diagram for this system.

5–5. Consider the system shown in Figure P5–5.
a. Find the system difference equation.
b. Find and graph the impulse response of this system.

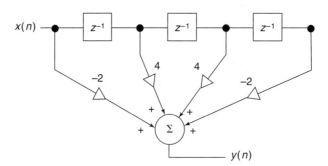

FIGURE P5–5

5–6. Consider the system shown in Figure P5–6.
a. Find the system difference equation.
b. Find and graph the impulse response of this system.

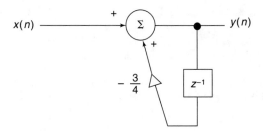

FIGURE P5–6

5–7. Consider two sequences:

$$h(n) = \delta(n) + 2\delta(n-1) - \delta(n-2) + \delta(n-3)$$
$$x(n) = 2\delta(n) + \delta(n-1) + \delta(n-3)$$

Let $y(n) = x(n) * h(n)$, where $*$ denotes convolution.
Evaluate $y(n)$ using the method illustrated by Figure 5–9.

5–8. Consider two sequences

$$h(n) = \delta(n) + \delta(n-1)$$
$$x(n) = 10\delta(n) - 5\delta(n-1) + 2\delta(n-2)$$

Let $y(n) = x(n) * h(n)$, where $*$ denotes convolution.
Evaluate $y(n)$ using the method illustrated by Figure 5–9.

5–9. Given the following system difference equation:

$$y(n) = \tfrac{1}{4}x(n) + \tfrac{1}{2}x(n-1) + x(n-2) + \tfrac{1}{2}x(n-3) + \tfrac{1}{4}x(n-4)$$

Write a pseudocode program to realize this system, in the style suggested in Section 5–1.

5–10. Let

$$x(n) = 5\delta(n) + 10\delta(n-1) - 3\delta(n-2)$$

Find $X(z)$.

5–11. Let

$$Y(z) = 10 + \frac{1}{2}z^{-1} - 6z^{-4} + \frac{1}{4}z^{-6}$$

Find $y(n)$.

5–12. Let

$$x(n) = 5u(n) + \left(\frac{1}{2}\right)^{n} u(n) + 6\left(-\frac{1}{3}\right)^{n} u(n)$$

Find $X(z)$.

5–13. Let

$$Y(z) = \frac{10z^{-3}}{1 + z^{-1}}$$

Find $y(n)$.

HINT: Start by finding the inverse Z transform of

$$\frac{10}{1 + z^{-1}}$$

then use the time shifting property.

5–14. Let

$$x(n) = \begin{cases} \alpha^n, n = 0,1,2,\dots, N-1 \\ 0, \text{otherwise} \end{cases}$$

Find $X(z)$.

HINT: This is a finite geometric series.

5–15. Let

$$y(n) = \left[a^n u(n) \right] * \left[b^n u(n) \right]$$

where $*$ denotes convolution.

Find $Y(z)$.

5–16. Solve Problem 5–7 using Z transforms.

5–17. Solve Problem 5–8 using Z transforms.

5–18. Given:

$$H(z) = -\tfrac{1}{2} + 2z^{-1} - \tfrac{1}{4}z^{-2} + 3z^{-3}$$

a. Find the system difference equation.
b. Find the system impulse response.
c. Is this system BIBO stable? Why or why not?
d. How many zeros does this system have?

5–19. Given

$$H(z) = \frac{1}{1 + 0.9z^{-1}}$$

a. Find the system difference equation.
b. Find the system impulse response.
c. Find the poles of this system and plot them on the z plane.
d. Is this system BIBO stable? Why or why not?

5–20. Given

$$H(z) = \frac{1 + z^{-1}}{1 - 2z^{-1}}$$

a. Find the system difference equation.
b. Find the poles and zeros of this system and plot them on the z plane.
c. Is this system BIBO stable? Why or why not?

5–21. Given

$$H(z) = \frac{1 + z^{-2}}{1 - \left(\dfrac{1}{\sqrt{2}} \right)z^{-1} + \dfrac{1}{4}z^{-2}}$$

a. Find the system difference equation.
b. Find the poles and zeros of this system and plot them on the z plane.
c. Is this system BIBO stable? Why or why not?

5–22. Given

$$h(n) = \left(\frac{5}{6} \right)^n \cos\left(\frac{\pi}{10} n \right) u(n)$$

a. Find the system transfer function $H(z)$.

HINT: Look back at Table 5–1.

b. Find the system difference equation.

c. Find the poles and zeros of this system and plot them on the z plane.

d. Is this system BIBO stable? Why or why not?

5–23. Given

$$H(z) = \left(1 - \lambda z^{-1}\right)\left(1 - \lambda^* z^{-1}\right)$$

where

$$\lambda = e^{j\frac{\pi}{4}}$$

a. Find the system difference equation. (The coefficients should turn out to be real numbers.)

b. Find the system impulse response.

c. Find the zeros of this system and plot them on the z plane.

d. Is this system BIBO stable? Why or why not?

5–24. Consider a system with the following impulse response:

$$h(n) = u(n)$$

a. Show that the system pole is *on* the unit circle.

b. Let $x(n) = u(n)$ be the input to this system. Show that

$$y(n) = x(n) * h(n) = u(n) * u(n) = (n + 1)u(n)$$

NOTE: Observe that this system is not BIBO stable; the input signal is amplitude bounded but the output blows up.

5–25. Given the following system difference equation:

$$y(n) = x(n) + ay(n - 1)$$

where $a \neq 1$. Suppose the input signal is $x(n) = u(n)$.
Use Z transforms to find $y(n)$.

5–26. Given the following system difference equation:

$$y(n) = \frac{1}{2}x(n) - x(n - 1) + \frac{1}{2}x(n - 2)$$

Suppose the input signal is

$$x(n) = \begin{cases} 1, n = 0, 1, 2 \\ 0, otherwise \end{cases}$$

Use Z transforms to find $y(n)$.

5–27. Consider a system described by the following difference equation:

$$y(n) = \frac{1}{N} \sum_{k=0}^{N-1} x(n - k)$$

This is an N point *moving average* filter, since $y(n)$ is the average value of $\{x(n), x(n - 1), x(n - 2), \ldots, x[n - (N - 1)]\}$.

a. Find the impulse response of this filter.

b. Find the transfer function of this filter.

5–28. Show that

$$\sum_{n=0}^{\infty} |h(n)| < \infty$$

is a sufficient condition for the BIBO stability of a causal system.

HINT:

$$\left|y(n)\right| = \left|\sum_{k=0}^{\infty} h(k)x(n-k)\right| \le \sum_{k=0}^{\infty} \left|h(k)\right|\left|x(n-k)\right|$$

Now assume that $x(n)$ is bounded, that is, $\left|x(n)\right| \le \alpha$ for all n, where α is a finite positive number.

5–29. Show that $x(n) * h(n) \leftrightarrow X(z)H(z)$ by directly showing that

$$\mathbf{Z}\{x(n) * h(n)\} = \sum_{n=-\infty}^{\infty} \left[\sum_{k=-\infty}^{\infty} x(k)h(n-k)\right]z^{-n} = X(z)H(z)$$

HINT: Write the double summation as

$$\mathbf{Z}\{x(n) * h(n)\} = \sum_{k=-\infty}^{\infty} x(k) \sum_{n=-\infty}^{\infty} h(n-k)z^{-n}$$

then change variables on the inner summation (let $m = n - k$, and sum over the ms).

5–30. Consider the following system difference equation:

$$y(n) = b_0 x(n) + b_1 x(n-1) + b_2 x(n-2) - a_1 y(n-1) - a_2 y(n-2)$$

a. Find the transfer function for this system.

b. Now consider the following set of two difference equations used to calculate an output signal $y(n)$ given an input signal $x(n)$. The first difference equation calculates an intermediate result called $p(n)$; the second difference equation calculates $y(n)$ from $p(n)$:

$$p(n) = x(n) - a_1 p(n-1) - a_2 p(n-2)$$
$$y(n) = b_0 p(n) + b_1 p(n-1) + b_2 p(n-2)$$

The transfer function for this system is

$$H(z) = \frac{Y(z)}{X(z)}$$

Show that this transfer function is the same as that found for Problem 5–30a.

c. Draw the block diagram for the system in Problem 5–30b using only one shift register that holds past values of $p(n)$. Note that the system in Problem 5–30a requires two of these shift registers for past values of both $x(n)$ and $y(n)$.

5–31. When considering the real-time realization of a difference equation, try to reduce the number of multiplications if possible, since multiplications are time-consuming operations. For example, this is possible in an FIR filter having a symmetric impulse response (which is usually the case). Suppose, for example, the FIR filter difference equation is

$$y(n) = \frac{1}{4}x(n) - \frac{1}{2}x(n-1) + \frac{3}{4}x(n-2) - \frac{1}{2}x(n-3) + \frac{1}{4}x(n-4)$$

If realized directly (as shown in Figure 5–10), five multiplications are required. By exploiting the symmetry, find a difference equation for this system requiring only three multiplications, and draw a block diagram for this realization.

HINT: Combine terms having the same coefficient:

$$y(n) = \frac{1}{4}\left[x(n) + x(n-1)\right] + \cdots$$

5-32. Consider the cascaded system shown in Figure P5-32.

    a. Suppose

$$H_2(z) = \frac{1}{H_1(z)}$$

    In this case, $H_2(z)$ is called the *inverse system* with respect to $H_1(z)$. Show that $g(n) = x(n)$.

    b. What must be true of $H_1(z)$ in order for its inverse system to be BIBO stable?

    HINT: What is the relationship between the zeros of $H_1(z)$ and the poles of $H_2(z)$?

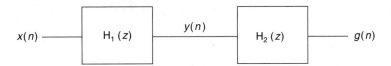

FIGURE P5-32

5-33. Consider a system having the transfer function

$$H(z) = 1 + 2z^{-1} + 3z^{-2}$$

    Suppose the input signal is $x(n) = (-1)^n u(n)$. Find $y(n)$ (that is, find the output signal).

5-34. Consider a system having the transfer function

$$H(z) = \frac{1}{1 + \frac{9}{10} z^{-1}}$$

    a. Suppose the input signal is $x(n) = \delta(n) + \delta(n - 1)$. Find $y(n)$ (that is, find the output signal).

    b. Find the step response of this system.

5-35. Consider a system having the transfer function

$$H(z) = \frac{10}{\left(1 - 0.2z^{-1}\right)\left(1 - 0.95z^{-1}\right)}$$

    a. Find the system impulse response.

    b. Find the system step response.

5-36. Consider a system having the transfer function

$$H(z) = \frac{1 - z^{-1}}{\left(1 - p_1 z^{-1}\right)\left(1 - p_1^* z^{-1}\right)}$$

where

$$p_1 = 0.9e^{j\frac{\pi}{5}}$$

    a. Find the system inpulse response.

    b. Find the system step response.

    c. Find the *form* of the output signal (A_k and α_k unevaluated) if the input signal is

$$x(n) = \cos\left(\frac{\pi}{2}n\right)u(n)$$

5–37. Consider a system having the transfer function

$$H(z) = \frac{b_0 + b_1 z^{-1} + b_2 z^{-2}}{1 + a_1 z^{-1} + a_2 z^{-2} + a_3 z^{-3} + a_4 z^{-4} + a_5 z^{-5}}$$

Assume the system poles are

$$p_1 = 0.95$$

$$p_2 = \frac{1}{2} e^{j0.58}$$

$$p_3 = p_2{}^*$$

$$p_4 = \frac{9}{10} e^{j2.1}$$

$$p_5 = p_4{}^*$$

Suppose that the input signal is $x(n) = \cos(1.5n)u(n)$.
a. Find the *form* of the output signal.
b. Identify the component of the output signal that is the transient response.
c. Identify the component of the output signal that is the steady-state response.

6

FREQUENCY DOMAIN ANALYSIS OF DISCRETE TIME SYSTEMS

6–1 FREQUENCY RESPONSE FUNCTION

We begin by considering an input signal $x(n)$ that is a complex exponential sequence:

$$x(n) = ce^{j\omega n} \tag{6–1}$$

Equation (6–1) can also be written as $x(n) = c\exp(j\omega n)$, and we will employ the exp( ) notation in places where it will improve the appearance of the text. The complex exponential sequence is convenient to deal with mathematically but the connection to the real world needs to be emphasized. The sequence $c\cos(\omega n)$ is related to the complex exponential sequence as follows:

$$c\cos(\omega n) = \frac{c}{2} e^{j\omega n} + \frac{c}{2} e^{-j\omega n} \tag{6–2}$$

This relationship follows from Euler's theorem. Basically, $c\cos(\omega n)$ can be thought of as consisting of two complex exponentials, one with frequency ω, and one with frequency $-\omega$. Another relationship worth noting is $c\cos(\omega n) = \mathrm{Re}\{c\exp(j\omega n)\}$.

The discrete time frequency ω is related to the continuous time frequency variable f(Hz) in the following way. Suppose that the complex exponential sequence $x(n)$, as in Equation (6–1), is created by sampling a continuous time complex exponential signal given by

$$x_a(t) = ce^{j2\pi ft} \tag{6–3}$$

Since $x(n) = x_a(nT)$, where the sampling frequency is $f_s = 1/T$, we have

$$x_n(t) = ce^{j2\pi fTn} = ce^{j\left(\frac{2\pi f}{f_s}\right)n} \tag{6–4}$$

If we define

$$\omega = 2\pi fT = \frac{2\pi f}{f_s} \tag{6-5}$$

then we see that the resulting sequence is $x(n) = c\exp(j\omega n)$. The quantity ωn has units of *radians*. If we think of the variable n as having units of samples, then the variable ω has units of radians per sample, and can be thought of as the frequency of the discrete time complex exponential signal.

Let $x(n) = c\exp(j\omega n)$ be the input to an LSI discrete time system with impulse response $h(n)$. Assume that this input is applied at time $n = \infty$ (since we are after a steady-state result). Let $y(n)$ be the system output. In Chapter 5, it was shown that the output $y(n)$ is equal to the convolution of $x(n)$ and $h(n)$, that is,

$$y(n) = h(n) * x(n) = \sum_{k=-\infty}^{\infty} h(k)x(n-k) \tag{6-6}$$

But $x(n-k) = c\exp[j\omega(n-k)]$ in this problem. Therefore,

$$y(n) = \sum_{k=-\infty}^{\infty} h(k)ce^{j\omega(n-k)} = \sum_{k=-\infty}^{\infty} h(k)ce^{j\omega n}e^{-j\omega k} \tag{6-7}$$

The complex exponential $c\exp(j\omega n)$ can be brought to the outside of the summation since it does not involve the summation index k:

$$y(n) = ce^{j\omega n} \left[\sum_{k=-\infty}^{\infty} h(k)e^{-j\omega k} \right] \tag{6-8}$$

The function in brackets in Equation (6–8) is the *frequency response function* of the system. We can use any letter we want as the summation index; let us use n instead of k. The frequency response function is thus defined in the following way:

$$H\left(e^{j\omega}\right) = \sum_{n=-\infty}^{\infty} h(n)e^{-j\omega n} \tag{6-9}$$

Let us summarize the result. If $x(n) = c\exp(j\omega n)$ is the steady-state input to an LSI discrete time system, then the steady-state output $y(n)$ is

$$y(n) = H\left(e^{j\omega}\right)ce^{j\omega n} \tag{6-10}$$

where $H[\exp(j\omega)]$ is the frequency response function, given by Equation (6–9). Of course, the frequency response function needs to be evaluated at the frequency of interest. For example, if $x(n) = \exp(j(\pi/8)n)$, then

$$y(n) = H\left(e^{j\omega}\right)\Big|_{\omega=\pi/8} \times e^{j\left(\frac{\pi}{8}\right)n} = H\left(e^{j\frac{\pi}{8}}\right)e^{j\left(\frac{\pi}{8}\right)n} \tag{6-11}$$

The frequency response function $H[\exp(j\omega)]$ is a complex function of the real variable ω. Therefore, it can be expressed in polar form as

$$H\left(e^{j\omega}\right) = \left|H\left(e^{j\omega}\right)\right|e^{j\theta_H(\omega)} \tag{6–12}$$

where $\left|H[\exp(j\omega)]\right|$ is the *amplitude response function* and $\theta_H(\omega)$ is the *phase response function*. These are given by

$$\left|H\left(e^{j\omega}\right)\right| = \sqrt{\mathrm{Re}^2\left\{H\left(e^{j\omega}\right)\right\} + \mathrm{Im}^2\left\{H\left(e^{j\omega}\right)\right\}} \tag{6–13}$$

$$\theta_H(\omega) = \angle\left\{H\left(e^{j\omega}\right)\right\} = \tan^{-1}\left[\frac{\mathrm{Im}\left\{H\left(e^{j\omega}\right)\right\}}{\mathrm{Re}\left\{H\left(e^{j\omega}\right)\right\}}\right] \tag{6–14}$$

The steady-state output due to input $x(n) = c\exp(j\omega n)$, as given in Equation (6–10), can be expressed in terms of the amplitude and phase response functions:

$$y(n) = c\left|H\left(e^{j\omega}\right)\right|e^{j(\omega n + \theta_H(\omega))} \tag{6–15}$$

Thus the steady-state output is also a complex exponential sequence with frequency ω; however, the amplitude and phase have been altered by the frequency response function of the system. Using this result in conjunction with Euler's theorem, it can be shown that if the system input is

$$x(n) = c\cos(\omega n) \tag{6–16}$$

then the output sequence will be

$$y(n) = c\left|H\left(e^{j\omega}\right)\right|\cos\left(\omega n + \theta_H(\omega)\right) \tag{6–17}$$

That is, if the input is a sinusoidal sequence, then the steady-state output will be a sinusoidal sequence with the same frequency ω; however, the amplitude and phase will be altered by the frequency response function of the system. (The results shown thus far in this section are analogous to the results shown in Chapter 2 for continuous time systems.)

The traditional notation for the frequency response function, $H[\exp(j\omega)]$, seems a bit cumbersome. Why use it? The notation was originally employed because of the relationship to the Z transform. Recall that the transfer function of a system with impulse response $h(n)$ is

$$H(z) = \sum_{n=-\infty}^{\infty} h(n)z^{-n} \tag{6–18}$$

Observe that if we replace z with $\exp(j\omega)$ in Equation (6–18), we obtain the summation formula for the frequency response function (Equation [6–9]). In other words,

$$H(z)\Big|_{z=\exp(j\omega)} = \sum_{n=-\infty}^{\infty} h(n)\left(e^{j\omega}\right)^{-n} = \sum_{n=-\infty}^{\infty} h(n)e^{-j\omega n} = H\left(e^{j\omega}\right) \tag{6–19}$$

In fact, it can be shown that as long as the system is BIBO stable (see Section 5–5), the frequency response function can be obtained from the transfer function by replacing z with $\exp(j\omega)$. Strictly speaking, the frequency response function for an unstable system is not defined. This is intuitive, since an unstable system will not produce a steady-state sinusoidal output in response to a sinusoidal input. The output of an unstable system will blow up instead. Mathematically, it can be shown that if a discrete time system is not BIBO stable, then the region of convergence for $H(z)$ does *not* include the unit circle, that is, it does not include $z = \exp(j\omega)$.

EXAMPLE 6–1.1

In Chapter 5, we considered a BIBO-stable IIR system with the difference equation

$$y(n) = x(n) + \frac{1}{2} y(n-1) \tag{6-20}$$

For this system the transfer function is (from Section 5–4)

$$H(z) = \frac{1}{1 - \frac{1}{2} z^{-1}} \tag{6-21}$$

By replacing z with $\exp(j\omega)$, we obtain the frequency response function

$$H\left(e^{j\omega}\right) = \frac{1}{1 - \frac{1}{2} e^{-j\omega}} \tag{6-22}$$

Using Euler's theorem, this can be expressed as

$$H\left(e^{j\omega}\right) = \frac{1}{1 - \frac{1}{2} \cos(\omega) + j \frac{1}{2} \sin(\omega)} \tag{6-23}$$

Now suppose that the input to this system is

$$x(n) = \cos\left(\frac{\pi}{2} n\right) \tag{6-24}$$

To evaluate the output signal $y(n)$, we can start by evaluating $H[\exp(j\omega)]$ at $\omega = \pi/2$:

$$H\left(e^{j\omega}\right)\Big|_{\omega = \frac{\pi}{2}} = \frac{1}{1 - \frac{1}{2} \cos\left(\frac{\pi}{2}\right) + j \frac{1}{2} \sin\left(\frac{\pi}{2}\right)} = \frac{1}{1 + j \frac{1}{2}} \tag{6-25}$$

Therefore,

$$\left| H\left(e^{j\frac{\pi}{2}}\right) \right| = \frac{1}{\sqrt{1^2 + \left(\frac{1}{2}\right)^2}} = 0.894 \tag{6-26}$$

$$\angle\left\{H\left(e^{j\frac{\pi}{2}}\right)\right\} = \theta_H(\omega) = \angle 1 - \angle\left(1 + j\frac{1}{2}\right)$$

(6–27)

$$= 0 - \tan^{-1}\left(\frac{1/2}{1}\right) = -0.4636 \text{ radians}$$

Thus the steady-state output signal is

$$y(n) = 0.894\cos\left(\frac{\pi}{2}n - 0.4636\right)$$

(6–28)

6–2   STEADY-STATE RESPONSE TO PERIODIC SIGNALS

Let $x(n)$ be a periodic sequence, that is, $x(n) = x(n + N)$, where N is the period. It can be shown that $x(n)$ can be represented as a weighted sum of N complex exponentials at frequencies $2\pi k/N$, $k = 0, 1, \ldots, N - 1$:

$$x(n) = \frac{1}{N}\sum_{k=0}^{N-1} c_k e^{j\left(\frac{2\pi k}{N}\right)n}$$

(6–29)

This is the *discrete Fourier series* (DFS) representation of $x(n)$. The DFS coefficients c_k can be calculated using the following formula:

$$c_k = \sum_{n=0}^{N-1} x(n)e^{-j\left(\frac{2\pi k}{N}\right)n}, k = 0, 1, \ldots N - 1$$

(6–30)

Coefficients c_k are in general complex numbers. (The DFS is analogous to the continuous time Fourier series. See Section 2–5.)

Suppose a periodic signal $x(n)$, represented by its DFS, is the steady-state input to an LSI discrete time system with frequency response function $H[\exp(j\omega)]$. By applying the principle of superposition, we obtain the steady-state output signal:

$$y(n) = \frac{1}{N}\sum_{k=0}^{N-1} c_k H\left(e^{j\frac{2\pi k}{N}}\right)e^{j\left(\frac{2\pi k}{N}\right)n}$$

(6–31)

That is, each component complex exponential at frequency $\omega = 2\pi k/N$ is multiplied by the frequency response function evaluated at $\omega = 2\pi k/N$.

We would like to consider discrete time signals that are more general, that is, not necessarily periodic. This brings us to the subject of the *discrete time Fourier transform*, or DTFT. In the process of studying the DTFT, we will also learn more about the properties of the frequency response function, which is actually the DTFT of the system impulse response $h(n)$.

6-3 THE DISCRETE TIME FOURIER TRANSFORM

The *discrete time Fourier transform* (DTFT) of a discrete time signal $x(n)$ is defined as

$$\mathbf{DTFT}\{x(n)\} = X\left(e^{j\omega}\right) = \sum_{n=-\infty}^{\infty} x(n)e^{-j\omega n} \qquad (6\text{-}32)$$

$x(n)$ and $X[\exp(j\omega)]$ form a DTFT pair, denoted by

$$x(n) \leftrightarrow X\left(e^{j\omega}\right) \qquad (6\text{-}33)$$

where $x(n)$ is in the time domain, and $X[\exp(j\omega)]$ is in the frequency domain. $X[\exp(j\omega)]$ is in general a complex function of ω, and is defined for both positive and negative values of ω. Note that although $x(n)$ is defined only for integer values of n, the frequency domain is continuous. That is, $X[\exp(j\omega)]$ is defined for any real value of ω. The DTFT of a signal is interpreted as the *spectrum*, *spectral content*, or *frequency content* of the signal.

The DTFT is often referred to as the Fourier transform of a discrete time signal. However, it is clearly not the same thing as the Fourier transform of a continuous time signal, as discussed in Chapter 3 (although there is an extremely important relationship between the two transforms that will be covered later). In most of the literature, it is easy to tell from the context of the discussion and the notation being employed which Fourier transform is under consideration.

If $X[\exp(j\omega)]$ is given, the corresponding time domain signal $x(n)$ can be obtained with the inverse DTFT (IDTFT) integral:

$$x(n) = \frac{1}{2\pi} \int_{-\pi}^{\pi} X\left(e^{j\omega}\right) e^{j\omega n} d\omega \qquad (6\text{-}34)$$

(Actually, the integral can be over any interval of length 2π.)

Since $X[\exp(j\omega)]$ is in general complex, it can be represented in rectangular or polar form:

$$X\left(e^{j\omega}\right) = \text{Re}\left\{X\left(e^{j\omega}\right)\right\} + j\text{Im}\left\{X\left(e^{j\omega}\right)\right\}$$

$$X\left(e^{j\omega}\right) = \left|X\left(e^{j\omega}\right)\right| e^{j\theta_X(\omega)} \qquad (6\text{-}35)$$

where

$$\left|X\left(e^{j\omega}\right)\right| = \sqrt{\text{Re}^2\left\{X\left(e^{j\omega}\right)\right\} + \text{Im}^2\left\{X\left(e^{j\omega}\right)\right\}}$$

$$\theta_X(\omega) = \angle\left\{X\left(e^{j\omega}\right)\right\} = \tan^{-1}\left[\frac{\text{Im}\left\{X\left(e^{j\omega}\right)\right\}}{\text{Re}\left\{X\left(e^{j\omega}\right)\right\}}\right] \qquad (6\text{-}36)$$

$\left|X[\exp(j\omega)]\right|$ could be defined as the spectral amplitude function, and $\theta_X(\omega)$ as the phase function.

If you compare Equations (6–9) and (6–32) you can see that *the frequency response function of an LSI discrete time system is the DTFT of the system impulse response:*

$$\mathbf{DTFT}\{h(n)\} = H\left(e^{j\omega}\right) \qquad (6\text{–}37)$$

Or using the "transform pair" notation,

$$h(n) \leftrightarrow H\left(e^{j\omega}\right) \qquad (6\text{–}38)$$

Thus the properties of the DTFT are also the properties of the frequency response function. The DTFT is a linear operator. That is,

$$ax(n) + bg(n) \leftrightarrow aX\left(e^{j\omega}\right) + bG\left(e^{j\omega}\right) \qquad (6\text{–}39)$$

The proof of this *linearity property* follows directly from the definition of the DTFT:

$$
\begin{aligned}
\mathbf{DTFT}\{ax(n) + bg(n)\} &= \sum_{n=-\infty}^{\infty}\left[ax(n) + bg(n)\right]e^{-j\omega n} \\
&= a\sum_{n=-\infty}^{\infty}x(n)e^{-j\omega n} + b\sum_{n=-\infty}^{\infty}g(n)e^{-j\omega n} \qquad (6\text{–}40)\\
&= a \times \mathbf{DTFT}\{x(n)\} + b \times \mathbf{DTFT}\{g(n)\}
\end{aligned}
$$

A list of DTFT theorems is shown in Table 6–1. These theorems will be discussed at various places throughout this chapter.

The relationship between the Z transform and the DTFT has already been hinted at rather strongly (Equation [6–19]). The relationship is as follows: let $x(n)$ be a sequence

TABLE 6–1
DTFT Theorems

$$x(n) \leftrightarrow X\left(e^{j\omega}\right)$$

$$ax(n) + bg(n) \leftrightarrow aX\left(e^{j\omega}\right) + bG\left(e^{j\omega}\right)$$

$$x(n - k) \leftrightarrow e^{-j\omega k} X\left(e^{j\omega}\right)$$

$$x(n)e^{j\phi n} \leftrightarrow X\left(e^{j(\omega-\phi)}\right)$$

$$x(n)\cos(\phi n) \leftrightarrow \frac{1}{2} X\left(e^{j(\omega-\phi)}\right) + \frac{1}{2} X\left(e^{j(\omega+\phi)}\right)$$

$$x(n)\sin(\phi n) \leftrightarrow \frac{1}{2j} X\left(e^{j(\omega-\phi)}\right) - \frac{1}{2j} X\left(e^{j(\omega+\phi)}\right)$$

$$x(n) * g(n) \leftrightarrow X\left(e^{j\omega}\right)G\left(e^{j\omega}\right)$$

$$x(n)g(n) \leftrightarrow \frac{1}{2\pi} X\left(e^{j\omega}\right) * G\left(e^{j\omega}\right)$$

having the Z transform $X(z)$. You can obtain the DTFT of $x(n)$ by letting $z = \exp(j\omega)$, that is, by evaluating $X(z)$ on the unit circle in the z plane, *provided* $X(z)$ is defined on the unit circle in the z plane. In other words, the substitution $z = \exp(j\omega)$ is valid *provided* the region of convergence for $X(z)$ includes the unit circle. For most sequences of interest that die out as $n \to \pm\infty$, this substitution is valid. Included in this valid category are all finite time duration sequences. Also included are those sequences in Table 5–1 (the Z transform table on page 128) that have coefficient values that cause the sequence to satisfy the property $x(n) \to 0$ as $n \to \infty$.

Given the close relationship between the Z transform and the DTFT and the fact that a system transfer function relates the Z transforms of the input and output signals as $Y(z) = H(z)X(z)$, the following DTFT property should come as no surprise. Let $X[\exp(j\omega)]$ and $Y[\exp(j\omega)]$ be the DTFTs of LSI system input and output signals $x(n)$ and $y(n)$, respectively. Let $H[\exp(j\omega)]$ be the frequency response function of this system. It can be shown that the input and output spectra are related by

$$Y\left(e^{j\omega}\right) = H\left(e^{j\omega}\right)X\left(e^{j\omega}\right) \tag{6–41}$$

That is, the spectrum of the output signal is found by multiplying the spectrum of the input signal by the frequency response function of the system. Thus the system alters the frequency content of the input signal. Note that Equation (6–41) is analogous to the relationship $Y_a(f) = H_a(f)X_a(f)$ for continuous time systems.

6–4 DTFT SYMMETRY AND PERIODICITY

The DTFT has symmetry properties similar to those of the Fourier transform (Chapter 3). Assuming that sequence $x(n)$ is real (and in this book, all time domain sequences are real unless stated otherwise), then the amplitude spectrum is an even function of ω, and the phase function is an odd function of ω. To show this, we begin with the following relationship:

$$X^*\left(e^{j\omega}\right) = X\left(e^{j(-\omega)}\right) \tag{6–42}$$

where $*$ denotes complex conjugate. Equation (6–42) is relatively easy to demonstrate, although it may help the reader to refer to the table of properties of complex numbers (Table 2–1 on page 25):

$$X^*\left(e^{j\omega}\right) = \left(\sum_{n=-\infty}^{\infty} x(n)e^{-j\omega n} \right)^* = \sum_{n=-\infty}^{\infty} \left(x(n)\right)^* \left(e^{-j\omega n}\right)^*$$

$$= \sum_{n=-\infty}^{\infty} x(n)e^{j\omega n} = X\left(e^{j(-\omega)}\right) \tag{6–43}$$

Now observe that the amplitude spectrum can be written as

$$\left|X\left(e^{j\omega}\right)\right| = \sqrt{\left|X\left(e^{j\omega}\right)\right|^2} = \sqrt{X\left(e^{j\omega}\right)X^*\left(e^{j\omega}\right)} = \sqrt{X\left(e^{j\omega}\right)X\left(e^{j(-\omega)}\right)} \tag{6–44}$$

It should be clear from inspecting the right side of Equation (6–44) that ω can be replaced by $-\omega$ without changing the function. Therefore, the amplitude spectrum is an even function of ω (that is, it has even symmetry about the point $\omega = 0$). In other words,

$$\left|X\left(e^{j\omega}\right)\right| = \left|X\left(e^{j(-\omega)}\right)\right| \tag{6–45}$$

The phase function $\theta_X(\omega)$ is given by

$$\theta_X(\omega) = \tan^{-1}\left(\frac{\operatorname{Im}\left\{X\left(e^{j\omega}\right)\right\}}{\operatorname{Re}\left\{X\left(e^{j\omega}\right)\right\}}\right) \tag{6–46}$$

To show that the phase function is an odd function of ω, we must first show that the imaginary part of $X[\exp(j\omega)]$ is an odd function of ω, and the real part of $X[\exp(j\omega)]$ is an even function of ω. Observe that Euler's theorem can be used to express the DTFT as

$$X\left(e^{j\omega}\right) = \sum_{n=-\infty}^{\infty} x(n)\cos(\omega n) - j \sum_{n=-\infty}^{\infty} x(n)\sin(\omega n) \tag{6–47}$$

It follows that, since $x(n)$ is assumed to be real,

$$\operatorname{Re}\left\{X\left(e^{j\omega}\right)\right\} = \sum_{n=-\infty}^{\infty} x(n)\cos(\omega n) \tag{6–48}$$

$$\operatorname{Im}\left\{X\left(e^{j\omega}\right)\right\} = -\sum_{n=-\infty}^{\infty} x(n)\sin(\omega n) \tag{6–49}$$

Since $\cos(\omega n) = \cos[(-\omega)n]$, it follows that $\operatorname{Re}\{X[\exp(j\omega)]\}$ is an even function of ω. Similarly, since $\sin(\omega n) = -\sin[(-\omega)n]$, it follows that $\operatorname{Im}\{X[\exp(j\omega)]\}$ is an odd function of ω. If we divide an odd function of ω by an even function of ω, the result is an odd function of ω. Therefore, the phase function can be expressed as

$$\theta_X(\omega) = \tan^{-1}\left(F(\omega)\right) \tag{6–50}$$

where $F(\omega)$ is odd, that is, $F(\omega) = -F(-\omega)$. Therefore,

$$\theta_X(\omega) = \tan^{-1}\left[F(\omega)\right] = \tan^{-1}\left[-F(-\omega)\right] = -\tan^{-1}\left[F(-\omega)\right] \tag{6–51}$$

From this we conclude that the phase function is an odd function of ω, that is,

$$\theta_X(\omega) = -\theta_X(-\omega) \tag{6–52}$$

Another symmetry property of interest is the following: If $x(n) = x(-n)$, that is, if $x(n)$ has even symmetry about the point $n = 0$, then $X[\exp(j\omega)]$ is a *real* function of ω. To show this, we begin with the DTFT of $x(n)$:

$$X\left(e^{j\omega}\right) = \sum_{n=-\infty}^{\infty} x(n)e^{-j\omega n} \tag{6–53}$$

If we change variables in the summation, and let $n = -m$, then the DTFT can be written as

$$X(e^{j\omega}) = \sum_{m=-\infty}^{\infty} x(-m)e^{j\omega m} \qquad (6\text{-}54)$$

Of course, we can use any letter for the summation index in Equation (6–54). Let's use n instead of m:

$$X(e^{j\omega}) = \sum_{n=-\infty}^{\infty} x(-n)e^{j\omega m} \qquad (6\text{-}55)$$

Now compare Equations (6–53) and (6–55). If $x(n) = x(-n)$, then

$$X(e^{j\omega}) = X(e^{j(-\omega)}) \qquad (6\text{-}56a)$$

But according to Equation (6–42), if this is the case then it must also be true that

$$X(e^{j\omega}) = X^*(e^{j\omega}) \qquad (6\text{-}56b)$$

In order for Equation (6–56) to be true, the imaginary part of $X[\exp(j\omega)]$ must be equal to zero. That is, $X[\exp(j\omega)]$ must be a real function of ω if $x(n) = x(-n)$.

The DTFT symmetry properties just discussed are very similar to the Fourier transform symmetry properties discussed in Chapter 3. However, there is one very significant difference between the DTFT and the Fourier transform: the DTFT is *periodic* in ω, with a period of 2π. In other words, if you add any integer multiple of 2π to the variable ω, the value of the DTFT is unchanged. This can be expressed in the following way:

$$X(e^{j\omega}) = X(e^{j(\omega+2\pi k)}) \qquad (6\text{-}57)$$

where k is any integer. This periodicity is a consequence of the fact that

$$e^{j(\omega+2\pi k)} = e^{j\omega}e^{j2\pi k} = e^{j\omega}[\cos(2\pi k) + j\sin(2\pi k)] = e^{j\omega}[1] = e^{j\omega} \qquad (6\text{-}58)$$

Of course, this means that the frequency response function $H[\exp(j\omega)]$ is also periodic.

Since the DTFT is periodic 2π, when plotting a function involving the DTFT such as an amplitude spectrum, there is usually no reason to plot more than one period. Most such plots show either the $[-\pi, \pi]$ interval or the $[0, 2\pi]$ interval. Furthermore, since $|X[\exp(j\omega)]|$ is an *even* function of ω (assuming $x(n)$ is real), amplitude spectra are often plotted on the interval $[0, \pi]$ only. Since the phase function $\theta_x(\omega)$ for a real sequence $x(n)$ is an *odd* function of ω, it can also be plotted on the interval $[0, \pi]$.

EXAMPLE 6–2

Given the exponential sequence

$$x(n) = \begin{cases} \alpha^n, & n \geq 0 \\ 0, & \text{otherwise} \end{cases} \qquad (6\text{-}59)$$

where $|\alpha| < 1$. (Assume α is real, so that $x(n)$ is a real sequence.) The DTFT of this sequence is

$$X\left(e^{j\omega}\right) = \sum_{n=0}^{\infty} \alpha^n e^{-j\omega n} = \sum_{n=0}^{\infty} \left(\alpha e^{-j\omega}\right)^n \qquad (6\text{–}60)$$

The summation on the right has the form of an infinite geometric series. (See Chapter 5, Equation [5–58]). Since $|\alpha| < 1$, it follows that $|\alpha \exp(-j\omega)| < 1$. Therefore,

$$X\left(e^{j\omega}\right) = \frac{1}{1 - \alpha e^{-j\omega}} \qquad (6\text{–}61)$$

which can also be written (applying Euler's theorem to the denominator) as

$$X\left(e^{j\omega}\right) = \frac{1}{1 - \alpha\cos(\omega) + j\alpha\sin(\omega)} \qquad (6\text{–}62)$$

The spectral amplitude function is:

$$\left|X\left(e^{j\omega}\right)\right| - \frac{|1|}{\left|1 - \alpha e^{-j\omega}\right|} = \frac{1}{\left|1 - \alpha\cos(\omega) + j\alpha\sin(\omega)\right|}$$

$$= \frac{1}{\sqrt{\left(1 - \alpha\cos(\omega)\right)^2 + \alpha^2\sin^2(\omega)}}$$

$$= \frac{1}{\sqrt{1 - 2\alpha\cos(\omega) + \alpha^2\cos^2(\omega) + \alpha^2\sin^2(\omega)}} \qquad (6\text{–}63)$$

$$= \frac{1}{\sqrt{1 + \alpha^2 - 2\alpha\cos(\omega)}}$$

To obtain the phase function $\theta_X(\omega)$, use the following property of complex numbers:

$$\angle\left\{\frac{z_1}{z_2}\right\} = \angle\{z_1\} - \angle\{z_2\} \qquad (6\text{–}64)$$

Applying this property to Equation (6–62) results in

$$\theta_X(\omega) = 0 - \tan^{-1}\left[\frac{\alpha\sin(\omega)}{1 - \alpha\cos(\omega)}\right] \qquad (6\text{–}65)$$

Figure 6–1 shows $|X[\exp(j\omega)]|$ and $\theta_X(\omega)$ plotted on the interval $-\pi < \omega \leq \pi$, for the specific case of $\alpha = 0.8$. In order to show that the DTFT is *periodic* 2π, the spectral amplitude and phase functions are also plotted on the interval $0 \leq \omega < 2\pi$ in Figure 6–2.

   NOTE: The spectral amplitude function is *even symmetric* about the point $\omega = \pi$, and the phase function is *odd symmetric* about the point $\omega = \pi$. This is true in general, and is a consequence of the combination of the following properties: $X[\exp(j\omega)]$ is periodic 2π, $X[\exp(j\omega)]$ is an even function of ω, and $\theta_X(\omega)$ is an odd function of ω.

FIGURE 6–1
(a) Spectral amplitude function. (b) Phase function.

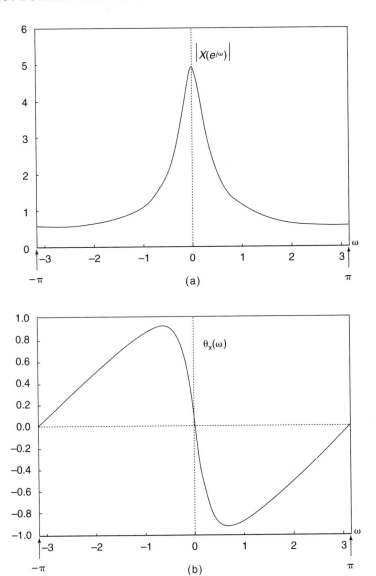

FIGURE 6–2
(a) Spectral amplitude
function. (b) Phase function.

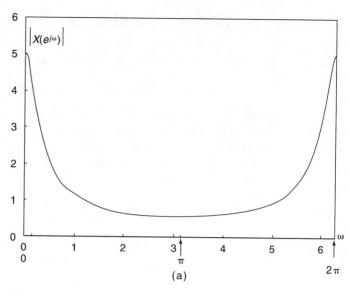

(a)

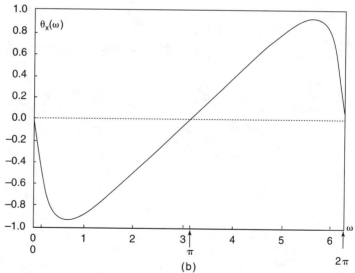

(b)

6–5   DTFT OF A SINUSOIDAL SEQUENCE

Consider the complex exponential sequence $x(n) = \exp(j\phi n)$, defined for all integer values of n, nonnegative and negative. The DTFT pair turns out to be

$$e^{j\phi n} \leftrightarrow 2\pi\delta(\omega - \phi), (\text{periodic} 2\pi) \tag{6–66}$$

where $\delta(\omega - \phi)$ is an impulse located at $\omega = \phi$, as shown in Figure 6–3 (with the periodic extension indicated). The impulse $\delta(\omega - \phi)$ is the Dirac delta function discussed in Chapters 2 and 3: a pulse with width approaching zero, height approaching infinity, and area equal to one. ($\alpha\delta(\omega - \phi)$ has an area of α.) This concept is illustrated in Figure 6–4.

The impulse described here satisfies the *sifting property* discussed in Chapter 2. In the present context, the sifting property is

$$\int_{-\infty}^{\infty} f(\omega)\delta(\omega - \phi)d\omega = f(\phi) \tag{6–67}$$

In order to verify the DTFT pair relationship in Equation (6–66), we can use the inverse DTFT integral in conjunction with the sifting property:

$$\mathbf{IDTFT}\{2\pi\delta(\omega - \phi)\} = \frac{1}{2\pi}\int_{-\pi}^{\pi} 2\pi\delta(\omega - \phi)e^{j\omega n}d\omega = e^{j\phi n} \tag{6–68}$$

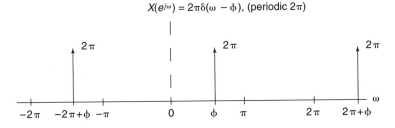

$X(e^{j\omega}) = 2\pi\delta(\omega - \phi), (\text{periodic } 2\pi)$

FIGURE 6–3
The DTFT of $x(n) = \exp(j\phi n)$ is an impulse located at $\omega = \phi$, periodically extended.

FIGURE 6–4
This frequency domain pulse becomes an impulse as $\Delta\omega \to 0$.

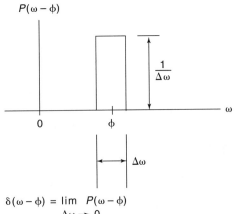

$P(\omega - \phi)$

$\delta(\omega - \phi) = \lim_{\Delta\omega \to 0} P(\omega - \phi)$

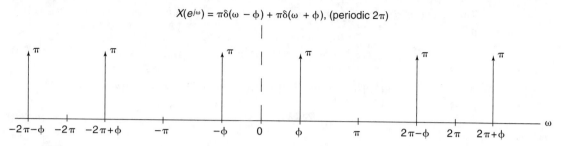

$$X(e^{j\omega}) = \pi\delta(\omega - \phi) + \pi\delta(\omega + \phi), \text{ (periodic } 2\pi)$$

FIGURE 6–5

The DTFT of $x(n) = \cos(\phi n)$ is a pair of impulses located at $\omega = \pm\phi$, periodically extended.

This result can be used to find the DTFT of a sinusoid. Note that

$$\cos(\phi n) = \frac{1}{2}e^{j\phi n} + \frac{1}{2}e^{-j\phi n} \tag{6–69}$$

Therefore, using the linearity property of the DTFT, along with Equation (6–66), we have the following DTFT pair:

$$\cos(\phi n) \leftrightarrow \pi\delta(\omega - \phi) + \pi\delta(\omega + \phi)$$
$$(\text{periodic} 2\pi) \tag{6–70}$$

Thus the DTFT of a sinusoid is a pair of impulses in the frequency domain, periodically extended. Figure 6–5 illustrates this result. Note that if this DTFT is displayed on the interval $0 \le \omega < 2\pi$, there will be two impulses: one at $\omega = \phi$ and another at $\omega = 2\pi - \phi$. It should be emphasized that the DTFT relationship of Equation (6–70) is for a sinusoid defined for all integer values of "time" n, both nonnegative and negative.

6–6 RELATIONSHIP BETWEEN THE FOURIER TRANSFORM AND THE DTFT

It stands to reason that if a discrete time signal $x(n)$ was obtained from a continuous time signal $x_a(t)$ by sampling, the respective spectra $X_a(f)$ and $X[\exp(j\omega)]$ must be related. The purpose of this section is to introduce that relationship on an "it can be shown" basis. The proof is deferred to Chapter 7, in which the relationship between $x_a(t)$ and $x(n)$ is explored in much greater detail.

Suppose there is a continuous time signal $x_a(t)$ with a corresponding spectrum $X_a(f)$. That is, we have the Fourier transform pair

$$x_a(t) \leftrightarrow X_a(f) \tag{6–71}$$

Suppose further that we create a discrete time signal $x(n)$ by sampling $x_a(t)$ at a sampling frequency $f_s = 1/T$:

$$x(n) = x_a(nT) \tag{6–72}$$

The sequence $x(n)$ has a corresponding spectrum $X[\exp(j\omega)]$. That is, we have the DTFT pair

$$x(n) \leftrightarrow X\left(e^{j\omega}\right) \tag{6-73}$$

It will be shown in Chapter 7 that the two spectra are related in the following way:

$$X\left(e^{j2\pi fT}\right) = \frac{1}{T} \sum_{n=-\infty}^{\infty} X_a\left(f - nf_s\right) \tag{6-74}$$

To understand Equation (6–74), it is first necessary to understand that $X[\exp(j2\pi fT)]$ is just $X[\exp(j\omega)]$ with the variable ω replaced by $2\pi fT$. In other words, the relationship between frequency variables ω and f that was first considered in Section 6–1 has appeared once again:

$$\omega = 2\pi fT = \frac{2\pi f}{f_s} \tag{6-75}$$

Therefore, $X[\exp(j2\pi fT)]$ is a *periodic* function of the frequency variable f (in Hz) that looks like $X[\exp(j\omega)]$; the only difference is that the frequency axis scaling is changed by Equation (6–75). The period of $X[\exp(j2\pi fT)]$ is $f = f_s$; this is easy to verify by letting $\omega = 2\pi$ in Equation (6–75) and solving for f. Recall that $\omega = 2\pi$ is the period of $X[\exp(j\omega)]$. Figure 6–6 illustrates the concept.

To understand the right side of Equation (6–74), observe that $X_a(f - nf_s)$, for some specific value of n, is just a shifted copy of spectrum $X_a(f)$. Figure 6–7 illustrates the concept. Thus the right side of Equation (6–74) says that we have a scaled $(1/T)$ copy of spectrum $X_a(f)$ centered at all integer multiples of the sampling frequency f_s, as shown in Figure 6–8. These shifted and scaled copies are *added* to form $X(\exp(j2\pi fT))$. Depending on the sampling

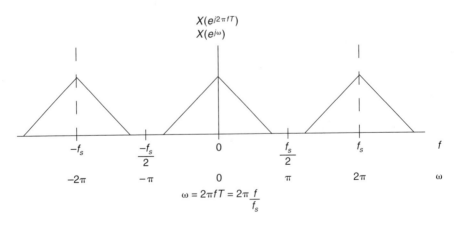

FIGURE 6–6
Hypothetical spectrum (DTFT) shown on the ω (radians per sample) frequency scale and the f (Hertz) frequency scale. The relationship between the two scales is $\omega = 2\pi f/fs$, where fs is the sampling frequency.

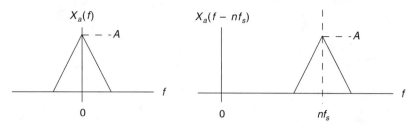

FIGURE 6–7
$X_a(f - nf_s)$ is a shifted copy of $X_a(f)$.

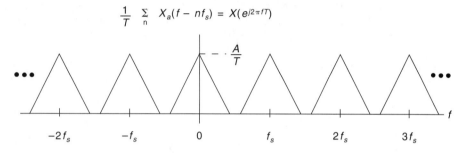

FIGURE 6–8
Scaled copies of $X_a(f)$ centered on integer multiples of the sampling frequency.

frequency and the bandwidth of the spectrum $X_a(f)$, these shifted copies might *overlap*. If the spectral copies overlap, the resulting spectrum is said to be *aliased*. Since aliasing is an extremely important concept, let's try to illustrate it with a simple example.

Consider a hypothetical signal $x_a(t)$ with a corresponding spectrum $X_a(f)$, as shown in Figure 6–9. Note that this signal is *bandlimited* to B Hz. Now suppose $x_a(t)$ is sampled at a sampling frequency f_s that is greater than two times the bandwidth B of the signal, that is, $f_s > 2B$. Figure 6–10 shows the resulting spectrum $X[\exp(j2\pi fT)]$; observe that the spectral copies do not overlap, which means that there is no aliasing. On the other hand, suppose $x_a(t)$ is sampled at a sampling frequency which is less than two times the bandwidth of the signal, that is, $f_s < 2B$. Figure 6–11 shows how the spectral copies are distributed in this case; note how they overlap. Spectrum $X[\exp(j2\pi fT)]$ is formed by adding these overlapped copies, as shown in Figure 6–12. In this case, the spectrum clearly suffers from aliasing.

FIGURE 6–9
Spectrum of a hypothetical continuous time signal bandlimited to B Hz. If this signal is sampled and the DTFT of the resulting sequence is computed, the way that $X[\exp(j2\pi fT)]$ relates to $X_a(f)$ depends on the sampling frequency, as shown in Figures 6–10, 6–11, and 6–12.

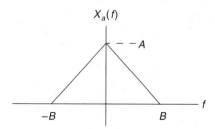

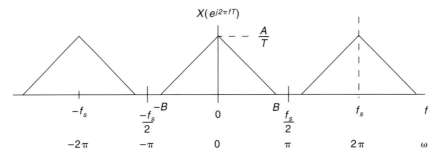

FIGURE 6–10

The DTFT can be formed by adding the shifted and scaled copies of the original spectrum $X_a(f)$ (see Figure 6–9). The copies are centered on all integer multiples of the sampling frequency. When $f_s > 2B$, the shifted and scaled copies do not overlap; consequently, within the interval $-f_s/2 < f < f_s/2$, the DTFT is a scaled duplicate of $X_a(f)$, as shown here. In this case, there is said to be no aliasing.

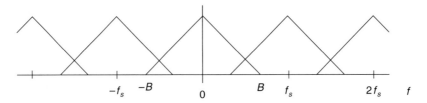

FIGURE 6–11

When $f_s < 2B$, the shifted and scaled copies of $X_a(f)$ will overlap as shown here. The DTFT is formed by adding these shifted and scaled copies, as shown in Figure 6–12.

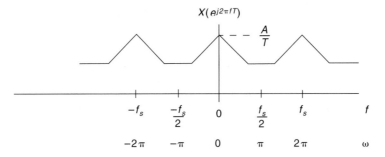

FIGURE 6–12

When $f_s < 2B$, the DTFT suffers from aliasing, as shown here. In this case, the DTFT is *not* a scaled duplicate of $X_a(f)$ within the interval $-f_s/2 < f < f_s/2$.

To summarize: $X[\exp(j2\pi fT)]$ is a *periodic* (f_s), *scaled* $(1/T)$, and possibly *aliased* version of $X_a(f)$. If $f_s > 2B$, where B is the bandwidth of the continuous time signal, there will be no aliasing.

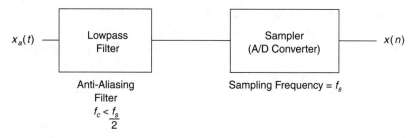

FIGURE 6–13
The purpose of the anti-aliasing filter is to bandlimit the continuous time signal before it is sampled.

A crucial observation is as follows: If $f_s > 2B$ (that is, if there is no aliasing), then within the frequency interval

$$\frac{-f_s}{2} < f < \frac{f_s}{2} \qquad (6\text{–}76)$$

$X[\exp(j2\pi fT)]$ is a *scaled duplicate* of $X_a(f)$. This is clearly illustrated in Figures 6–9 and 6–10. (On the ω scale, the interval is $-\pi < \omega < \pi$, as also shown in Figure 6–10.) Therefore, if there is no aliasing, it seems reasonable to suppose that the sampling process can be reversed (in theory, anyway). That is, it would seem that if there is no aliasing, a continuous time signal can be reconstructed from its samples. In fact, it is true that if there is no aliasing, an *ideal* digital-to-analog (D/A) converter would accomplish this feat. This subject will be covered in Chapter 7.

Real-world signals are never strictly bandlimited in the sense suggested by Figure 6–9. In the real world, bandwidth is defined as that range of frequencies within which the spectrum is significant. This means that aliasing cannot be completely eliminated, but the system designer should choose a sampling frequency that is large enough to make the aliasing negligible, so that for all practical purposes $X[\exp(j2\pi fT)]$ is a scaled duplicate of the original spectrum within the frequency interval $-f_s/2 < f < f_s/2$.

In many discrete time systems, the input continuous time signal is processed by an analog lowpass filter before the sampling operation. The cutoff frequency of this *anti-aliasing filter* is less than $f_s/2$. This is illustrated by Figure 6–13. Of course, the sampling frequency and the cutoff frequency of the anti-aliasing filter must be selected such that none of the significant and/or desirable spectral components are eliminated by the anti-aliasing filter. (The definition of *significant and/or desirable* depends on the specific application under consideration.)

EXAMPLE 6–6.1
It is interesting to see how sinusoids are affected by sampling and aliasing. Consider the continuous time sinusoidal signal:

$$x_a(t) = \cos(2\pi f_o t) \qquad (6\text{–}77)$$

where $f_o = 10$kHz. The spectrum $X_a(f)$ consists of a pair of impulses located at ± 10khz, as shown in Figure 6–14(a). This pair of impulses is associated as one unit, as indicated in the figure by the symbol ⌣. The amplitude scale factors will be ignored since they clutter up the discussion. The bandwidth of $x_a(t)$ is $B = f_o = 10$ kHz.

Suppose $x_a(t)$ is sampled at $f_s = 34$ kHz. The resulting spectrum, $X[\exp(j2\pi fT)]$, as shown in Figure 6–14(b), was drawn by applying the relationship given in Equation (6–74), with the amplitude scale factor ignored. Copies of the original spectrum (i.e., the pair of impulses associated as a unit) are centered at all integer multiples of the sampling frequency. Since $f_s > 2B$, $X[\exp(j2\pi fT)]$ is a scaled duplicate of $X_a(f)$ within the interval $-f_s/2 < f < f_s/2$.

Now suppose $x_a(t)$ is sampled at $f_s = 12$ kHz instead. Since f_s is now less than $2B$, the resulting spectrum $X[\exp(j2\pi fT)]$ will now be corrupted by aliasing, as shown in Figure 6–14(c), which was drawn by applying the relationship of Equation (6–74). Copies of the original spectrum (the pair of impulses associated as a unit) are centered at all integer multiples of the sampling frequency, but this time the copies overlap. Therefore, $X[\exp(j2\pi fT)]$ is *not* a scaled duplicate of the original spectrum within the interval $-f_s/2 < f < f_s/2$. When viewing the spectrum in this interval, it appears that the original sinusoid must have had a frequency of $f_o = 2$ kHz instead of $f_o = 10$ kHz. And in fact, if a 2 kHz sinusoid is sampled at $f_s = 12$ kHz, Figure 6–14(c) is what $X[\exp(j2\pi fT)]$ would look like. Let's see why this is so.

Consider a continuous time sinusoid $x_a(t) = \cos(2\pi f_o t)$, where $f_o = 2$ kHz. The spectrum $X_a(f)$ consists of a pair of impulses located at ± 2 kHz, as shown in Figure 6–15(a). The pair of impulses is associated as one unit. Now suppose $x_a(t)$ is sampled at $f_s = 12$ kHz. The resulting spectrum $X[\exp(j2\pi fT)]$, as shown in Figure 6–15(b), was obtained by applying the relationship of Equation (6–74). Note that $X[\exp(j2\pi fT)]$ is a scaled duplicate of $X_a(f)$ within the interval $-f_s/2 < f < f_s/2$. Note also that Figure 6–15(b) is the same as Figure 6–14(c) (ignoring the amplitude scale factors). The implication is that a 10 kHz sinusoid sampled at $f_s = 12$ kHz will result in the same sequence $x(n)$ that would be obtained by sampling a 2 kHz sinusoid at $f_s = 12$ kHz. Let us verify that this is true.

Let $x_a(t) = \cos(2\pi f_o t)$, where $f_o = 10$ kHz. If $x_a(t)$ is sampled at $f_s = 12$ kHz, the resulting sequence is

$$x(n) = \cos\left(\frac{2\pi f_o n}{f_s}\right) = \cos\left(\frac{2\pi \times 1000 \times n}{12000}\right) = \cos\left(\frac{5\pi n}{3}\right) \qquad (6\text{–}78)$$

Now consider another signal, $y_a(t) = \cos(2\pi f_o t)$, where $f_o = 2$ kHz. If $y_a(t)$ is sampled at $f_s = 12$ kHz, the resulting sequence is

$$y(n) = \cos\left(\frac{2\pi f_o n}{f_s}\right) = \cos\left(\frac{2\pi \times 2000 \times n}{12000}\right) = \cos\left(\frac{\pi n}{3}\right) \qquad (6\text{–}79)$$

Observe that since n is an integer,

$$\cos\left(\frac{5\pi n}{3}\right) = \cos\left(2\pi n - \frac{\pi n}{3}\right) = \cos\left(\frac{-\pi n}{3}\right) = \cos\left(\frac{\pi n}{3}\right) \qquad (6\text{–}80)$$

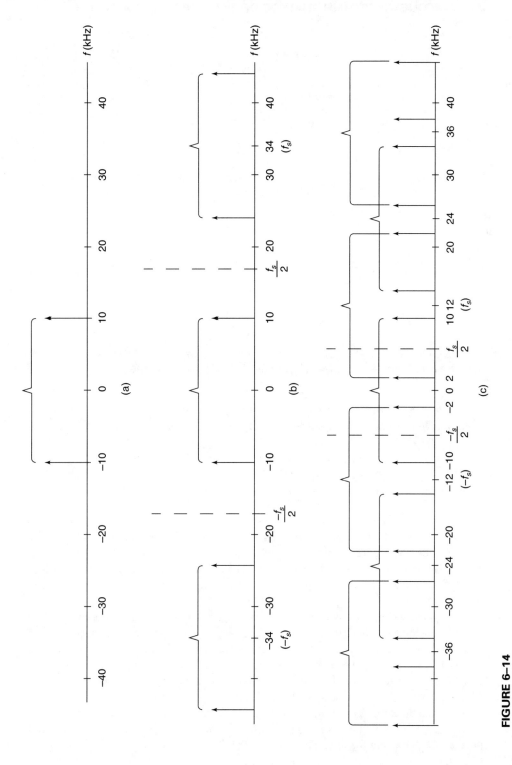

FIGURE 6–14

Illustrations for Example 6–6.1. (a) Original spectrum. (b) DTFT when $f_s = 34$ kHz. There is no aliasing. (c) DTFT when $f_s = 12$ kHz. The spectrum suffers from aliasing.

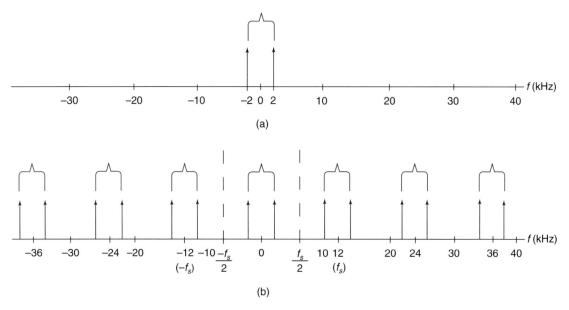

FIGURE 6–15
More illustrations for Example 6–6.1. (a) Original spectrum. (b) DTFT when f_s = 12 kHz. There is no aliasing. Note that the aliased spectrum shown in Figure 6–14(c) is just like this one.

Therefore, $x(n) = y(n)$. If aliasing is precluded by choosing a large enough sampling frequency and/or by using an anti-aliasing filter, this kind of ambiguity can be avoided.

6–7 DISCRETE TIME PROCESSING OF CONTINUOUS TIME SIGNALS

Consider the hypothetical continuous time system depicted in Figure 6–16(a). Suppose the spectrum $X_a(f)$ of the input signal and the frequency response function $H_a(f)$ of the system are as depicted in Figure 6–16(b).

NOTE: This system is a lowpass filter. The output spectrum $Y_a(f)$ is found by multiplying the input spectrum by the frequency response function; this is also depicted in Figure 6–16(b).

Suppose we want to achieve the same result using digital signal processing. We begin by replacing the continuous time system with a discrete time system having an A/D converter at the input and a D/A converter at the output, as depicted in Figure 6–17(a). The problem is to design the discrete time system, that is, find the impulse response $h(n)$ or the transfer function $H(z)$, such that the overall result, in terms of the spectra of the input and output continuous time signals, is the same as shown in Figure 6–16(b). For the purposes of this discussion, assume that the A/D and D/A converters are ideal, and that there is no aliasing.

The key observation is that in the absence of aliasing, $X[\exp(j2\pi fT)]$ is a scaled duplicate of $X_a(f)$ within the interval $-f_s/2 < f < f_s/2$. This suggests that if the overall result is to be the same, then the frequency response function $H[\exp(j2\pi fT)]$ of the discrete time

FIGURE 6–16
Continuous time filtering.

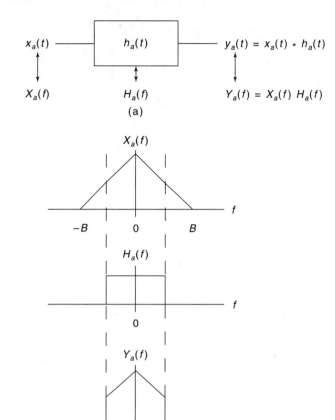

FIGURE 6–17
Discrete time processing of continuous time signals. Compare Figure 6–17(b) to Figure 6–16(b) and note that, with respect to the way in which the original spectrum is altered, the result at the output is the same. However, the discrete time system is more flexible since the characteristics of the filter can be changed just by altering the numbers in a computer program.

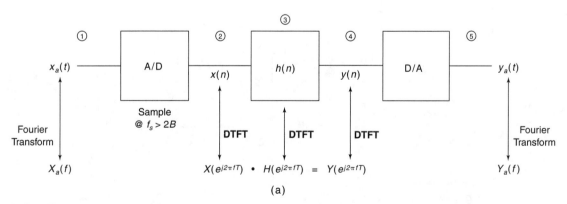

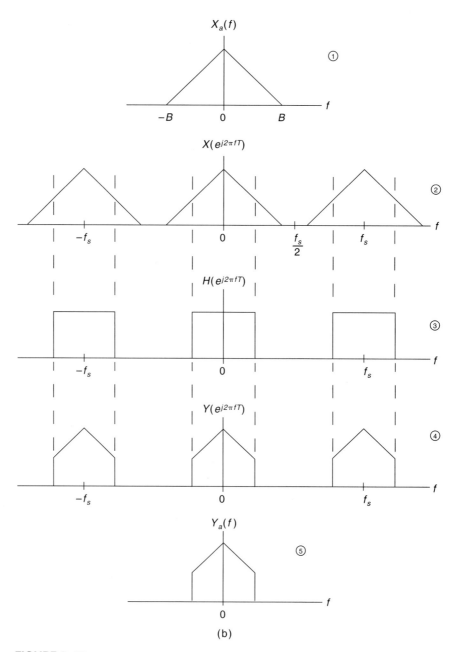

FIGURE 6–17
(continued)

system must be a duplicate of $H_a(f)$ within the same interval. (Of course, outside of this interval it will be periodically extended.) Figure 6–17(b) shows how the spectrum is altered as the signal passes through the system of Figure 6–17(a).

NOTE: The ideal D/A converter in effect reverses the sampling process. The scaling factor $1/T$ has been neglected in Figure 6–17(b).

Generally speaking, the purpose of designing a system such as the one depicted in Figure 6–17(a) is not to accomplish a signal processing task that could have been done just as well with a continuous time system. However, if the signal processing task requires that the characteristics of the filter be changed easily (or even adapted automatically) to fit the requirements imposed by different input signals, then digital signal processing is appropriate. To take a simple example, suppose the cutoff frequency of the lowpass filter shown in Figures 6–16 and 6–17 needs to be variable. In a continuous time lowpass filter, this requires changing resistor and/or capacitor values. In a discrete time system, this change can be accomplished by altering the numbers in a program. Note also that there are certain types of filters that are much easier to design and implement in the discrete time domain. Examples include very sharp lowpass filters having exact linear phase in the passband and a Hilbert transform filter (that shifts the phase of all frequency components by 90 degrees).

6–8 RELATIONSHIP BETWEEN FREQUENCY RESPONSE FUNCTION AND SYSTEM POLES AND ZEROS

As noted in Section 6–1, the frequency response function of a BIBO stable discrete time LSI system can be found by evaluating the transfer function at $z = \exp(j\omega)$. For any particular frequency ω, $z = \exp(j\omega)$ is a point on the unit circle on the z plane; it can also be thought of as a vector, as shown in Figure 6–18(a). This vector rotates as ω changes.

Recall from Chapter 5 that the transfer function of a discrete time LSI system can be expressed as

$$H(z) = \frac{G \prod_{k=1}^{M}\left(1 - \lambda_k z^{-1}\right)}{\prod_{k=1}^{N}\left(1 - p_k z^{-1}\right)} \tag{6–81}$$

where $p_k = \lambda_k$ are the system poles and zeros, respectively. The system poles and zeros can be plotted as points on the z plane, as shown in Figure 5–11 (Chapter 5, page 134). These points can also be thought of as vectors. For example, if the system has poles at $z = 0.8\exp(\pm j\pi/4)$ and a zero at $z = -0.5$, the location of these vectors is as shown in Figure 6–18(b).

The frequency response function for a system having the transfer function of Equation (6–81) is

$$H(e^{j\omega}) = \frac{G \prod_{k=1}^{M}\left(1 - \lambda_k e^{-j\omega}\right)}{\prod_{k=1}^{N}\left(1 - p_k e^{-j\omega}\right)} \tag{6–82}$$

Therefore, the amplitude response function is

$$\left|H(e^{j\omega})\right| = \frac{|G| \prod_{k=1}^{M}\left|1 - \lambda_k e^{-j\omega}\right|}{\prod_{k=1}^{N}\left|1 - p_k e^{-j\omega}\right|} \tag{6–83}$$

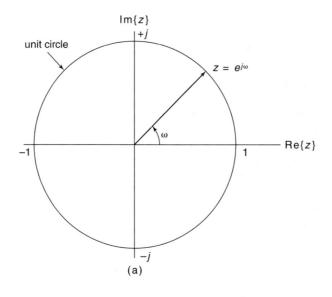

(a)

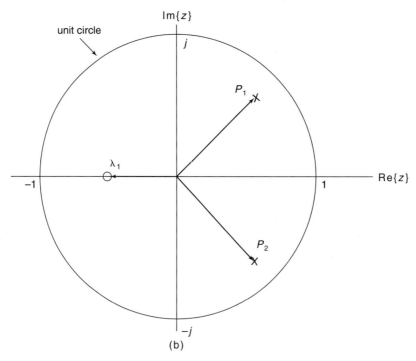

(b)

FIGURE 6–18
Geometric interpretation of points on the z plane.

Let us consider the kth factor in the numerator of Equation (6–82), which can be expressed as

$$\left(1 - \lambda_k e^{-j\omega}\right) = e^{-j\omega}\left(e^{j\omega} - \lambda_k\right) \tag{6–84}$$

Therefore, the kth term in the numerator of the amplitude response function (Equation [6–83]) can be expressed as

$$\begin{aligned}
\left|1 - \lambda_k e^{-j\omega}\right| &= \left|e^{-j\omega}\left(e^{j\omega} - \lambda_k\right)\right| = \left|e^{-j\omega}\right|\left|e^{j\omega} - \lambda_k\right| \\
&= \left|e^{j\omega} - \lambda_k\right|
\end{aligned} \tag{6–85}$$

Similarly, the kth term in the denominator of the amplitude response function can be expressed as

$$\left|1 - p_k e^{-j\omega}\right| = \left|e^{j\omega} - p_k\right| \tag{6–86}$$

The vectors $\exp(j\omega)$ and λ_k (for a particular value of ω) are shown in Figure 6–19(a). Also shown is the vector N_k, having its tail at the point $z = \lambda_k$ and its tip at the point $z = \exp(j\omega)$. Observe that using tip-to-tail vector addition, the relationship between these vectors is:

$$\lambda_k + N_k = e^{j\omega} \tag{6–87}$$

Therefore,

$$N_k = e^{j\omega} - \lambda_k \tag{6–88}$$

This means that the terms in the numerator of the amplitude response function can be expressed as

$$\left|1 - \lambda_k e^{-j\omega}\right| = \left|e^{j\omega} - \lambda_k\right| = \left|N_k\right| \tag{6–89}$$

where $\left|N_k\right|$ is the length of the vector N_k. As the frequency (ω) changes, so does the length of this vector.

Similar analysis of the kth term in the denominator of the amplitude response function results in

$$\left|1 - p_k e^{-j\omega}\right| = \left|e^{j\omega} - p_k\right| = \left|D_k\right| \tag{6–90}$$

where vector D_k has its tail at the point $z = p_k$ and its tip at the point $z = \exp(j\omega)$, as shown in Figure 6–19(b). $\left|D_k\right|$ is the length of this vector; as frequency (ω) changes, so does the length of this vector.

Using the above results, the amplitude response function can be expressed in terms of the lengths of vectors N_k and D_k:

$$\left|H\left(e^{j\omega}\right)\right| = \frac{|G|\prod\limits_{k=1}^{M}|N_k|}{\prod\limits_{k=1}^{N}|D_k|} \tag{6–91}$$

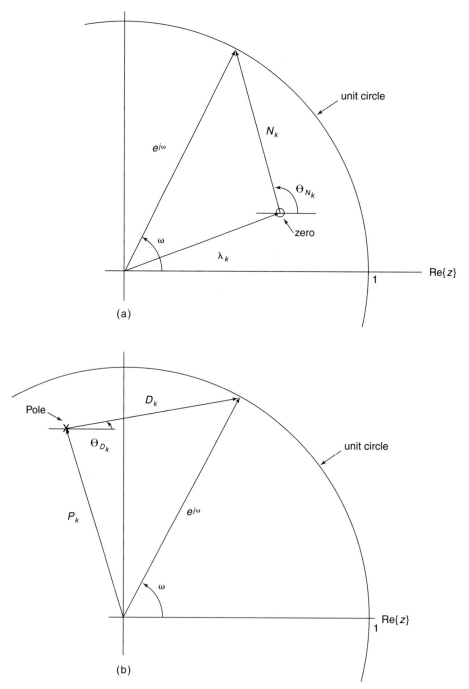

(a)

(b)

FIGURE 6–19

Geometric interpretation of the kth numerator and denominator terms in the factored form of the amplitude response function, using tip-to-tail vector addition.

Therefore, given the location of the poles and zeros, we can (in theory) graphically determine the value of the amplitude response function at any particular frequency.

EXAMPLE 6–8.1

Suppose we have a system with a pair of complex conjugate zeros at $z = \exp(\pm j\pi/4)$, and a pair of complex conjugate poles at $z = 0.9\exp(\pm j\pi/3)$. Figure 6–20(a) shows the lengths of $N_1, N_2, D_1,$ and D_2 when the frequency is $\omega = \pi/10$. Figure 6–20(b) shows these lengths when the frequency is close to (but not quite) $\omega = \pi/4$. Figure 6–20(c) shows these lengths when the frequency is $\omega = \pi/3$. Figure 6–20(d) shows these lengths when $\omega = 3\pi/4$. The reader is invited to use a ruler to measure these lengths and determine which of these four frequencies results in the largest value of $|H[\exp(j\omega)]|$. Figure 6–20(e), which is a plot of the actual amplitude response function for a system with these poles and zeros (and with $G = 1$), can be used to verify the results obtained graphically.

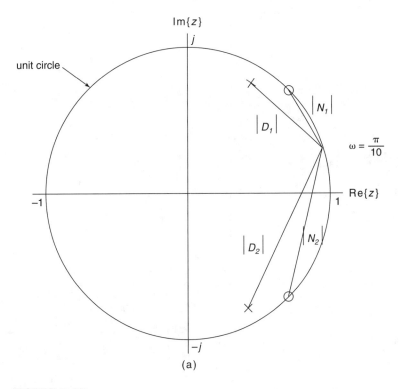

(a)

FIGURE 6–20

The distances from the poles and zeros to a point $z = \exp(j\omega)$ on the unit circle, for Example 6–8.1. These distances, which determine the shape of the amplitude response function, change as ω changes. (e) shows the amplitude response function for Example 6–8.1.

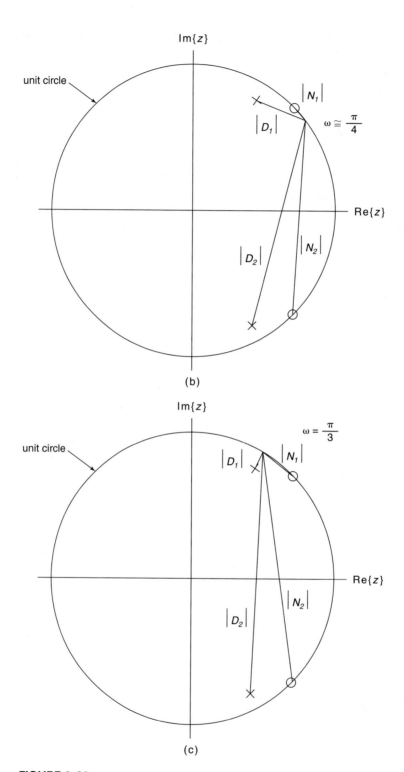

FIGURE 6–20
(continued)

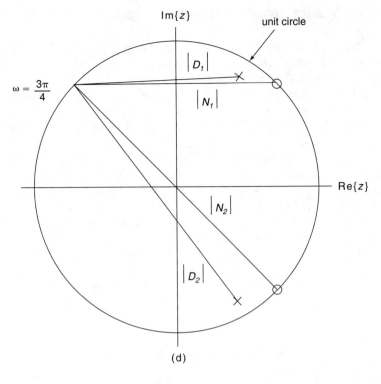

(d)

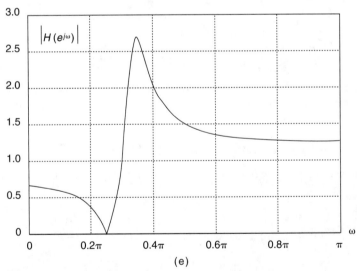

(e)

FIGURE 6–20
(continued)

The real utility of this geometric view of the amplitude response function is that it can give us an approximate idea of what to expect if the poles and zeros are in certain locations. For example, if there happens to be a zero on the unit circle of the z plane, that is,

$\lambda_k = \exp(j\omega_o)$, then when $\omega = \omega_o$, the amplitude response function is equal to zero. This is illustrated by Figure 6–20, in which the system being illustrated has a zero on the unit circle at $z = \exp(j\pi/4)$. In general, if there is a zero $\lambda_k = |\lambda_k \exp(j\omega_o)|$ which is close to the unit circle, then there will probably be a dip (that is, local minimum) in the amplitude response function at the frequency $\omega = \omega_o$, since corresponding length $|N_k|$ will be very small at that frequency Similarly, if there is a pole $p_k = |p_k| \exp(j\omega_o)$ that is close to the unit circle, there will probably be a peak (that is, local maximum) in the amplitude response function at the frequency $\omega = \omega_o$, since corresponding length $|D_k|$ will be very small at that frequency. This is also illustrated by Figure 6–20; there is a pole at $z = 0.9\exp(j\pi/3)$, and the amplitude response function has a definite peak at the frequency $\omega = \pi/3$. Of course, a BIBO-stable system will not have any poles located *on* the unit circle. We say there will *probably* be a peak or dip because the actual value of the amplitude response function depends on *all* of the lengths, and it is of course possible to have a case in which one very small length in the numerator of the amplitude response function is effectively cancelled out by another very small length in the denominator and vice versa.

Using complex number properties 3 and 4 from Table 2–1 (Chapter 2, page 25), it is easy to show that the phase response function for a system having the transfer function of Equation (6–81) is (assuming G is positive)

$$\theta_H(\omega) = \omega(N - M) + \sum_{k=1}^{M} \theta_{N_k} - \sum_{k=1}^{N} \theta_{D_k} \tag{6–92}$$

where

$$\begin{aligned}
\theta_{N_k} &= \angle\{N_k\} = \angle\{e^{j\omega} - \lambda_k\} \\
\theta_{D_k} &= \angle\{D_k\} = \angle\{e^{j\omega} - p_k\}
\end{aligned} \tag{6–93}$$

These angles are defined as usual for vectors. Figure 6–19(a) shows the angle θ_{Nk}. If G is negative, the phase response function of Equation (6–92) must be shifted by plus or minus π radians.

6–9 DTFT SHIFTING AND MODULATION THEOREMS

The *time shifting theorem* of the DTFT states that if $x(n) \leftrightarrow X[\exp(j\omega)]$, then

$$x(n - k) \leftrightarrow e^{-j\omega k} X(e^{j\omega}) \tag{6–94}$$

To prove this theorem, we take the DTFT of the sequence $x(n - k)$:

$$\textbf{DTFT}\big[x(n - k)\big] = \sum_{n=-\infty}^{\infty} x(n - k)e^{-j\omega n} \tag{6–95}$$

A change of variables in the summation expression will do the trick. Let $(n - k) = m$:

$$\textbf{DTFT}\big[x(n - k)\big] = \sum_{m=-\infty}^{\infty} x(m)e^{-j\omega(m+k)} = \sum_{m=-\infty}^{\infty} x(m)e^{-j\omega m} e^{-j\omega k} \tag{6–96}$$

Since exp(–jωk) does not depend on summation index m, it can be brought outside the summation:

$$\mathbf{DTFT}\big[x(n - k)\big] = e^{-j\omega k}\left[\sum_{m=-\infty}^{\infty} x(m)e^{-j\omega m}\right] \qquad (6\text{--}97)$$

The expression inside the brackets in Equation (6–97) is $X[\exp(j\omega)]$. (The summation index can be changed from m to any other letter, such as n.) Thus the time shifting theorem is demonstrated. Observe that since $|\exp(-j\omega k)| = 1$, the spectral amplitude function is not changed by a time domain shift.

The *modulation theorem* (or *frequency shifting theorem*) of the DTFT states that if $x(n) \leftrightarrow X[\exp(j\omega)]$, then:

$$e^{j\phi n}x(n) \leftrightarrow X\left(e^{j(\omega-\phi)}\right) \qquad (6\text{--}98)$$

$X\{\exp[j(\omega - \phi)]\}$ is a frequency shifted version of orignal spectrum $X[\exp(j\omega)]$. Figure 6–21 illustrates the concept: The original spectrum is shifted such that it is centered on $\omega = \phi$ instead of $\omega = 0$. Note that the *periodic extensions* are shifted by the same amount, so that the resulting spectrum is still periodic 2π.

The modulation theorem is easy to verify by taking the DTFT of $\exp(j\phi n)x(n)$:

$$\mathbf{DTFT}\big[e^{j\phi n}x(n)\big] = \sum_{n=-\infty}^{\infty}\left(e^{j\phi n}x(n)\right)e^{-j\omega n} = \sum_{n=-\infty}^{\infty}x(n)e^{-j(\omega-\phi)n} \qquad (6\text{--}99)$$

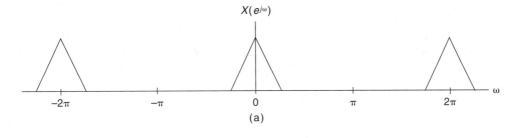
(a)

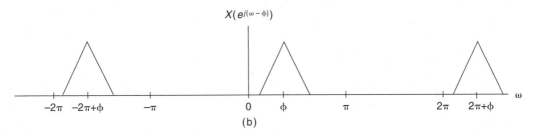
(b)

FIGURE 6–21
Frequency shifting.

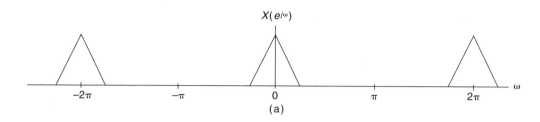

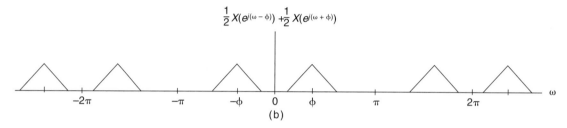

FIGURE 6–22
(a) Spectrum of a hypothetical signal $x(n)$. (b) Spectrum of $x(n)\cos(\phi n)$.

The far right side of Equation (6–99) is the DTFT summation formula with ω replaced by $(\omega - \phi)$. Thus the modulation theorem is demonstrated.

An important extension of the modulation theorem is

$$x(n)\cos(\phi n) \leftrightarrow \frac{1}{2} X\left(e^{j(\omega-\phi)}\right) + \frac{1}{2} X\left(e^{j(\omega+\phi)}\right) \tag{6–100}$$

Figure 6–22 illustrates the concept. Multiplying by $\cos(\phi n)$ in the time domain creates two frequency domain shifts: there are scaled (1/2) copies of the original spectrum centered at $\omega = \phi$ and $\omega = -\phi$. (And again, the periodic extensions are also shifted, such that the result is still periodic 2π.)

This extension of the modulation theorem is easy to verify using the identity

$$\cos(\phi n) = \frac{1}{2} e^{j\phi n} + \frac{1}{2} e^{-j\phi n} \tag{6–101}$$

which means that

$$x(n)\cos(\phi n) = \frac{1}{2} e^{j\phi n} x(n) + \frac{1}{2} e^{-j\phi n} x(n) \tag{6–102}$$

Now apply the modulation theorem (Equation [6–98]) to both terms on the right of Equation (6–102); this will demonstrate the validity of Equation (6–100).

6–10 DTFT CONVOLUTION THEOREMS

In Chapter 5 we observed that the output signal from an LSI discrete time system can be found by convolving the input signal with the system impulse response, that is,

$y(n) = h(n) * x(n)$. This fact, combined with Equation (6–41), implies the following DTFT pair:

$$h(n) * x(n) \leftrightarrow H\left(e^{j\omega}\right)X\left(e^{j\omega}\right) \qquad (6\text{–}103)$$

The Z transform pair is $h(n) * x(n) \leftrightarrow H(z)X(n)$.

NOTE: Equation (6–103) can be derived directly, without any reference to Z transforms, by taking the DTFT of $h(n) * x(n)$ directly.

Equation (6–103) says that *convolution in the time domain is multiplication in the frequency domain*. It can be shown that the *dual* of this relationship is also true: *Multiplication in the time domain is convolution in the frequency domain*. That is,

$$x(n)g(n) \leftrightarrow \frac{1}{2\pi} X\left(e^{j\omega}\right) * G\left(e^{j\omega}\right) \qquad (6\text{–}104)$$

where the convolution of the continuous functions $X[\exp(j\omega)]$ and $G[\exp(j\omega)]$ is given by

$$\frac{1}{2\pi} X\left(e^{j\omega}\right) * G\left(e^{j\omega}\right) = \frac{1}{2\pi} \int_{-\pi}^{\pi} X\left(e^{j\alpha}\right)G\left(e^{j(\omega-\alpha)}\right)d\alpha \qquad (6\text{–}105)$$

Equation (6–104) is important because the *truncation in time* of a signal can be mathematically represented as multiplying the signal by a rectangular window sequence. Suppose we have a signal $x(n)$ that is truncated to create a signal $\hat{x}(n)$ that is N points in time duration:

$$\hat{x}(n) = \begin{cases} x(n), n = 0, 1, \ldots, N-1 \\ 0, \text{otherwise} \end{cases} \qquad (6\text{–}106)$$

If we define a *rectangular window* $r(n)$ as

$$r(n) = \begin{cases} 1, n = 0, 1, \ldots, N-1 \\ 0, \text{otherwise} \end{cases} \qquad (6\text{–}107)$$

then clearly

$$\hat{x}(n) = r(n)x(n) \qquad (6\text{–}108)$$

The rectangular window $r(n)$ is just another sequence and thus it has a spectrum. That is,

$$r(n) \leftrightarrow R\left(e^{j\omega}\right) \qquad (6\text{–}109)$$

We are now in a position to answer the following question: If $x(n)$ is truncated in this manner, how is its spectrum affected? That is, what is the spectrum of the truncated sequence $\hat{x}(n)$? If we define

$$x(n) \leftrightarrow X\left(e^{j\omega}\right)$$
$$\hat{x}(n) \leftrightarrow \hat{X}\left(e^{j\omega}\right) \qquad (6\text{–}110)$$

then according to Equation (6–104), we have:

$$\hat{X}\left(e^{j\omega}\right) = \frac{1}{2\pi} R\left(e^{j\omega}\right) * X\left(e^{j\omega}\right) \tag{6–111}$$

Thus the original spectrum is convolved with the spectrum of the rectangular window. Roughly speaking, the original spectrum is blended in a complicated manner with the spectrum of the rectangular window; the resulting spectrum can be expected to inherit characteristics from both of its parents. Section 6–14 (A Simple FIR Lowpass Filter) will illustrate this concept. In Chapter 9, which covers the design of FIR filters, this concept plays an important role. In Section 6–11, we will learn how to derive the spectrum of the rectangular window.

6–11   THE RECTANGULAR WINDOW AND ITS SPECTRUM

As suggested in the previous section, the rectangular window plays an important role in signal processing theory because of its use as a truncation function. Let $r(n)$ be the rectangular window sequence of length N points:

$$r(n) = \begin{cases} 1, n = 0, 1, \ldots, N-1 \\ 0, otherwise \end{cases} \tag{6–112}$$

The DTFT of the rectangular window is

$$R\left(e^{j\omega}\right) = \sum_{n=0}^{N-1} (1)e^{-j\omega n} \tag{6–113}$$

Note that the summation limits do not need to cover values of n where $r(n) = 0$. In order to put Equation (6–113) in closed form, we appeal to the following formula (a geometric series with a finite number of terms, as covered in Chapter 5):

$$\sum_{n=0}^{N-1} a^n = \frac{1 - a^N}{1 - a} \tag{6–114}$$

In Equation 6–113, exp(–$j\omega$) played the role of a in Equation (6–114). Therefore,

$$R\left(e^{j\omega}\right) = \frac{1 - e^{-j\omega N}}{1 - e^{-j\omega}} \tag{6–115}$$

As a first step toward finding the spectral amplitude function, observe that Equation (6–115) can be written as

$$R\left(e^{j\omega}\right) = \frac{e^{-j\omega\frac{N}{2}}\left[e^{j\omega\frac{N}{2}} - e^{-j\omega\frac{N}{2}}\right]}{e^{-j\frac{\omega}{2}}\left[e^{j\frac{\omega}{2}} - e^{-j\frac{\omega}{2}}\right]} \tag{6–116}$$

Applying Euler's theorem to the expressions inside the brackets and combining the expressions outside the brackets results in

$$R\left(e^{j\omega}\right) = e^{-j\omega\left(\frac{N-1}{2}\right)} \left[\frac{\sin\left(\frac{\omega N}{2}\right)}{\sin\left(\frac{\omega}{2}\right)} \right] \tag{6–117}$$

Since $|\exp(j\theta)| = 1$, the spectral amplitude function for the rectangular window is

$$\left|R\left(e^{j\omega}\right)\right| = \left| \frac{\sin\left(\frac{\omega N}{2}\right)}{\sin\left(\frac{\omega}{2}\right)} \right| \tag{6–118}$$

$|R[\exp(j\omega)]|$ for a rectangular window of $N = 16$ points is shown in Figure 6–23. Note that this spectral amplitude function is characterized by a main lobe and sidelobes. The peak value of the main lobe is equal to N. The width of the main lobe is equal to $4\pi/N$. Thus as N increases (i.e., as the window gets longer), the main lobe of the spectral amplitude function becomes higher and narrower. (The fact that it becomes narrower as N increases is another manifestation of the rule of thumb that states the bandwidth is inversely proportional to

FIGURE 6–23
Amplitude spectrum of an $N = 16$ point rectangular window.

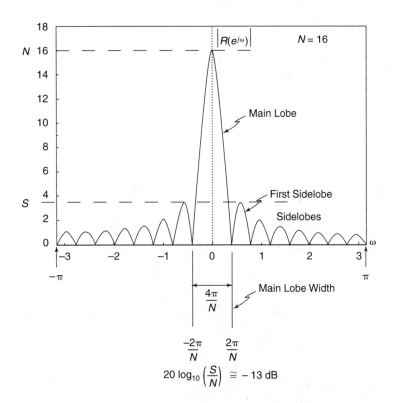

pulsewidth.) Note also that the amplitude of the sidelobes more or less increases along with the amplitude of the main lobe. "It can be shown" that the ratio (first sidelobe peak value)/(main lobe peak value) remains nearly constant as N increases. That is to say, with respect to the notation on Figure 6–23, $S/N \approx$ constant. In fact, it turns out that

$$20\log\left(\frac{S}{N}\right) \cong -13\text{dB} \qquad (6\text{–}119)$$

These qualities are illustrated by Figure 6–24 (a through d), which shows $|R[\exp(j\omega)]|$ for $N = 8, 16, 32,$ and $64,$ respectively.

FIGURE 6–24
Amplitude spectra of
rectangular windows of
various lengths: (a) $N = 8$; (b)
$N = 16$; (c) $N = 32$; (d) $N = 64$.

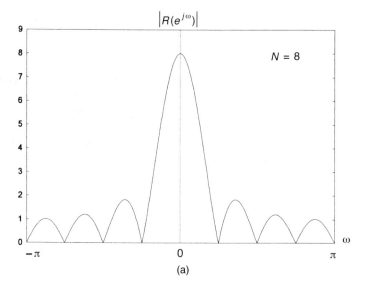

(a)

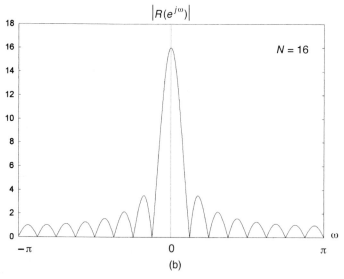

(b)

FIGURE 6–24
(continued)

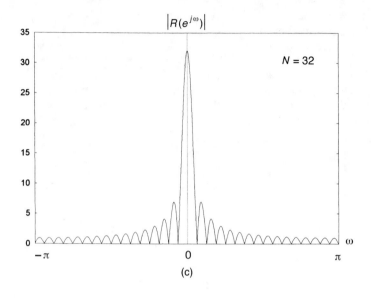

(c)

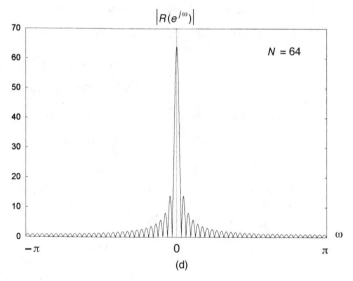

(d)

Figures 6–23 and 6–24 are plotted on the interval $-\pi < \omega < \pi$. Figure 6–25 shows $|R[\exp(j\omega)]|$ (for $N = 16$) on the interval $0 \leq \omega < 2\pi$, in order to emphasize once again that the DTFT is periodic 2π.

FIGURE 6–25
Amplitude spectrum of an
$N = 16$ point rectangular
window, shown on the
interval $0 \leq \omega \leq 2\pi$. When
compared to Figure 6–23,
the periodicity of the DTFT
is revealed.

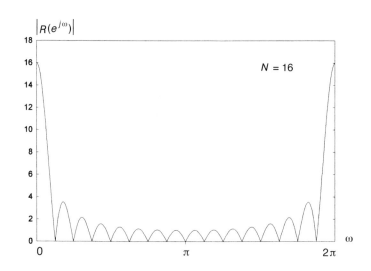

6–12 DTFT OF A TRUNCATED SINUSOID

Suppose $x(n)$ is a truncated (or *windowed*) sinusoid. Specifically, suppose

$$x(n) = \begin{cases} \cos(\phi n), n = 0, 1, \dots, N-1 \\ 0, otherwise \end{cases} \qquad (6\text{–}120)$$

This can also be expressed as

$$x(n) = r(n)\cos(\phi n) \qquad (6\text{–}121)$$

where $r(n)$ is an N-point rectangular window (see Equation [6–112]).

We would like to find the spectrum $X[\exp(j\omega)]$ of the truncated sinusoid. The modulation (or frequency shifting) theorem discussed in Section 6–9 (Equation [6–100]) is directly applicable to this problem. If

$$r(n) \leftrightarrow R\left(e^{j\omega}\right)$$

then

$$x(n) = r(n)\cos(\phi n) \leftrightarrow X\left(e^{j\omega}\right) = \frac{1}{2} R\left(e^{j(\omega-\phi)}\right) + \frac{1}{2} R\left(e^{j(\omega+\phi)}\right) \qquad (6\text{–}122)$$

The spectrum of the rectangular window, $R[\exp(j\omega)]$, was derived in Section 6–11 (see Equation [6–117]). The amplitude spectrum $\left| R[\exp(j\omega)] \right|$ is shown in Figures 6–23 and 6–24. When $r(n)$ is multiplied by $\cos(\phi n)$, the result in the frequency domain is that copies of $R[\exp(j\omega)]$ are shifted to $\omega = \pm\phi$ (and of course all of the periodic extensions are also shifted). Thus when the sinusoid is truncated, the result in the frequency domain is that instead of having impulses at $\omega = \pm\phi$, there are copies of the window spectrum at $\omega = \pm\phi$. As shown in Figures 6–23 and 6–24, the longer the window length (N), the higher and narrower the window spectrum main lobe will be.

The spectral amplitude function $|X[\exp(j\omega)]|$ of a truncated sinusoid is shown in Figure 6–26 for two cases: (a) $N = 16$ and (b) $N = 64$. In both cases, $\phi = \pi/3$. Note also that these spectral plots are shown for the interval $0 \le \omega < 2\pi$, which is why the main lobes appear centered at $\omega = \pi/3$ and $\omega = 2\pi - (\pi/3)$.

FIGURE 6–26
Amplitude spectrum of $x(n) =$ $r(n)\cos(\pi/3 n)$, where $r(n)$ is an N-point rectangular window. (a) $N = 16$; (b) $N = 64$.

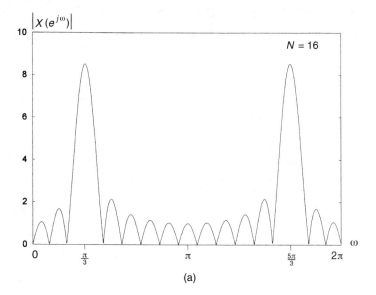

(a)

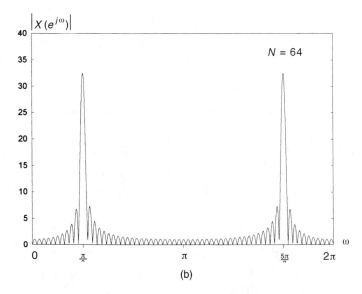

(b)

By appealing to the linearity property of the DTFT, this result can be extended to the case in which $x(n)$ is a truncated sum of two sinusoids at different frequencies, that is,

$$x(n) = r(n)\left[A_1\cos(\phi_1 n) + A_2\cos(\phi_2 n)\right] \qquad (6\text{–}123)$$

With no truncation, the spectrum would be characterized by impulses at $\omega = \pm\phi_1$ and $\omega \pm \phi_2$ (periodically extended). However, as a result of the multiplication by the rectangular window, the spectrum is characterized by copies of the rectangular window spectrum appearing at these frequencies instead of the impulses:

$$X\left(e^{j\omega}\right) = \frac{A_1}{2}R\left(e^{j(\omega-\phi_1)}\right) + \frac{A_1}{2}R\left(e^{j(\omega+\phi_1)}\right) + \frac{A_2}{2}R\left(e^{j(\omega-\phi_2)}\right) + \frac{A_2}{2}R\left(e^{j(\omega+\phi_2)}\right) \qquad (6\text{–}124)$$

Whether or not the amplitude spectrum has distinct peaks at $\omega = \phi_1$ and $\omega = \phi_2$ depends on the window length (N) and the spacing of the two frequencies. This is illustrated by the following example.

EXAMPLE 6–12.1

A continuous time signal consists of the sum of two sinusoids. The frequencies of the sinusoids are $f_1 = 113$ Hz and $f_2 = 120$ Hz. Both sinusoids have a peak amplitude of 1. This signal is sampled at $f_s = 400$ Hz and truncated by an N-point rectangular window. The resulting sequence is as shown in Equation (6–123), where $A_1 = 1$, $A_2 = 1$, and

$$\phi_1 = 2\pi\left(\frac{113}{400}\right)$$
$$ \qquad (6\text{–}125)$$
$$\phi_2 = 2\pi\left(\frac{120}{400}\right)$$

For this example, the amplitude spectrum will be shown as a function of f, that is, as $|X[\exp(j2\pi fT)]|$, and will be plotted within the interval $0 \leq f \leq f_s/2$. The spectrum is characterized by copies of the rectangular window spectrum at f_1 and f_2. These copies are added together. If the copies overlap to a small extent, the resulting amplitude spectrum will be characterized by distinct peaks at these two frequencies. On the other hand, if these copies overlap to a significant extent—in particular, if the main lobes overlap—the existence of two distinct peaks will be adversely affected.

On the ω frequency scale, the main lobe of $R[\exp(j\omega)]$ has a width of $\omega = 4\pi/N$. Since $\omega = 2\pi f/f_s$, the corresponding width on the f frequency scale is $f = 2f_s/N$. (For the example at hand, since $f_s = 400$ Hz, the main lobe width is $800/N$ Hz.) On either scale, the crucial observation is that the main lobe width increases as N decreases. Therefore, if N is not large enough, the main lobes will overlap to such an extent that the existence of two distinct *frequency components* in the signal will not be evident upon inspection of $|X[\exp(j2\pi fT)]|$.

The main lobes will overlap if the main lobe width is greater than the frequency spacing of the sinusoids. That is, the main lobes will overlap if $(2f_s/N) > |f_2 - f_1|$. Therefore, in order to avoid overlapping the main lobes, the rectangular window must have a length $N > (2f_s/|f_2 - f_1|)$. For the example at hand, this translates to $N > 115$.

Figure 6–27(a) shows the amplitude spectrum when $N = 200$ points. The main lobes do not overlap, so this spectrum has two very distinct peaks at frequencies f_1 and f_2. According to the formula given above, the main lobe width in this case is $2 \times 400/200 = 4$ Hz; this is consistent with the main lobe widths suggested in Figure 6–27(a).

Figure 6–27(b) shows the amplitude spectrum when $N = 50$ points. In this case, the main lobes overlap, but the overlap is not significant enough to eliminate the two distinct peaks.

Figure 6–27(c) shows the amplitude spectrum when $N = 40$ points. The main lobes overlap to such an extent that the two distinct peaks are almost eliminated.

FIGURE 6–27
Amplitude spectrum of the windowed sum of two sinusoids having frequencies of 113 Hz and 120 Hz. The sampling frequency is $f_s = 400$ Hz. An N-point rectangular window is used. (a) $N = 200$; (b) $N = 50$; (c) $N = 40$; (d) $N = 25$. In (d), the existence of two distinct *frequency components* in the signal is obscured because the window is too short.

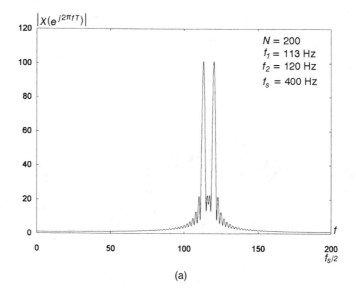

(a)

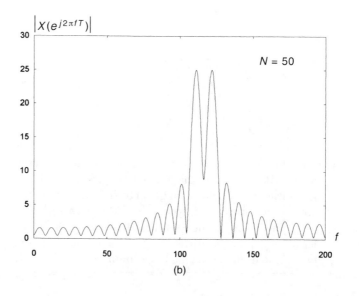

(b)

FIGURE 6–27
(continued)

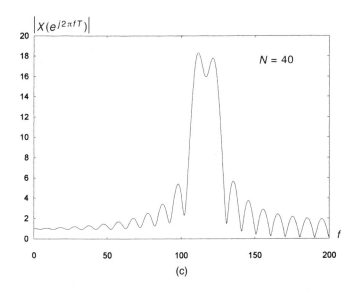

(c)

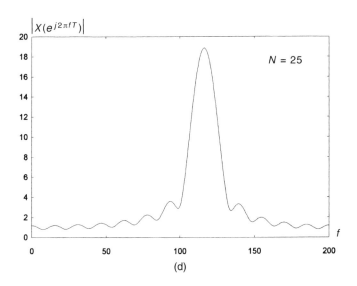

(d)

Finally, Figure 6–27(d) shows the amplitude spectrum when $N = 25$ points. The main lobe overlap is so great that when the copies centered at f_1 and f_2 are added, only one peak is evident. The window length is not sufficient to allow us to resolve the two frequency components of $x(n)$ by inspection of $|X[\exp(j2\pi fT)]|$.

This example suggests the following rough rule of thumb: A rectangular window length of approximately $N > f_s/|f_2 - f_1|$ is required in order to resolve the frequencies of two sinusoids in the amplitude spectrum.

Equations (6–123) and (6–124) can obviously be extended to more than two sinusoids. For example, suppose $g_a(t)$ is a *periodic* continuous time signal with the Fourier series:

$$g_a(t) = \sum_{k=-\infty}^{\infty} c_k e^{j2\pi k f_o t} \qquad (6\text{–}126)$$

where f_o is the fundamental frequency (and therefore also the spacing between harmonics on the frequency scale) and c_k are the Fourier series coefficients. Suppose this signal is sampled and then truncated with a rectangular window. If there is no aliasing (meaning, as a practical matter, that the sampling frequency is greater than two times the highest *significant* harmonic frequency), then the spectrum will be characterized by a weighted (c_k) copy of the rectangular window spectrum located at each *frequency component* kf_o. If the rectangular window is long enough such that the main lobe width of the rectangular window spectrum is less than the spacing of the harmonics, $|X[\exp(j2\pi fT)]|$ will exhibit definite peaks at the harmonic frequencies, and the height of the peak at frequency kf_o will be approximately proportional to $|c_k|$. If the window length is less than the sampling frequency divided by the harmonic spacing, the distinct peaks will be obscured or lost.

6–13 IDEAL LOWPASS FILTER

An *ideal lowpass filter* with cutoff frequency ω_c is defined in the frequency domain as follows:

$$H_i(e^{j\omega}) = \begin{cases} 1, |\omega| \leq \omega_c \\ 0, \omega_c < |\omega| \leq \pi \end{cases} \qquad (6\text{–}127)$$

$$\text{periodic } 2\pi$$

The subscript i denotes *ideal*. The frequency response function $H_i[\exp(j\omega)]$ is shown in Figure 6–28, which also shows the periodic (2π) extensions.

To find the impulse response $h_i(n)$ of an ideal lowpass filter, we must employ the inverse DTFT integral:

$$h_i(n) = \frac{1}{2\pi} \int_{-\pi}^{\pi} H_i(e^{j\omega}) e^{j\omega n} d\alpha \qquad (6\text{–}128)$$

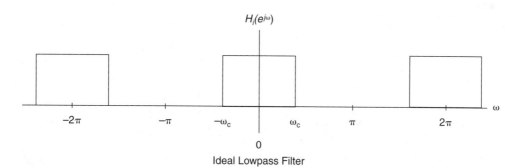

Ideal Lowpass Filter

FIGURE 6–28
Frequency response function of an ideal lowpass filter.

For the specific problem at hand, this becomes:

$$h_i(n) = \frac{1}{2\pi} \int_{-\omega_c}^{\omega_c} (1)e^{j\omega n} d\omega$$

$$= \frac{e^{j\omega n}}{j2\pi n} \bigg|_{-\omega_c}^{\omega_c} = \frac{e^{j\omega_c n} - e^{-j\omega_c n}}{j2\pi n} \tag{6-129}$$

But we can use Euler's theorem to obtain

$$e^{j\omega_c n} - e^{-j\omega_c n} = 2j\sin(\omega_c n) \tag{6-130}$$

By substituting Equation (6–130) into Equation (6–129), we obtain

$$h_i(n) = \frac{\sin(\omega_c n)}{\pi n} \tag{6-131}$$

This impulse response is defined for all integer values of n ($-\infty < n < \infty$); it dies out as n gets larger in either the positive or negative direction. The impulse response of an ideal lowpass filter with cutoff frequency $\omega_c = \pi/4$ is shown in Figure 6–29. Two plotting styles are shown for $h_i(n)$; Figure 6–29(b) is a connect-the-dots version of Figure 6–29(a).

NOTE: L'Hopital's rule must be used to establish the fact that $h_i(0) = \omega_c = \pi$, since direct substitution of $n = 0$ into Equation (6–131) results in the indeterminate form 0/0.

Observe that the impulse response of this system starts before the impulse is applied to the input! (Recall that the impulse is defined as $\delta(n) = 1$ when $n = 0$, but has zero value for all other ns.) As discussed in Chapter 5, a system like this one is said to be *noncausal*. *Since an ideal lowpass filter is a noncausal system, it cannot be realized in the real world.* However, the ideal lowpass filter impulse response (Equation [6–131]) does serve as a starting point for the design of discrete time FIR lowpass filters that *can* be realized, as will be shown in detail in Chapter 9. In Section 6–14, we will look at a simple FIR lowpass filter design obtained by truncating and shifting the impulse response of an ideal lowpass filter.

FIGURE 6–29

Impulse response of an ideal lowpass filter with cutoff frequency $\omega_c = \pi/4$, showing two styles of sequence plots.

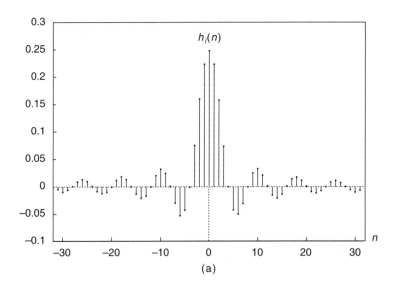

FIGURE 6–29
Impulse response of an ideal lowpass filter with cutoff frequency $\omega_c = \pi/4$, showing two styles of sequence plots.

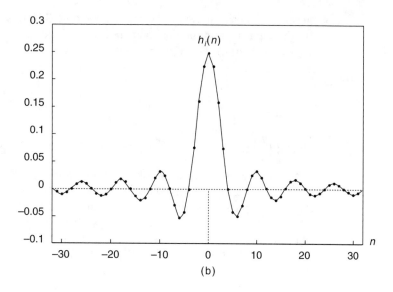

(b)

6–14 A SIMPLE FIR LOWPASS FILTER

The ideal lowpass filter was introduced in Section 6–13. The impulse response of this filter is given in Equation (6–131). Suppose we create a new impulse response, denoted $h(n)$, by altering $h_i(n)$ in the following manner:

1. Shift $h_i(n)$ so that it is centered at $n = (N-1)/2$ instead of $n = 0$ (N is an integer).
2. Truncate the resulting sequence so that it is zero outside of the interval $0 \le n \le N-1$.

The resulting sequence is:

$$h(n) = \begin{cases} h_i\left[n - \left(\dfrac{N-1}{2}\right)\right] = \dfrac{\sin\left\{\omega_c\left[n - \left(\dfrac{N-1}{2}\right)\right]\right\}}{\pi\left[n - \left(\dfrac{N-1}{2}\right)\right]}, & n = 0,1,\ldots,N-1 \\[4mm] 0, \text{otherwise} \end{cases}$$

(6–132)

Impulse response $h(n)$ is shown in Figure 6–30(a) for the case $N = 29$ and $\omega_c = 0.46\pi$. Note that we can also express $h(n)$ as

$$h(n) = h_i\left[n - \left(\frac{N-1}{2}\right)\right]r(n)$$

(6–133)

where $r(n)$ is an N-point rectangular window (see Section 6–11). Note that $h(n)$ is the impulse response of a causal FIR system, and is therefore realizable.

Let $H[\exp(j\omega)]$ be the frequency response function of the system with impulse response $h(n)$. Let $H_i[\exp(j\omega)]$ be the frequency response function of the ideal lowpass filter (see Figure 6–28). How are these two frequency response functions related? To answer this question, two principles must be applied: the time shifting theorem (Equation [6–94]) and the "multiplication in the time domain is convolution in the frequency domain" theorem (Equation [6–104]). The resulting relationship is

$$H\left(e^{j\omega}\right) = \frac{1}{2\pi}\left[e^{-j\omega\left(\frac{N-1}{2}\right)}H_i\left(e^{j\omega}\right)\right] * R\left(e^{j\omega}\right) \tag{6-134}$$

where $R[\exp(j\omega)]$ is the DTFT of the rectangular window (Section 6–11), and $*$ denotes convolution.

When two spectra are convolved, the resulting spectrum will in some sense have qualities, or characteristics, inherited from each of its parents. In this particular case, the resulting amplitude response function $|H[\exp(j\omega)]|$ still looks like a lowpass filter, but since $R[\exp(j\omega)]$ is smooth (has no jump discontinuities), the sudden transitions in the frequency response function of the ideal lowpass filter are rounded off, creating a transition band between the stopband and passband. Furthermore, the sidelobes in the spectrum of the rectangular window manifest themselves as ripples in both the passband and stopband of the resulting frequency response function. Figure 6–30(b) illustrates this result for the case $N = 29$ and $\omega_c = 0.46\pi$. The terminology *passband*, *stopband*, and *transition band* is explained in this figure.

In Section 6–11, we noted that the main lobe of the spectrum of a rectangular window is inversely proportional to window length N. The wider this main lobe is, the wider the transition band will be in $|H[\exp(j\omega)]|$. (That is, the smaller N is, the wider the transition band will be.) This is illustrated in Figures 6–31 and 6–32, which show $h(n)$ and $|H[\exp(j\omega)]|$ for $N = 9$ and $N = 79$. In fact, the transition band has a width approximately equal to $(1.81\pi)/(N - 1)$.

Figures 6–30, 6–31, and 6–32 also illustrate the fact that the peak approximation error, or peak ripple value, δ in Figure 6–30(b), is approximately 0.09, and does *not* significantly

FIGURE 6–30
A simple $N = 29$ point FIR filter formed by truncating and shifting the ideal lowpass filter impulse response. (a) FIR filter impulse response. (b) FIR filter amplitude response function.

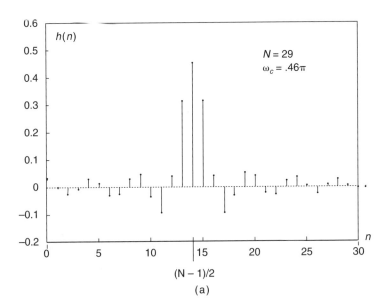

(a)

FIGURE 6–30
(continued)

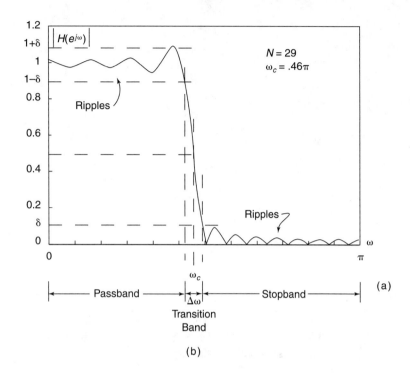

N = 29
$\omega_c = .46\pi$

Ripples

Ripples

ω

π

Passband

ω_c

$\Delta\omega$

Stopband (a)

Transition
Band

(b)

change when N is changed. Since $20\log(0.09) \cong -21$, this simple FIR lowpass filter is said
to have 21 dB of attenuation in the stopband.

The examples presented to this point all have an odd length (N is an odd integer),
such that the midpoint of $h(n)$, at $n = (N - 1)/2$, is also an integer. In these cases, $h(n)$ is a

FIGURE 6–31
A simple $N = 9$ point FIR filter
formed by truncating and
shifting the ideal lowpass
filter impulse response.
Compare Figures 6–30(b) and
6–31(b) to see how reducing
N changes the amplitude
response function.

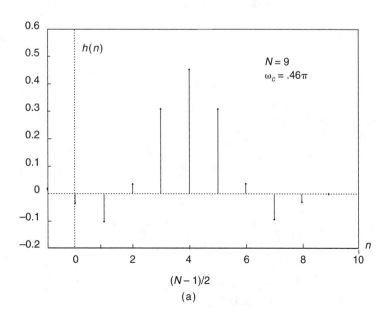

$h(n)$

N = 9
$\omega_c = .46\pi$

n

$(N-1)/2$

(a)

FIGURE 6–31
(continued)

FIGURE 6–31
(continued)

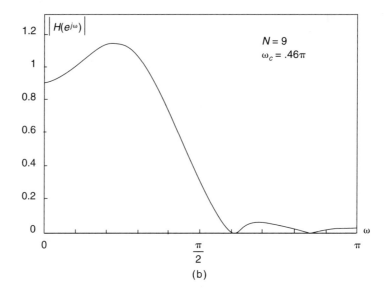

(b)

shifted (and truncated) version of $h_i(n)$ in the sense originally presented in Chapter 5, Figure 5–3 (on page 112). However, it is possible to consider shifting a sequence by a *non-integer* amount. For example, if

$$x(n) = \cos(\phi n) \tag{6–135}$$

then

$$x(n - \alpha) = \cos(\phi(n - \alpha)) \tag{6–136}$$

FIGURE 6–32

A simple $N = 79$ point FIR filter formed by truncating and shifting the ideal lowpass filter impulse response. Compare Figures 6–30(b) and 6–32(b) to see how increasing N changes the amplitude response function.

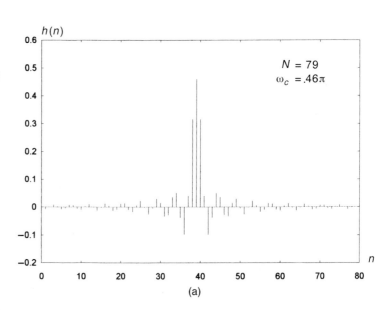

(a)

FIGURE 6–32
(continued)

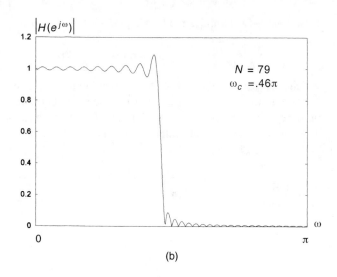

(b)

where α is any real number. If α is not an integer, then $x(n - \alpha)$ is not just a copy of $x(n)$ that has been moved to a new location on the n axis. However, $x(n - \alpha)$ is a sequence, the values of which can be obtained by plugging in integer values of n into the right side of Equation (6–136). One possible interpretation is that the underlying continuous time function $x_a(t)$ has been shifted by τ seconds, where τ is *not* an integer multiple of sampling interval T, and then resampled. In other words,

$$x(n) = x_a(t)\big|_{t=nT}$$
$$x(n - \alpha) = x_a(t - \tau)\big|_{t=nT} \tag{6-137}$$

The time shifting theorem of the DTFT can be generalized to allow for a noninteger shift by appealing to the inverse DTFT integral:

$$x(n) = \frac{1}{2\pi} \int_{-\pi}^{\pi} X(e^{j\omega}) e^{j\omega n} d\omega \tag{6-138}$$

If we replace n with $n - \alpha$ on the left side of Equation (6–138), we must also do so on the right side:

$$x(n - \alpha) = \frac{1}{2\pi} \int_{-\pi}^{\pi} X(e^{j\omega}) e^{j\omega(n-\alpha)} d\omega \tag{6-139}$$

The terms inside the integral can be rearranged, leading to

$$x(n - \alpha) = \frac{1}{2\pi} \int_{-\pi}^{\pi} \left[e^{-j\omega\alpha} X(e^{j\omega}) \right] e^{j\omega n} d\omega \tag{6-140}$$

Equation (6–140) implies the DTFT pair

$$x(n - \alpha) \leftrightarrow e^{-j\omega\alpha} X(e^{j\omega})$$ (6–141)

which is the same as the shifting theorem presented earlier, except that α is any real number.

With this in mind, let us return to the shifted and truncated ideal lowpass filter impulse response and consider *even* lengths N. That is, in Equation (6–132) we will let N be an even integer, such that midpoint $(N - 1)/2$ is an integer plus $1/2$. Figure 6–33(a) shows

FIGURE 6–33
Even length ($N = 30$) FIR filter. The ideal lowpass filter impulse response is shifted by a noninteger factor before being truncated. (See text for explanation.)

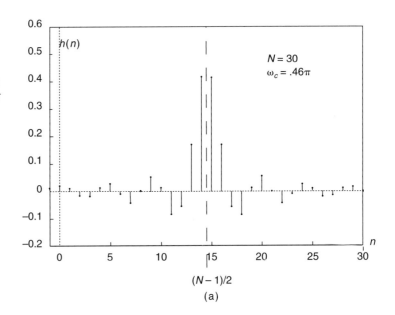

(a)

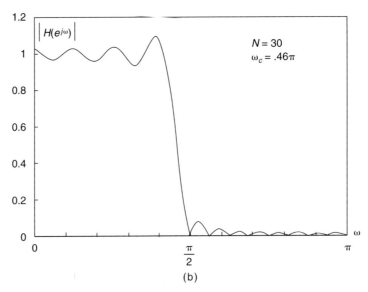

(b)

the resulting impulse response $h(n)$ for the case $N = 30$ and $\omega_c = 0.46\pi$; the midpoint in this case is 14.5. Figure 6–33(b) shows the resulting amplitude response function $|H[\exp(j\omega)]|$. Note that qualitatively it looks very similar to Figure 6–30(b) (the $N = 29$ case).

Note that impulse response $h(n)$ obtained in this section is symmetric about its midpoint whether N is even or odd. This impulse response symmetry is characteristic of a system that has a *generalized linear phase* phase response function $\theta_H(\omega)$. One way to view the significance of this characteristic is to observe that if the input signal to this system, $x(n)$, is symmetric about its midpoint, the output signal, $y(n)$, will also have symmetry about its midpoint (although its shape will be altered in most cases). The reason for this is that the output sequence $y(n)$ is equal to the input sequence $x(n)$ convolved with impulse response $h(n)$ and if two symmetric sequences are convolved, the result will also be symmetric. This preservation of symmetry can be important in some applications. (*Linear phase* will be considered in more detail in Section 6–15.)

The window method for designing FIR lowpass filters, which is discussed in detail in Chapter 9, is based on the results presented in this section. The main problem with the results shown here is the objectionable size of peak approximation error δ, as shown in Figure 6–30(b). The peak approximation error can be reduced by employing a window function with tapered edges, as opposed to the rectangular window that has a sudden transition from 0 to 1 on each edge. (These *tapered windows* are symmetric about their midpoints, so that the resulting impulse response is still symmetric about its midpoint.) As the reader can probably guess, there are design tradeoffs involved; there is a price to be paid for reducing δ.

6–15 SYSTEMS HAVING GENERALIZED LINEAR PHASE

Consider an LSI discrete time system having a frequency response function of the form

$$H\left(e^{j\omega}\right) = A(\omega)e^{-j\alpha\omega} \tag{6–142}$$

where $A(\omega)$ is a *real* function of ω, and α is a positive real number. Let us find the phase response function $\theta_H(\omega)$ of this system.

There are two cases to consider:

1. $A(\omega) \geq 0$
2. $A(\omega) < 0$

If $A(\omega) \geq 0$, then Equation (6–142) is already in polar form, where $|H[\exp(j\omega)]| = A(\omega)$ and $\theta_H(\omega) = -\alpha\omega$. On the other hand, if $A(\omega) < 0$, Equation (6–142) can be written as

$$H\left(e^{j\omega}\right) = -|A(\omega)|e^{-j\alpha\omega} \tag{6–143}$$

At this point we appeal to Euler's theorem and note that

$$e^{\pm j\pi} = \cos(\pi) \pm j\sin(\pi) = -1 \tag{6–144}$$

This allows us to take care of the (–) sign in Equation (6–143) and write

$$H\left(e^{j\omega}\right) = |A(\omega)|e^{j(-a\omega\pm\pi)} \tag{6–145}$$

Thus if $A(\omega) < 0$, we have $|H[\exp(j\omega)]| = |A(\omega)|$, and $\theta_H(\omega) = -\alpha\omega \pm \pi$.

To summarize, if $H[\exp(j\omega)]$ has the form indicated in Equation (6–142), where $A(\omega)$ is real, then the amplitude response function is $|H[\exp(j\omega)]| = |A(\omega)|$ and the phase response function is

$$\theta_H(\omega) = \begin{cases} -\alpha\omega, & A(\omega) \geq 0 \\ -\alpha\omega \pm \pi, & A(\omega) < 0 \end{cases} \tag{6–146}$$

A system with a phase response function of this form is said to have *generalized linear phase*, or just linear phase. Figure 6–34 shows a hypothetical example.

NOTE: The phase response plot is a straight line with a constant slope of $-\alpha$, with jump discontinuities of plus or minus π wherever $A(\omega)$ changes sign.

To understand the significance of linear phase, consider a complex exponential input signal $x(n)$:

$$x(n) = c e^{j\phi n} \tag{6–147}$$

(Remember that any periodic sequence can be represented as a weighted sum of complex exponentials at different frequencies, as explained in Section 6–2.) Let $x(n)$ be the input to a system having generalized linear phase. The output $y(n)$ is given by

$$y(n) = cH\left(e^{j\phi}\right)e^{j\phi n} = cA(\phi)e^{-j\alpha\phi}e^{j\phi n} \tag{6–148}$$

where $A(\phi)$ is real. If $A(\phi)$ is nonnegative, then

$$y(n) = c|A(\phi)|e^{j\phi(n-\alpha)} \tag{6–149}$$

On the other hand, if $A(\phi)$ is negative, then

$$y(n) = -c|A(\phi)|e^{j\phi(n-\alpha)} \tag{6–150}$$

Observe that *the time delay* (α) *does not depend on the frequency* (ϕ). Thus for an input signal consisting of many frequency components (for example, a periodic signal as represented by a discrete Fourier series), if the system has generalized linear phase, *all of the frequency components will be time delayed by the same amount*. (Components with frequency ϕ, where $A(\phi)$ is negative, will also be inverted.)

Any discrete time system having a real impulse response $h(n)$ that is symmetric about its midpoint is a system with generalized linear phase. (The FIR lowpass filters considered in Section 6–14 have this property.) To see why this is so, consider a hypothetical impulse response $\hat{h}(n)$ which is an even function of n, that is, $\hat{h}(n) = \hat{h}(-n)$ (the midpoint here is $n = 0$). As shown in Section 6–4, the DTFT of a sequence having this kind of even symmetry is a real function of ω. Therefore, the frequency response function of such a system is a real function of ω. That is, using the notation employed earlier,

$$\hat{H}\left(e^{j\omega}\right) = A(\omega) \tag{6–151}$$

where $A(\omega)$ is a real function of ω.

FIGURE 6–34

Hypothetical system having generalized linear phase.

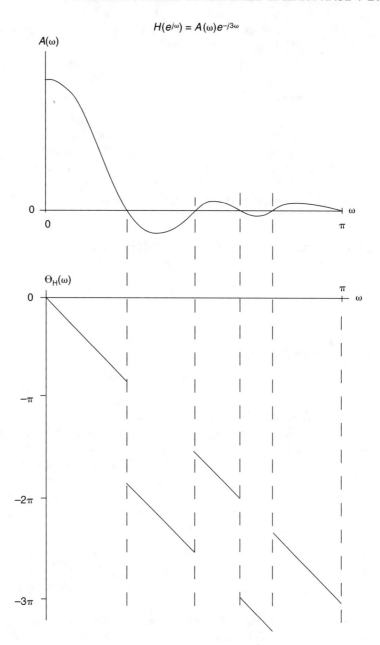

$$H(e^{j\omega}) = A(\omega)e^{-j3\omega}$$

Next we consider an impulse response $h(n)$ that is symmetric about some midpoint α. (As shown in Section 6–14, this midpoint can be an integer or an integer + 1/2.) Any such impulse response can be thought of as a shifted version of an even impulse response $\hat{h}(n)$. That is,

$$h(n) = \hat{h}(n - \alpha) \tag{6–152}$$

The resulting frequency response function $H[\exp(j\omega)]$ can be found by appealing to the shifting theorem:

$$H\left(e^{j\omega}\right) = e^{-j\alpha\omega}\hat{H}\left(e^{j\omega}\right) \tag{6-153}$$

That is,

$$H\left(e^{j\omega}\right) = A(\omega)e^{-j\alpha\omega} \tag{6-154}$$

As shown earlier, a frequency response function with this form represents a system having generalized linear phase.

See Section 6–14 for a discussion of the significance of linear phase with respect to the preservation of input signal symmetry.

6–16 PHASE FUNCTIONS: PRINCIPAL VALUE VERSUS UNWRAPPED PHASE

The DTFT can be expressed in polar form as

$$X\left(e^{j\omega}\right) = \left|X\left(e^{j\omega}\right)\right|e^{j\theta_x(\omega)} \tag{6-155}$$

But since $\exp(jk2\pi) = 1$ for any integer value of k (positive or negative), we can also write

$$X\left(e^{j\omega}\right) = \left|X\left(e^{j\omega}\right)\right|e^{j\theta_x(\omega)}e^{jk2\pi} = \left|X\left(e^{j\omega}\right)\right|e^{j(\theta_x(\omega)+k2\pi)} \tag{6-156}$$

This means that integer multiples of 2π radians can be added to or subtracted from the phase function $\theta_x(\omega)$ without changing $X[\exp(j\omega)]$. Therefore, strictly speaking, $\theta_x(\omega)$ is not a single-valued function; however, it is almost always plotted as a single-valued function. There are two basic styles of single-valued phase plots: the *principal value* phase plot and the *unwrapped* phase plot.

The *principal value* of $\theta_x(\omega)$, when evaluated at some particular frequency ω, is a number (in radians) within the range $-\pi < \theta \leq \pi$. When phase is calculated at a particular frequency ω using a calculator or a computer program, based on the formula

$$\theta_x(\omega) = \tan^{-1}\left(\frac{\mathrm{Im}\left\{X\left(e^{j\omega}\right)\right\}}{\mathrm{Re}\left\{X\left(e^{j\omega}\right)\right\}}\right) \tag{6-157}$$

the result will always be within the principal value range. Figure 6–35 shows a principal value version of the phase plot of Figure 6–34.

An *unwrapped* phase plot is a single-valued phase plot that is allowed to extend beyond the principal value range. An unwrapped phase plot is drawn such that within the frequency range $-\pi < \omega \leq \pi$, the plot has a monotone decreasing characteristic in a global sense. It does not wrap around from $\theta = -\pi$ to $\theta = \pi$ like a principal value plot sometimes does. Sign changes may force the inclusion of jump discontinuities of plus or minus π radians; thus the

FIGURE 6–35
Principal value phase plot for
the phase response function
of Figure 6–34.

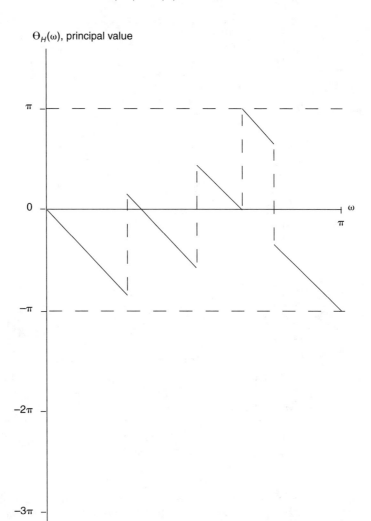

$$H(e^{j\omega}) = A(\omega)e^{-j3\omega}$$

$\Theta_H(\omega)$, principal value

monotone decreasing characteristic may not be true locally. The phase plot shown in Figure 6–34 can be thought of as an unwrapped phase plot, although a purist might argue for changing all of the $+\pi$ jump discontinuities to $-\pi$ jump discontinuities so that the desired monotone decreasing characteristic would be obtained.

A simple example can be used to illustrate the basic difference between principal value phase plots and unwrapped phase plots. Consider the simple linear phase function $\theta_x(\omega) = -5\omega$. Figure 6–36(a) shows the unwrapped phase plot for this phase function. Figure 6–36(b) shows the principal value phase plot for this phase function.

$\Theta_x(\omega) = -5\omega$, unwrapped

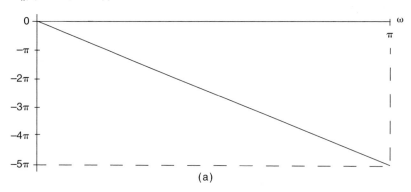

(a)

$\Theta_x(\omega) = -5\omega$, principal value

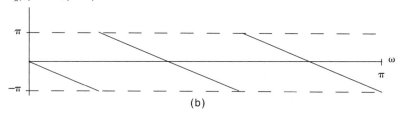

(b)

FIGURE 6–36
Hypothetical linear phase function. (a) Unwrapped phase plot. (b) Principal value phase plot.

PROBLEMS

6–1. Let $x_a(t) = \cos(2\pi f_o t)$, where $f_o = 100$ Hz. Suppose this signal is sampled to produce the sequence $x(n) = \cos(\omega_o n)$. Find the frequency ω_o if
    a. $f_s = 1000$ Hz
    b. $f_s = 10000$ Hz

6–2. Use superposition and Euler's theorem to demonstrate that if $x(n) = A\cos(\omega n + \phi)$ is the steady-state input to a BIBO stable discrete time system with the frequency response function $H[\exp(j\omega)]$, then the steady-state output signal is:

$$y(n) = A\left|H\left(e^{j\omega}\right)\right|\cos\left[\omega n + \phi + \theta_H(\omega)\right]$$

HINT: Look back at Equation (6–2). Equation (6–42) will also help.

6–3. Suppose $h(n) = \delta(n - 10)$ is the impulse response of a discrete time system.
    a. Find $H(e^{j\omega})$.
    b. Find and sketch $\left|H(e^{j\omega})\right|$ on the interval $-\pi < \omega < \pi$.
    c. Find and sketch $\theta_H(\omega)$ on the interval $-\pi < \omega < \pi$.
    d. Let

$$x(n) = \cos\left(\frac{\pi}{6}n\right)$$

be the steady-state input to this system. Find the steady-state output signal.

    e. If $x(n)$ was obtained by sampling $x_a(t) = \cos(2\pi f_o t)$ at $f_s = 10$ kHz, what is f_o?

6–4. Suppose $h(n) = \delta(n) + \delta(n-10)$ is the impulse response of a discrete time system.

    a. Find $H(e^{j\omega})$.

    b. Find and sketch $|H(e^{j\omega})|$ on the interval $0 \le \omega < 2$. (Be sure to find the values of ω where $|H(e^{j\omega})| = 0$).

    c. Find $\theta_H(\omega)$.

    d. Let $x(n)$ be the steady-state input to this system, obtained by sampling $x_a(t) = \cos(2\pi f_o t)$ at $f_s = 5$ kHz. If the steady-state output is $y(n) = 0$, what is f_o? Assume there is no aliasing. (There is more than one solution, even if there is no aliasing.)

    e. Let

$$x(n) = \cos\left(\frac{\pi}{6}n\right)$$

    be the steady-state input to this system. Find the steady-state output signal.

    f. Let

$$x(n) = \cos\left(\frac{\pi}{3}n\right) + \frac{1}{2}\cos\left(\frac{\pi}{9}n\right)$$

    be the steady-state input to this system. Find the steady-state output signal.

6–5. Given the following transfer function:

$$H(z) = \left(1 - \lambda z^{-1}\right)\left(1 - \lambda * z^{-1}\right)$$

where

$$\lambda = e^{j\frac{\pi}{4}}$$

    a. Find $H(e^{j\omega})$.

    b. This transfer function has zeros on the unit circle at

$$z = e^{\pm j\frac{\pi}{4}}$$

    What does that tell you about the amplitude response function $|H(e^{j\omega})|$ when

$$\omega = \frac{\pi}{4}$$

    c. Sketch $|H(e^{j\omega})|$ on the following intervals:

      i. $0 \le \omega \, 2\pi$

      ii. $-\pi < \omega < \pi$

    d. Let $x(n)$ be the steady-state input to this system, obtained by sampling $x_a(t) = A\cos(2\pi f_o t)$ at $f_s = 480$ Hz, where $f_o = 60$ Hz. Find the steady-state output signal.

6–6. For the system having the transfer function $H(z) = (1 - \lambda z^{-1})(1 - \lambda^* z^{-1})$, choose the value of λ such that if the sampling frequency is $f_s = 10$ kHz and the steady-state input signal is obtained by sampling $x_a(t) = A\cos(2\pi f_o t)$, where $f_o = 60$ Hz, the steady-state output will be $y(n) = 0$.

6–7. Given the transfer function

$$H(z) = \frac{1 - z^{-1}}{1 + 0.8z^{-1}}$$

    a. Find the poles and zeros of this system.

    b. Find $H(e^{j\omega})$.

    c. Find and sketch $|H(e^{j\omega})|$ on the interval $0 \le \omega \le \pi$.

    NOTE: Problem 6–7(a) should help you locate the frequencies (ω) at which $|H(e^{j\omega})|$ has its minimum and maximum values.

d. Find $\theta_H(\omega)$.

e. Let $x(n)$ be the steady-state input to this system, obtained by sampling $x_a(t) = 10\cos(2\pi f_o t)$ at $f_s = 10$ kHz. Let $f_o = 2$ kHz. Find the steady-state output.

6–8. Given the transfer function

$$H(z) = \frac{1 + z^{-1}}{1 - 0.9z^{-1}}$$

a. Find the poles and zeros of this system.

b. Find and sketch $|H(e^{j\omega})|$ on the interval $0 \leq \omega \leq \pi$.

NOTE: Problem 6–8(a) should help you locate the frequencies (ω) where $|H(e^{j\omega})|$ has its minimum and maximum values.

6–9. Given the transfer function

$$H(z) = \frac{\left(1 - az^{-1}\right)\left(1 - a^* z^{-1}\right)}{\left(1 - pz^{-1}\right)\left(1 - p^* z^{-1}\right)}$$

where

$$a = e^{j\frac{\pi}{4}}$$

and

$$p = 0.9e^{j\frac{3\pi}{4}}$$

Sketch $|H(e^{j\omega})|$ on the interval $0 \leq \omega \leq \pi$. (Use the information about the poles and zeros to obtain a rough sketch showing the location of the peaks and valleys of $|H(e^{j\omega})|$.)

6–10. The sequence $x(n) = A$ (for all values of n) can be thought of as $x(n) = A\cos(\omega_o n)$ where $\omega_o = 0$. Suppose this sequence is the input to a system having the transfer function $H(z) = 1 - z^{-1}$. Find the output sequence.

6–11. Show that if $x(n)$ is real and $x(n) \leftrightarrow X(e^{j\omega})$, then

$$x(-n) \leftrightarrow X^*\left(e^{j\omega}\right)$$

(This is the time reversal property of the DTFT.)

6–12. Show that if $x(n) = -x(-n)$, that is, $x(n)$ has odd symmetry about the point $n = 0$, then $X(e^{j\omega})$ is an imaginary function of ω, that is, $\text{Re}\{X(e^{j\omega})\} = 0$.

6–13. Show that

$$\sin(\omega_o n) \leftrightarrow \frac{\pi}{j}\delta(\omega - \omega_o) - \frac{\pi}{j}\delta(\omega + \omega_o)\,(\text{periodic } 2\pi)$$

6–14. Show that

$$x(n)\sin(\omega_o n) \leftrightarrow \frac{1}{2j}X\left(e^{j(\omega - \omega_o)}\right) - \frac{1}{2j}X\left(e^{j(\omega + \omega_o)}\right)$$

6–15. Show that

$$A\delta(n - k) \leftrightarrow Ae^{-j\omega k}$$

6–16. Show that if $x(n) = A$ for all values of n, then $X(e^{j\omega}) = 2\pi A\delta(\omega)$ (periodic 2π).

6–17. The frequency response function of an *ideal highpass filter* is shown in Chapter 9 (Figure 9–20 on page 272), where it is shown on the interval $-\pi < \omega < \pi$. Sketch the frequency response function of an ideal highpass filter on the interval $0 < \omega < 2\pi$.

6–18. The frequency response function of an *ideal bandpass filter* is shown in Chapter 9 (Figure 9–15 on page 269), where it is shown on the interval $-\pi < \omega < \pi$. Sketch the frequency response function of an ideal bandpass filter on the interval $0 \le \omega < 2\pi$.

6–19. The frequency response function of an *ideal bandstop filter* is shown in Chapter 9 (Figure 9–24 on page 275), where it is shown on the interval $-\pi < \omega < \pi$. Sketch the frequency response function of an ideal bandstop filter on the interval $0 \le \omega < 2\pi$.

6–20. Let

$$x(n) = \cos\left(\frac{\pi}{3}n\right)$$

for all n. Sketch $X(e^{j\omega})$ on the following intervals:
  a. $0 \le \omega < \pi$
  b. $-\pi < \omega < \pi$
  c. $0 \le \omega < 2\pi$

6–21. Let

$$x(n) = \cos\left(\frac{\pi}{3}n\right) + \frac{1}{2}\cos\left(\frac{\pi}{2}n\right) + \frac{1}{4}\cos\left(\frac{3\pi}{4}n\right)$$

for all n. Sketch $X(e^{j\omega})$ on the following intervals:
  a. $0 \le \omega < \pi$
  b. $-\pi < \omega < \pi$
  c. $0 \le \omega < 2\pi$

6–22. Let $x_a(t) = A\cos(2\pi f_o t)$ with $f_o = 10$ Hz, to create the sequence $x(n)$. Sketch $X(e^{j2\pi fT})$ on the interval $0 \le f < f_s$ for each of the three sampling frequencies listed below. Also indicate on each sketch the corresponding frequencies on the ω scale.
  a. $f_s = 30$ Hz
  b. $f_s = 200$ Hz
  c. $f_s = 15$ Hz

6–23. Let

$$x_a(t) = \cos(2\pi f_1 t) + \frac{1}{2}\cos(2\pi f_2 t)$$

with $f_1 = 8$ Hz and $f_2 = 15$ Hz, be sampled to create the sequence $x(n)$. Sketch $X(e^{j2\pi fT})$ on the interval $0 \le f < f_s$ for each of the three sampling frequencies listed below. Also indicate on each sketch the corresponding frequencies on the ω scale.
  a. $f_s = 40$ Hz
  b. $f_s = 200$ Hz
  c. $f_s = 20$ Hz

6–24. Let

$$x_a(t) = \frac{\sin(2\pi f_c t)}{\pi t}$$

(defined for all values of t), where $f_c = 100$ Hz. Suppose this signal is sampled to create $x(n)$. Sketch $X(e^{j\omega})$ on the interval $-\pi < \omega < \pi$ for each of the three sampling frequencies listed below.

HINT: $x_a(t)$ is the same function as the impulse response of an ideal lowpass filter in the continuous time domain (see Chapter 3).
  a. $f_s = 1000$ Hz
  b. $f_s = 300$ Hz
  c. $f_s = 150$ Hz

6–25. Let $x_a(t) = (2\pi)(100)e^{-(2\pi)(100)t}u(t)$. Suppose this signal is sampled to create $x(n)$. Sketch $X(e^{j2\pi f/T})$ on the interval $0 \leq f < f_s$ for each of the two sampling frequencies listed below.

HINT: $x_a(t)$ is the same function as the impulse response of a simple first-order lowpass filter with a cutoff frequency of 100 Hz, as discussed in Chapter 2 (see Figure 2–3 on page 000 and Equation [2–55]). Example 6–2 is also applicable to this problem.
  a. $f_s = 2000$ Hz
  b. $f_s = 1000$ Hz

6–26. Consider a hypothetical signal $x_a(t)$ having the spectrum $X_a(f)$ shown in Figure P6–26. Suppose this signal is processed by the system depicted in Figure 6–17(a) (assume the A/D and D/A converters are ideal). Suppose the discrete time filter in Figure 6–17(a) is an ideal lowpass filter with a cutoff frequency of $\omega_c = \pi/4$. Sketch the spectrum at points 2, 3, 4, and 5 with respect to Figure 6–17(a) for each of the four cases listed below. These sketches should be in the style of Figure 6–17(b).
  a. $B = 100$ Hz; $f_s = 200$ Hz
  b. $B = 100$ Hz; $f_s = 400$ Hz
  c. $B = 100$ Hz; $f_s = 800$ Hz
  d. $B = 20$ Hz; $f_s = 160$ Hz

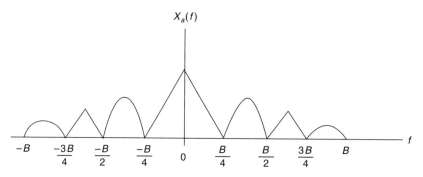

$X_a(f)$

$-B$ $\quad \frac{-3B}{4}$ $\quad \frac{-B}{2}$ $\quad \frac{-B}{4}$ $\quad 0 \quad$ $\frac{B}{4}$ $\quad \frac{B}{2}$ $\quad \frac{3B}{4}$ $\quad B$ $\qquad f$

FIGURE P6–26

6–27. Let $x_a(t) = \cos(2\pi f_1 t) + \cos(2\pi f_2 t)$, where $f_1 = 10$ Hz and $f_2 = 15$ Hz. Suppose this signal is processed by the system depicted in Figure 6–17(a). Suppose the discrete time filter in Figure 6–17(a) is an ideal bandpass filter with an upper cutoff frequency of $\omega_u = 5\pi/8$ and a lower cutoff frequency of $\omega_\ell = 3\pi/8$. (See Figure 9–15 on page 269 in Chapter 9 for a description of the ideal bandpass filter.) Find the continuous time steady-state output signal $y_a(t)$ for each of the four cases listed below.

HINT: For each case, find the upper and lower cutoff frequencies in Hertz.
  a. $f_s = 40$ Hz
  b. $f_s = 50$ Hz
  c. $f_s = 60$ Hz
  d. $f_s = 100$ Hz

6–28. Consider the system shown in Figure 6–17(a). Suppose the discrete time filter in this system has the amplitude and phase response functions shown in Figure 6–1 and Equations (6–63) and (6–65) (with $\alpha = 0.8$). Suppose $x_a(t) = \cos(2\pi f_1 t) + \cos(2\pi f_2 t)$ is the input to the system, with $f_1 = 10$ Hz and $f_2 = 15$ Hz. Find the continuous time steady-state output signal for each of these cases:
  a. $f_s = 40$ Hz
  b. $f_s = 60$ Hz

6–29. Let

$$x(n) = r(n)\cos\left(\frac{\pi}{2}n\right)$$

where $r(n)$ is an N-point rectangular window (see Equation [6–112]). Sketch $|X(e^{j\omega})|$ on the interval $0 \leq \omega < 2\pi$ for each of the cases listed here. The sketch must be accurate with respect to the location and width of the main lobe.

   a. $N = 20$

   b. $N = 100$

6–30. Suppose the following signal is sampled at $f_s = 120$ Hz to create a sequence $x(n)$:

$$x_a(t) = 4\cos\left[2\pi(20)t\right] + 2\cos\left[2\pi(40)t\right]$$

Suppose $x(n)$ is multiplied by an $N = 24$ point rectangular window (Equation [6–112]) to create the sequence $g(n)$:

$$g(n) = r(n)x(n)$$

Sketch the amplitude spectrum $|G(e^{j\omega})|$ on the interval $0 \leq \omega < 2\pi$. The sketch must be accurate with respect the main lobe widths, heights (in a proportional sense), and locations.

6–31. Let $x_a(t) = \cos(2\pi f_1 t) + \cos(2\pi f_2 t)$, where $f_1 = 40$ Hz and $f_2 = 55$ Hz. This signal is to be sampled and truncated by an N-point rectangular window; the resulting sequence is $g(n) = r(n)x(n)$. If the sampling frequency is $f_s = 10$ kHz, what is the approximate minimum required value of N such that the amplitude spectrum $|G(e^{j\omega})|$ is characterized by two distinct peaks corresponding to the two frequency components of the signal?

6–32. Let $H(e^{j\omega})$ be the frequency response function of the N-point moving average filter described by the difference equation

$$y(n) = \frac{1}{N}\sum_{k=0}^{N-1} x(n-k)$$

   a. Find $H(e^{j\omega})$ and $|H(e^{j\omega})|$.

   HINT: The impulse response of this system looks like a rectangular window multiplied by $1/N$.

   b. Use sketches to show how $|H(e^{j\omega})|$ depends on the moving average length N.

   c. Intuitively, averaging is a smoothing out process, and thus is related to lowpass filtering. Does $|H(e^{j\omega})|$ have the qualities of a lowpass filter? If so, what is the approximate cutoff frequency of a moving average filter? How does it compare with an ideal lowpass filter?

6–33. Suppose $x(n)$ has the spectrum shown in Figure P6–33(a). Now consider the discrete time system shown in Figure P6–33(b).

   a. Sketch the spectrum at points 1 and 2 (with respect to Figure P6–33[b]) on the interval $-\pi < \omega < \pi$.

   b. Sketch the spectrum at points 1 and 2 (with respect to Figure P6–33[b]) on the interval $0 \leq \omega < 2\pi$.

6–34. Consider a discrete time *ideal Hilbert transformer*, which is defined by the following frequency response function:

$$H(e^{j\omega}) = \begin{cases} +j, -\pi < \omega < 0 \\ 0, \omega = 0 \\ -j, 0 < \omega \leq \pi \end{cases} \quad (\text{periodic } 2\pi)$$

   a. Sketch $|H(e^{j\omega})|$ on the interval $-\pi < \omega < \pi$.

   b. Sketch $\theta_H(\omega)$ on the interval $-\pi < \omega < \pi$.

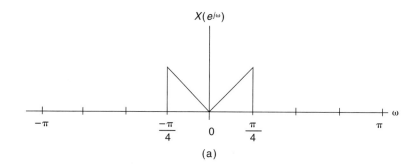

(a)

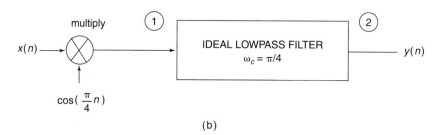

(b)

FIGURE P6–33

c. Let

$$x(n) = \cos\left(\frac{\pi}{10}n\right) + \frac{1}{2}\cos\left(\frac{\pi}{6}n - \frac{\pi}{4}\right) + \frac{1}{3}\cos\left(\frac{\pi}{2}n + \frac{\pi}{10}\right)$$

be the input signal to an ideal Hilbert transformer. Find the steady-state output signal.

d. Use the inverse DTFT to show that the impulse response of an ideal Hilbert transformer is

$$h(n) = \frac{1 - \cos(\pi n)}{\pi n}$$

defined for $-\infty < n < \infty$.

e. Is an ideal Hilbert transformer a causal system?

f. Use l'Hopital's rule to show that $h(0) = 0$.

g. Plot $h(n)$. Is the symmetry of $h(n)$ what you would expect to observe, given problem 6–12?

h. Show that the impulse response can also be expressed as

$$h(n) = \frac{1 - \sin\left(\frac{\pi}{2}n\right)}{\pi n}$$

6–35. The general form of an *all-pass filter* is

$$H(z) = G\prod_{k=1}^{M}\frac{\left(z^{-1} - d_k^*\right)}{\left(1 - d_k z^{-1}\right)}$$

where $|d_k| < 1$ for all k (for stability).
There are M poles d_k, $k = 1, 2, \ldots, M$
There are M zeros

$$\frac{1}{d_k^*}, \; k = 1, 2, \ldots, M$$

Show that $|H(e^{j\omega})| = |G|$.

HINT: Use $|H(e^{j\omega})|^2 = H(e^{j\omega})H^*(e^{j\omega})$

6–36. Consider the following FIR filters:

$$H_1(z) = G\left(1 - \frac{1}{4}z^{-1}\right)\left(1 - \frac{1}{2}z^{-1}\right)$$

$$H_2(z) = \frac{1}{4}G\left(1 - 4z^{-1}\right)\left(1 - \frac{1}{2}z^{-1}\right)$$

$$H_3(z) = \frac{1}{2}G\left(1 - \frac{1}{4}z^{-1}\right)\left(1 - 2z^{-1}\right)$$

$$H_4(z) = \left(\frac{1}{4}\right)\left(\frac{1}{2}\right)G\left(1 - 4z^{-1}\right)\left(1 - 2z^{-1}\right)$$

a. Find the impulse response of each of these four filters.

b. Show that all four filters have the *same* amplitude response function, $|H(e^{j\omega})|$, even though their impulse responses are different.

NOTE: Each filter has a difference phase response function. This problem illustrates the fact that the amplitude response function ($|H(e^{j\omega})|$) does not uniquely define the system.

HINT: Start by showing that $|1 - ae^{-j\omega}|^2 = |a - e^{-j\omega}|^2$

c. A *minimum phase* system is one having all of its zeros inside the unit circle. (This definition applies to both FIR and IIR filters.) Which of the four filters listed above has minimum phase? A BIBO stable minimum phase system has the property that its inverse system $1/H(z)$ is also BIBO stable. Why? If a system is not minimum phase, its inverse system is not BIBO stable. Why not?

d. Consider a simple linear phase FIR filter: $H(z) = 1 + 4z^{-1} + z^{-2}$. (The impulse response has the required symmetry.) Use this example to demonstrate that a "linear phase" system is *not* the same thing as a "minimum phase" system.

6–37. Consider a system with the linear phase response function $\theta_H(\omega) = -20\omega$.
Sketch the *principal value* phase response plot for this system on the interval $-\pi < \omega < \pi$.

6–38. Consider the following FIR filter:

$$h(n) = \frac{1}{4}\delta(n) - \frac{1}{2}\delta(n-1) + \delta(n-2) - \frac{1}{2}\delta(n-3) + \frac{1}{4}\delta(n-4)$$

Does this filter have generalized linear phase? Why or why not?

7

SAMPLING THEOREM AND REAL-WORLD D/A CONVERSION

7–1 RELATIONSHIP BETWEEN THE FOURIER TRANSFORM AND THE DTFT

Let $x(n)$ be a discrete time signal obtained by sampling a continuous time signal $x_a(t)$ at sampling frequency $f_s = 1/T$. That is, let $x(n) = x_a(nT)$. Let $X_a(f)$ be the Fourier transform of $x_a(t)$; let $X[\exp(j2\pi fT)]$ be the DTFT of $x(n)$ (with $\omega = 2\pi fT$). The purpose of this section is to derive the relationship originally presented in Section 6–6, shown here as Equation (7–1). In the process of deriving this result, other theoretical relationships will be obtained that will be used in Sections 7–2 and 7–3 to help explain the difference between ideal D/A conversion and real-world D/A conversion. In this discussion, the quantization errors produced by a real-world sampling process (i.e., by a practical A/D converter) will not be addressed.

$$X\left(e^{j2\pi fT}\right) = \frac{1}{T} \sum_{n=-\infty}^{\infty} X_a\left(f - nf_s\right) \tag{7–1}$$

To demonstrate the validity of Equation (7–1), we begin by considering the following impulse train in the continuous time domain:

$$s(t) = \sum_{n=-\infty}^{\infty} \delta(t - nT) \tag{7–2}$$

$s(t)$ is shown in Figure 7–1. (Note that the impulses are spaced by the sampling interval T.) Now suppose $s(t)$ and $x_a(t)$ are multiplied together to create a weighted impulse train that we will call $x_s(t)$:

$$\begin{aligned} x_s(t) &= x_a(t)s(t) = x_a(t) \sum_{n=-\infty}^{\infty} \delta(t - nT) \\ &= \sum_{n=-\infty}^{\infty} x_a(t)\delta(t - nT) \end{aligned} \tag{7–3}$$

227

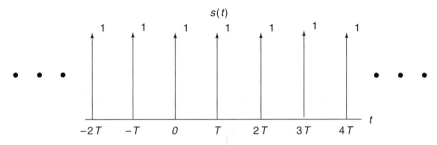

FIGURE 7–1
Impulse train $s(t)$ in the continuous time domain $(f_s = 1/T)$.

Since the impulse train is zero except at the specific times $t = nT$, this can be expressed as

$$x_s(t) = \sum_{n=-\infty}^{\infty} x_a(nT)\delta(t - nT) \tag{7–4}$$

Furthermore, since we have defined $x(n) = x_a(nT)$, we have

$$x_s(t) = \sum_{n=-\infty}^{\infty} x(n)\delta(t - nT) \tag{7–5}$$

$x_s(t)$ is not a discrete time signal; it is a weighted impulse train that (theoretically) exists in the continuous time domain. However, the weights (or areas) of the impulses are equal to the values of the sequence $x(n)$.

In order to demonstrate the validity of Equation (7–1), we will find the Fourier transform of $x_s(t)$, denoted $X_s(f)$, using two different approaches. The first approach will result in the right side of Equation (7–1); the second approach will result in the left side of Equation (7–1). Since both sides will be shown to be equal to $X_s(f)$, they are obviously equal to each other. Of course, the left side of Equation (7–1) is also the DTFT of sequence $x(n)$ (with $\omega = 2\pi fT$); therefore, from a mathematical perspective, the Fourier transform of $x_s(t)$ and the DTFT of $x(n)$ turn out to be the same thing, even though $x(n)$ is a signal in the discrete time domain and $x_s(t)$ is a signal in the continuous time domain.

To show that $X_s(f)$ is equal to the right side of Equation (7–1), we begin by noting that the impulse train $s(t)$ is a periodic function (period $= T$; fundamental frequency $= f_s$). Therefore, $s(t)$ can be represented as a Fourier series:

$$s(t) = \sum_{n=-\infty}^{\infty} c_n e^{j2\pi n f_s t} \tag{7–6}$$

where the Fourier series coefficients are given by

$$c_n = \frac{1}{T} \int_{-T/2}^{T/2} s(t) e^{-j2\pi n f_s t} dt \tag{7–7}$$

Within the interval $-T/2 < t < T/2$, only one impulse, $\delta(t)$, appears in $s(t)$. Thus Equation (7–7) reduces to

$$c_n = \frac{1}{T} \int_{-T/2}^{T/2} \delta(t) e^{-j2\pi n f_s t} dt \tag{7–8}$$

Using the sifting property of the impulse, Equation (7–8) clearly reduces to

$$c_n = \frac{1}{T} e^{-j2\pi n f_s(0)} = \frac{1}{T} \tag{7-9}$$

Therefore, $s(t)$ can be represented by the following Fourier series:

$$s(t) = \frac{1}{T} \sum_{n=-\infty}^{\infty} e^{j2\pi n f_s t} \tag{7-10}$$

Let $S_a(f)$ be the Fourier transform of $s(t)$. We have

$$S_a(f) = \mathbf{F}\{s(t)\}$$
$$= \mathbf{F}\left\{\frac{1}{T} \sum_{n=-\infty}^{\infty} e^{j2\pi n f_s t}\right\} \tag{7-11}$$
$$= \frac{1}{T} \sum_{n=-\infty}^{\infty} \mathbf{F}\left\{e^{j2\pi n f_s t}\right\}$$

Using the Fourier transform pair

$$e^{j2\pi f_o t} \leftrightarrow \delta(f - f_o) \tag{7-12}$$

we see that Equation (7–11) reduces to

$$S_a(f) = \frac{1}{T} \sum_{n=-\infty}^{\infty} \delta(f - n f_s) \tag{7-13}$$

Since $x_s(t) = x_a(t)s(t)$, we know that $X_s(f)$ can be found by convolving $X_a(f)$ and $S_a(f)$:

$$X_s(f) = S_a(f) * X_a(f) = \int_{-\infty}^{\infty} S_a(\lambda) X_a(f - \lambda) d\lambda$$
$$= \int_{-\infty}^{\infty} \frac{1}{T} \sum_{n=-\infty}^{\infty} \delta(\lambda - n f_s) X_a(f - \lambda) d\lambda \tag{7-14}$$
$$= \frac{1}{T} \sum_{n=-\infty}^{\infty} \int_{-\infty}^{\infty} \delta(\lambda - n f_s) X_a(f - \lambda) d\lambda$$

Using the sifting property of the impulse, Equation (7–14) reduces to

$$X_s(f) = \frac{1}{T} \sum_{n=-\infty}^{\infty} X_a(f - n f_s) \tag{7-15}$$

which is, in fact, the right side of Equation (7–1). Note that if $X_a(f)$ is the spectrum depicted in Figure 7–2, and if there is no aliasing, then $X_s(f)$ is as depicted in Figure 7–3. On the other hand, if there is aliasing, then $X_s(f)$ is as shown in Figure 7–4. Let's emphasize again that $X_s(f)$ is the spectrum of a signal $x_s(t)$ that exists in the continuous time domain. If for some reason we wish to consider filtering $x_s(t)$, we must do the job with a continuous time (i.e., analog) type filter.

Next we consider a different (and more direct) approach for finding $X_s(f)$. We have

$$x_s(t) = \sum_{n=-\infty}^{\infty} x(n)\delta(t - nT) \tag{7-16}$$

FIGURE 7–2
Spectrum of a hypothetical
bandlimited continuous time
signal $x_a(t)$.

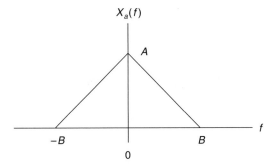

FIGURE 7–3
Spectrum of $x_s(t) = x_a(t)s(t)$ when $fs > 2B$. There is no aliasing. $X_s(f)$ is a scaled duplicate of $X_a(f)$ in the interval $-f_s/2 < f < f_s/2$.

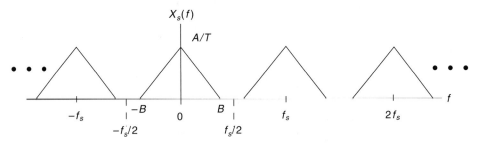

(a)

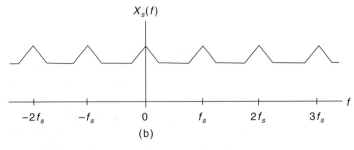

(b)

FIGURE 7–4
Spectrum of $x_s(t) = x_a(t)s(t)$ when $fs < 2B$. This spectrum suffers from aliasing; note that $X_s(f)$ is *not* a scaled duplicate of $X_a(f)$ in the interval $-f_s/2 < f < f_s/2$.

Therefore,

$$X_a(f) = \mathbf{F}\left\{ \sum_{n=-\infty}^{\infty} x(n)\delta(t - nT) \right\}$$

$$= \sum_{n=-\infty}^{\infty} x(n) \cdot \mathbf{F}\{\delta(t - nT)\}$$

(7–17)

Using the Fourier transform pair

$$\delta(t - \xi) \leftrightarrow e^{-j2\pi f\xi}$$

(7–18)

we see that Equation (7–17) reduces to

$$X_s(f) = \sum_{n=-\infty}^{\infty} x(n)e^{-j2\pi fTn}$$

(7–19)

which is the same thing as the DTFT of $x(n)$, $X[\exp(j\omega)]$, with $\omega = 2\pi fT$. That is,

$$X_s(f) = X\left(e^{j2\pi fT}\right)$$

(7–20)

Equations (7–15), (7–19), and (7–20) demonstrate the validity of Equation (7–1).

7–2 THE SAMPLING THEOREM AND IDEAL D/A CONVERSION

The *sampling theorem* can be stated as follows: A continuous time signal $x_a(t)$ that is band-limited to B Hz can (in theory) be uniquely recovered from its samples $x_a(nT)$, where $f_s = 1/T$, *provided* $f_s > 2B$. The results presented in Section 7–1 will provide the necessary ammunition for proving the sampling theorem, which was first hinted at in Section 6–6.

Figures 7–2 and 7–3 show that if the sampling frequency is greater than two times the bandwidth of $x_a(t)$ (and therefore there is no aliasing), then $X_s(f)$ is a scaled duplicate of $X_a(f)$ within the interval $-f_s/2 < f < f_s/2$. Therefore, $x_a(t)$ can (in theory) be recovered from $x_s(t)$ by processing $x_s(t)$ with an ideal lowpass filter having a cutoff frequency of $f_s/2$ Hz and a passband gain of $T = 1/f_s$. The ideal lowpass filter will chop off the unwanted spectral copies in $X_s(f)$ to recover the original spectrum $X_a(f)$. Since $x_s(t)$ is related to the sequence $x(n)$ as shown by Equation (7–16), then $x_a(t)$ can (in theory) be recovered from the sequence $x(n)$ if this sequence can somehow be converted to the impulse train $x_s(t)$. In other words, we can (in theory) consider an *ideal digital-to-analog (D/A) converter*, which would first create $x_s(t)$ from $x(n)$, and then process $x_s(t)$ with an ideal lowpass filter. The ideal D/A converter, which is shown in Figure 7–5, is one theoretical realization of the sampling theorem.

There are two obvious reasons that an *ideal* D/A converter cannot be realized in practice. First, an ideal lowpass filter cannot be realized. Second, although we could certainly create a weighted pulse train in which the pulses are very narrow (relative to the sampling interval T) and have heights proportional to $x(n)$, and thus create an *approximation* to $x_s(t)$, we cannot obtain the Dirac delta functions that are actually required. In addition, one should keep in mind that even if an ideal D/A converter could be built, there are

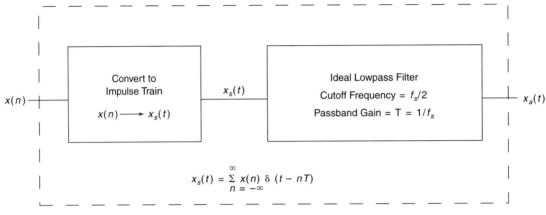

FIGURE 7–5
Ideal digital-to-analog (D/A) converter.

two additional factors that prevent the sampling theorem from being *exactly* realized: quantization errors, $x(n)$ is not precisely equal to $x_a(nT)$, and a small (though hopefully negligible) amount of aliasing that is present because no real-world signal is *perfectly* bandlimited. Thus there are at least four factors that prevent a continuous time signal from being *exactly* recovered from its samples.

There is another theoretical realization of the sampling theorem: the *ideal interpolation formula*, which can be developed as follows. As stated above, $x_a(t)$ can be recovered from $x_s(t)$ by processing $x_s(t)$ with an ideal lowpass filter with a cutoff frequency of $f_c = f_s/2$ Hz and a passband gain of T. In the time domain, this processing involves convolving $x_s(t)$ with the impulse response of the ideal lowpass filter:

$$x_a(t) = x_s(t) * h_a(t) \qquad (7\text{–}21)$$

where $h_a(t)$ is the impulse response of the required continuous time domain ideal lowpass filter:

$$h_a(t) = \frac{T\sin(2\pi f_c t)}{\pi t} = \frac{\sin(\pi f_s t)}{\left(\dfrac{\pi}{T}\right)t} = \frac{\sin\left[\left(\dfrac{\pi}{T}\right)t\right]}{\left(\dfrac{\pi}{T}\right)t} \qquad (7\text{–}22)$$

Therefore,

$$x_a(t) = \int_{-\infty}^{\infty} x_s(\tau) h_a(t-\tau)d\tau = \int_{-\infty}^{\infty} x_s(\tau)\frac{\sin\left(\dfrac{\pi}{T}(t-\tau)\right)}{\dfrac{\pi}{T}(t-\tau)}d\tau \qquad (7\text{–}23)$$

where (from Equation [7–3])

$$x_s(\tau) = \sum_{n=-\infty}^{\infty} x_a(nT)\delta(\tau - nT) \tag{7-24}$$

We now substitute Equation (7–24) into Equation (7–23):

$$x_a(t) = \int_{-\infty}^{\infty} \left[\sum_{n=-\infty}^{\infty} x_a(nT)\delta(\tau - nT) \right] \frac{\sin\left(\frac{\pi}{T}(t - \tau)\right)}{\frac{\pi}{T}(t - \tau)} d\tau \tag{7-25}$$

Equation (7–25) can be rearranged as follows:

$$x_a(t) = \sum_{n=-\infty}^{\infty} x_a(nT) \int_{-\infty}^{\infty} \delta(\tau - nT) \frac{\sin\left(\frac{\pi}{T}(t - \tau)\right)}{\frac{\pi}{T}(t - \tau)} d\tau \tag{7-26}$$

The integral can be evaluated using the sifting property of the impulse:

$$x_a(t) = \sum_{n=-\infty}^{\infty} x_a(nT) \frac{\sin\left(\frac{\pi}{T}(t - nT)\right)}{\frac{\pi}{T}(t - nT)} \tag{7-27}$$

Equation (7–27) is the ideal interpolation formula.

7–3   REAL-WORLD D/A CONVERSION

As noted in Section 7–2, an ideal D/A converter would create the weighted impulse train $x_s(t)$ from sequence $x(n)$ and then process $x_s(t)$ with an ideal lowpass filter. A practical, off-the-shelf D/A converter creates a weighted pulse train with a staircase appearance, as shown in Figure 7–6. This pulse train is here denoted $x_p(t)$. (The ideal result, $x_a(t)$, is also shown in Figure 7–6.) At each sampling interval (nT), a pulse with height $x(n)$ and width T is in effect created. The resulting pulse train can be represented mathematically as

$$x_p(t) = \sum_{n=-\infty}^{\infty} x(n)g_a(t - nT) \tag{7-28}$$

where $g_a(t)$ is the single pulse shown in Figure 7–7. Most D/A converters do not incorporate any additional lowpass filtering; a lowpass filter must be placed at the output of the D/A converter if it is desirable to smooth off the sudden transitions in $x_p(t)$. This postfilter will be considered later.

The qualitative difference between $x_a(t)$ and $x_p(t)$ in the time domain is clear from inspection of Figure 7–6(b). However, it is also interesting to compare these signals in the frequency domain, where the ideal D/A converter can be represented by

$$X_a(f) = H_a(f)X_s(f) \tag{7-29}$$

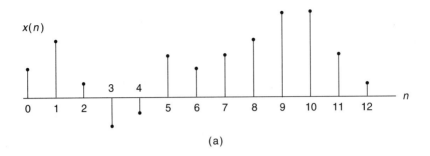

(a)

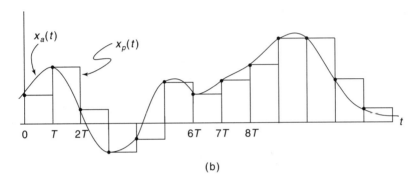

(b)

FIGURE 7–6

A real-world D/A converter converts $x(n)$ to a weighted pulse train in the continuous time domain, which approximates the actual continuous time signal $x_a(t)$. The approximation will be improved if the sampling frequency is at least several times greater than $2B$, since this would make the individual pulses narrower and the steps in the waveform smaller.

FIGURE 7–7

The pulse $g_a(t)$ is in effect the impulse response of a real-world D/A converter. Mathematically, the output of a real-world D/A converter can be obtained by convolving $g_a(t)$ with impulse train $s(t)$. (See text for explanation.)

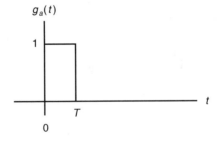

where $H_a(f)$ is an ideal lowpass filter with cutoff frequency $f_c = f_s/2$, as discussed in the previous section. On the other hand, it will now be shown that the practical D/A converter gives us

$$X_p(f) = G_a(f)X_s(f) \qquad (7\text{--}30)$$

where $X_p(f)$ is the Fourier transform of $x_p(t)$ and $G_a(f)$ is in effect the frequency response function of the practical D/A converter. By comparing $H_a(f)$ and $G_a(f)$, we can gain additional appreciation of the difference between ideal and practical D/A conversion.

To find $G_a(f)$, we begin by observing that mathematically, $x_p(t)$ can be obtained by *convolving $x_s(t)$* with the single pulse $g_a(t)$:

$$x_s(t) * g_a(t) = \int_{-\infty}^{\infty} x_s(\tau)g_a(t-\tau)d\tau$$

$$= \int_{-\infty}^{\infty} \left[\sum_{n=-\infty}^{\infty} x(n)\delta(\tau - nT) \right] g_a(t-\tau)d\tau$$

$$= \sum_{n=-\infty}^{\infty} x(n)\int_{-\infty}^{\infty} \delta(\tau - nT)g_a(t-\tau)d\tau \qquad \text{(7–31)}$$

$$= \sum_{n=-\infty}^{\infty} x(n)g_a(t-nT) = x_p(t)$$

Therefore, in the frequency domain we have the relationship shown in Equation (7–30), where $G_a(f)$ is the Fourier transform of $g_a(t)$. $G_a(f)$ can be determined by reviewing Example 3–7.1 (Chapter 3, page 50), where the Fourier transform of a rectangular pulse was first considered. The time shifting theorem (Equation [3–26]) also comes into play. The result is

$$G_a(f) = e^{-j2\pi f \frac{T}{2}} \frac{\sin(\pi fT)}{\pi f} \qquad \text{(7–32)}$$

The amplitude response function is therefore

$$|G_a(f)| = \frac{|\sin(\pi fT)|}{|\pi f|} \qquad \text{(7–33)}$$

The ideal amplitude response function $|H_a(f)|$ and the practical amplitude response function $|G_a(f)|$ are shown plotted together in Figure 7–8. Keep in mind that the hypothetical signal being processed by these filters is $x_s(t)$ and that the ideal purpose is to preserve without alteration that part of the spectrum in the interval $-f_s/2 < f < f_s/2$, where $X_s(f)$ is a

FIGURE 7–8
In the frequency domain, the difference between an ideal D/A converter and a real-world D/A converter is essentially the difference between amplitude response functions $|H_a(f)|$ (ideal) and $|G_a(f)|$ (real-world). $G_a(f)$ is the Fourier transform of the pulse $g_a(t)$ shown in Figure 7–7. See text for explanation.

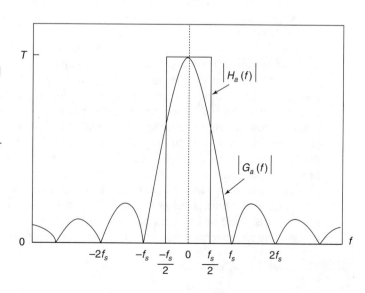

FIGURE 7–9
If the sampling frequency is at least several times greater than 2B, then within the frequency interval of interest, $|H_a(f)|$ and $|G_a(f)|$ are very similar. This means that in this case the real-world D/A converter may be reasonably close to an ideal D/A converter.

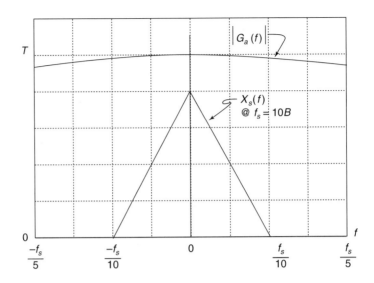

scaled duplicate of $X_a(f)$ (assuming there is no aliasing), and eliminate the part of the spectrum outside of this interval, where the extra copies of $X_a(f)$ are distributed. Figure 7–8 shows that $G_a(f)$ is a relatively crude lowpass filter having an approximate cutoff frequency of $f_c = f_s/2$. The undesirable copies of $X_a(f)$ will be attenuated by this filter, but not completely eliminated. Additional attenuation can be obtained by placing a lowpass filter with a cutoff frequency of $f_c = f_s/2$ at the output of the D/A converter. As noted above, in the time domain this postfilter will tend to smooth off the sharp staircase transitions of $x_p(t)$.

It is important to note that the copy of $X_a(f)$ within the interval $-f_s/2 < f < f_s/2$ will be distorted to some extent by $G_a(f)$, since $|G_a(f)|$ is not flat (i.e., does not have constant value) in this interval. A conventional postfilter (e.g., Butterworth or Chebyshev lowpass filter) will not compensate for this distortion.

The distortion created by $G_a(f)$ within the desired interval on the frequency axis will be less pronounced if the sampling frequency is several times greater than 2B (where B is the bandwidth of the signal being reconstructed). This is because if $f_s \gg 2B$, the desired copy of $X_a(f)$ will be contained in a smaller subinterval (relative to $f_s/2$) centered at $f = 0$, as shown in Figure 7–9 (where $f_s = 10B$ for the purposes of illustration). Within this smaller interval, $G_a(f)$ more closely approximates the desired flat response, as also shown in Figure 7–9. In the time domain, the larger sampling frequency will result in $x_p(t)$ having narrower, shorter steps, which means that a relatively low order (perhaps even first-order) lowpass postfilter might be sufficient to obtain the desired smoothing. However, this improvement comes with a price tag: If the processing is to be done in real time, increasing the sampling frequency will obviously increase the number of mathematical operations the processor must carry out per unit time. Obviously, the sampling frequency cannot be increased to a point at which the maximum processing rate of the processor is exceeded.

Decimation/interpolation oversampling schemes exist that allow the sampling frequency to be increased without proportionally increasing the processing rate. (The processing rate must increase because of the added complexity inherit in these schemes, but not by as great a factor as would otherwise be the case.) In addition to reducing D/A conversion distortion, oversampling can be used to improve the performance of the A/D conversion process with respect to quantization errors (i.e., quantization noise). This topic will be considered in Chapter 12.

PROBLEMS

7–1. Consider $|G_a(f)|$, as shown in Figure 7–8. As discussed in Section 7–3, the curvature of this amplitude response function within the interval $-f_s/2 < f < f_s/2$ will distort the spectral copy within this interval. An ideal $|G_a(f)|$ would have a constant value (T) within this interval, thus causing no distortion. One possible distortion measure is the difference in decibels between the ideal and the actual at one-half the sampling frequency, that is,

$$D_{dB} = 20\log(T) - 20\log\left|G_a\left(\frac{f_s}{2}\right)\right|$$

Calculate D_{dB}, assuming $|G_a(f)|$ is as given in Equation (7–33).

7–2. Suppose a D/A converter has been designed such that the pulse $g_a(t)$, instead of being as shown in Figure 7–7, is taller and narrower than the sampling interval. That is, suppose

$$g_a(t) = \begin{cases} \left(\dfrac{T}{\tau}\right), 0 \le t \le \tau \\ 0, otherwise \end{cases}$$

where $\tau < T$, and $T = 1/f_s$ is the sampling interval.

a. Show that in this case, the effective amplitude response function of the D/A converter is

$$|G_a(f)| = \left(\frac{T}{\tau}\right)\frac{|\sin(\pi f\tau)|}{|\pi f|}$$

b. Suppose $\tau = 0.25T$. Sketch $|G_a(f)|$ on the interval $-8f_s \le f \le 8f_s$. In order to compare this result with that discussed in Section 7–3, sketch $|G_a(f)|$ from Equation (7–33) on the same interval.

c. Suppose $\tau = 0.25T$. Calculate D_{dB} (as defined above in Problem 7–1).

d. Let $\tau = \alpha T$, where $\alpha < 1$. Choose α such that $D_{dB} = 0.1$ dB.

8

DISCRETE FOURIER TRANSFORM (DFT) AND FAST FOURIER TRANSFORM (FFT)

8–1 THE DFT AND ITS RELATIONSHIP TO THE DTFT

The discrete Fourier transform (DFT) is defined only for sequences of finite duration, for example,

$$x(n) = \begin{cases} \text{possible nonzero values,} & n = 0, 1, ..., N-1 \\ 0, & \text{otherwise} \end{cases} \tag{8–1}$$

The sequence $x(n)$ in Equation (8–1) is said to be a finite duration sequence of N points in length. Figure 8–1 illustrates the concept.

Of course, any sequence $x(n)$ has a DTFT, given by

$$X\left(e^{j\omega}\right) = \sum_{n=-\infty}^{\infty} x(n)e^{-j\omega n} \tag{8–2}$$

FIGURE 8–1
N-point finite duration sequence. The discrete Fourier transform (DFT) is only defined for this kind of sequence. The fast Fourier Transform (FFT) is an efficient algorithm for calculating the DFT.

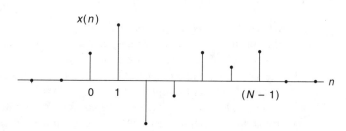

But since we are considering an N-point finite duration sequence in this case (as defined by Equation [8–1]), the summation limits in the DTFT need only cover that range of values of n where $x(n)$ can be nonzero:

$$X\left(e^{j\omega}\right) = \sum_{n=0}^{N-1} x(n)e^{-j\omega n} \tag{8–3}$$

Recall that $X[\exp(j\omega)]$ is a continuous function of ω.

The discrete Fourier transform (DFT) of an N-point finite duration sequence consists of *samples* of $X[\exp(j\omega)]$, evaluated at the following values of ω:

$$\omega = 0, \frac{2\pi(1)}{N}, \frac{2\pi(2)}{N}, \frac{2\pi(3)}{N}, \dots, \frac{2\pi(N-1)}{N} \tag{8–4}$$

Note that there are N samples in the frequency domain. In other words,

$$\mathbf{DFT}\{x(n)\} = X\left(e^{j\omega}\right)\Big|_{\omega=\frac{2\pi k}{N}, \; k=0,1,\dots,N-1} \tag{8–5}$$

That is,

$$\mathbf{DFT}\{x(n)\} = X\left(e^{j\left(\frac{2\pi k}{N}\right)}\right) = \sum_{n=0}^{N-1} x(n)e^{-j\left(\frac{2\pi k}{N}\right)n}, \quad k = 0,1,\dots, N-1 \tag{8–6}$$

Traditionally, the samples are referred to as $X(k)$ instead of $X[\exp(j2\pi k/N)]$, which trades mathematical rigor for notational simplicity. Thus the finite duration sequence $x(n)$ has a DFT $X(k)$:

$$\begin{array}{ccc} x(n) & \leftrightarrow & X(k) \\ {\scriptstyle n \,-\, 0,1,\,\dots,\,N\,-\,1} & & {\scriptstyle k \,-\, 0,1,\,\dots,\,N\,-\,1} \end{array} \tag{8–7}$$

where

$$X(k) = \sum_{n=0}^{N-1} x(n)e^{-j\frac{2\pi}{N}kn}, \quad k = 0,1,\dots, N-1 \tag{8–8}$$

The following points should be emphasized:

1. $X(k)$ consists of N samples of $X[\exp(j\omega)]$. The numbers $X(k)$ are in general complex.
2. The spacing between the samples is $2\pi/N$. Thus the longer $x(n)$ is, the more closely spaced the samples are in the frequency domain.
3. The samples are in the interval $0 \le \omega < 2\pi$, not $-\pi \le \omega < \pi$. The first sample ($k = 0$) is at $\omega = 0$. The last sample ($k = N-1$) is at $\omega = 2\pi(N-1)/N$; there is no sample taken at $\omega = 2\pi$. Figure 8–2 illustrates the concept for $N = 8$.
4. If you *plot* the points $(k, |X(k)|)$ using a computer plotting package, you have in effect plotted $|X[\exp(j\omega)]|$ on the interval $0 \le \omega < 2\pi$. However, to obtain a decent plot, N needs to be large so that the plot will be smooth.

5. The relationship between the frequency variable ω and the DFT index (k) is

$$\omega = \frac{2\pi k}{N} \qquad (8\text{–}9)$$

6. The relationship between ω and the frequency in hertz (f) is (from Section 6–1):

$$\omega = \frac{2\pi f}{f_s} \qquad (8\text{–}10)$$

7. By combining Equations (8–9) and (8–10), the following relationship between frequency in Hertz (f) and the DFT index (k) can be obtained:

$$f = \left(\frac{f_s}{N}\right)k \qquad (8\text{–}11)$$

Therefore, in terms of frequency in Hertz, the spacing between frequency domain samples is $\Delta f = f_s/N$; this is sometimes referred to as the *frequency resolution*, or *bin width*, of the DFT. The relationship between ω, k, and f is shown in Figure 8–2.

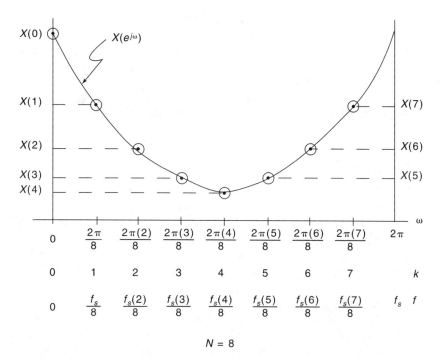

$N = 8$

FIGURE 8–2

The DFT $X(k)$ consists of N samples of $X(\exp(j\omega))$. (The case $N = 8$ is shown here.) Note the relationship between the DFT index (k) and the frequency variables (ω and f). The fast Fourier transform (FFT) is an efficient algorithm for calculating the DFT.

If the DFT $X(k)$, $k = 0,1, \ldots, N - 1$ is given, the corresponding finite duration sequence $x(n)$, n = 0, 1, ..., $N - 1$ can be found using the inverse DFT (IDFT) formula:

$$x(n) = \frac{1}{N} \sum_{k=0}^{N-1} X(k) e^{j\frac{2\pi}{N}kn}, \quad n = 0,1,...,N-1 \tag{8-12}$$

8-2 THE FAST FOURIER TRANSFORM (FFT)

Evaluating $X(k)$ directly from Equation (8–8) requires on the order of N^2 complex multiplications. In 1965, Cooley and Tukey discovered that $X(k)$ can actually be evaluated with approximately $N\log_2(N)$ complex multiplications using an algorithm that came to be known as the fast Fourier transform (FFT). Thus the FFT is an efficient method of calculating the DFT. In terms of results, the FFT of a finite duration sequence is the *same thing* as the DFT of that sequence.

To see why the FFT is such a big deal, consider the number of complex multiplications required if $N = 1024$:

$$(1024) \log_2 (1024) = 10,240 \tag{8-13}$$

$$(1024)^2 = 1,048,576 \tag{8-14}$$

The details of the FFT algorithm are beyond the scope of this book. For our purposes, it is enough to think of the FFT as an efficient black box that calculates the DFT. FFT subroutines written in FORTRAN and C are available.

The inverse DFT can be efficiently calculated using the inverse FFT algorithm, which is similar in structure to the forward FFT algorithm.

8-3 ZERO PADDING

Most FFT algorithms require the length (N) of the sequence to be a power of 2, that is, 2, 4, 8, 16, 32, 64, ... As we shall soon see, this constraint imposes no difficulty even though the sequences we want to take the DFT of generally don't have a length that is a power of 2. We need only *append zeros* to the sequence in question such that the overall length, including the zeros, is a power of 2. In fact, we can use the same trick, known as *zero padding*, to obtain a large number of closely spaced samples in the frequency domain (i.e., improve the frequency resolution) even if $x(n)$ has a small length: append zeros to $x(n)$ and take the FFT of the zero-padded sequence.

To understand zero padding, it is important to observe that appending zeros to a sequence does *not* change the DTFT $X[\exp(j\omega)]$. Consider a finite duration sequence $x(n)$ with L points, where $L < N$. The DTFT of this sequence can be expressed as

$$X\left(e^{j\omega}\right) = \sum_{n=0}^{L-1} x(n) e^{-j\omega n} \tag{8-15}$$

But since $x(n) = 0$ for $n > L - 1$, nothing is changed if we extend the upper summation limit to $N - 1$:

$$X\left(e^{j\omega}\right) = \sum_{n=0}^{N-1} x(n)e^{-j\omega n} \qquad (8\text{--}16)$$

This is the same thing as taking the DTFT of the zero-padded sequence, denoted as $x_{zp}(n)$:

$$x_{zp}(n) = \begin{cases} x(n), n = 0, 1, ..., L - 1 \\ 0, n = L, L + 1, ..., N - 1 \end{cases} \qquad (8\text{--}17)$$

Thus the N point DFT of the zero-padded sequence, given by

$$X(k) = \sum_{n=0}^{N-1} x_{zp}(n)e^{-j\frac{2\pi}{N}kn}, k = 0, 1, ..., N - 1 \qquad (8\text{--}18)$$

consists of N samples of the DTFT $X[\exp(j\omega)]$, spaced at an interval of $2\pi/N$ in the interval $0 \le \omega < 2\pi$.

Let us consider an example. Suppose we have an FIR filter having a 15-point impulse response:

$$h(n) = \begin{cases} \dfrac{\sin\left(\frac{2\pi}{3}(n - 7)\right)}{\pi(n - 7)}, & n = 0, 1, ..., 14 \\ 0, & \text{otherwise} \end{cases} \qquad (8\text{--}19)$$

The frequency response function of this filter, $H[\exp(j\omega)]$, is of course the DTFT of $h(n)$. Suppose we want to plot the amplitude response function $|H[\exp(j\omega)]|$. The FFT can be used to find samples, $H(k)$ of the frequency response function:

$$H(k) = H\left(e^{j\omega}\right)\Big|_{\omega = \frac{2\pi k}{N}}, k = 0, 1, ..., N - 1 \qquad (8\text{--}20)$$

where N is the number of samples in the frequency domain in the interval $0 \le \omega < 2\pi$. To obtain these N samples, $h(n)$ must be zero-padded to a total length of N points before using the FFT to find $H(k)$. As noted above, N must be a power of 2.

Let's start by choosing $N = 16$. We zero-pad $h(n)$ to a length of 16 points and then use the FFT to find $H(k)$, $k = 0, 1, . . ., 15$, as illustrated in Figure 8–3. To plot $|H[\exp(j\omega)]|$, we plot the points $(k, |H(k)|)$ using a convenient plotting package. Figure 8–4(a) shows the resulting plot using only points (no connecting lines). Since there are only 16 points, when the plotting package connects these points with straight lines in an effort to show the continuous function $|H[\exp(j\omega)]|$, as shown in Figure 8–4(b), the result is very misleading; it is definitely *not* a good picture of $|H[\exp(j\omega)]|$. To obtain a good picture, more zero padding is required.

To help illustrate the effects of zero padding, let's choose $N = 32$ as the next step, even though this is still not enough zero padding to get a good picture. Figure 8–5 illustrates the

FIGURE 8–3
Zero padding a 15-point
sequence $h(n)$ to a length of
16 points before taking the
FFT. The FFT yields 16
samples of $H[\exp(j\omega)]$.

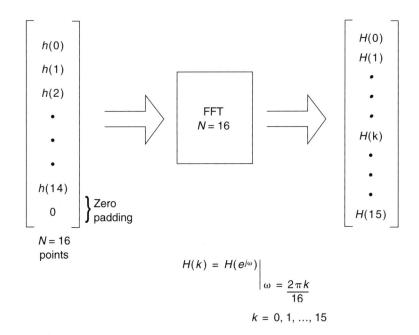

$$H(k) = H(e^{j\omega})\Big|_{\omega = \frac{2\pi k}{16}}$$

$$k = 0, 1, ..., 15$$

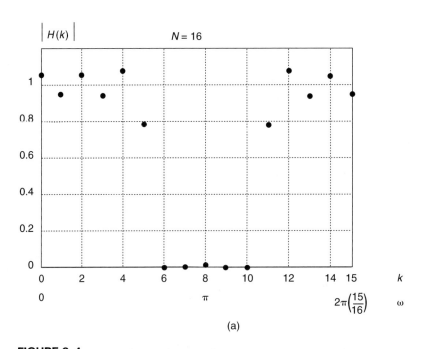

(a)

FIGURE 8–4
Amplitude response plot based on the 16-point FFT. Compare with Figure 8–8, in which a
512-point FFT is used instead.

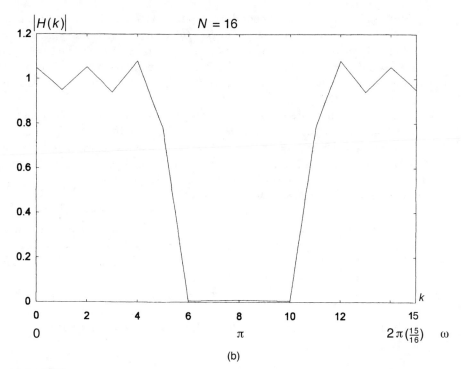

$N = 16$

(b)

FIGURE 8–4
(continued)

FIGURE 8–5
Zero padding a 15-point
sequence $h(n)$ to a length of
32 points before taking the
FFT. The FFT yields 32
samples of $H[\exp(j\omega)]$.

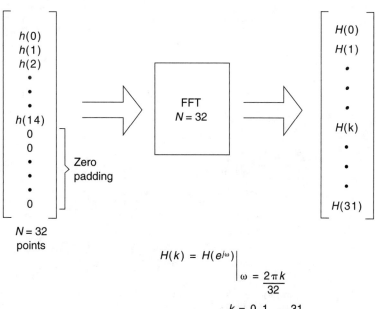

process in which frequency domain samples $H(k)$, $k = 0, 1, \ldots, 31$ are obtained with the FFT. Once again, we plot the points $(k, |H(k)|)$ using the plotting package of our choice. Figure 8–6(a) shows the resulting plot using only points; this should be compared to Figure 8–4(a). Figure 8–6(b) shows the plot that results when the points are connected by straight lines; this should be compared to Figure 8–4(b). It is clear that we still don't have a good picture of $|H[\exp(j\omega)]|$.

To obtain a good picture of $|H[\exp(j\omega)]|$, we can zero-pad $h(n)$ out to a total length of $N = 512$ points and then take the FFT, as illustrated by Figure 8–7. Once again, we plot the points $(k, |H(k)|)$. Figure 8–8 shows the plot that results when the points are connected by straight lines; this time the points are so close together that the result is a smooth looking curve, which is the good picture of $|H[\exp(j\omega)]|$ that we are after. (Compare Figure 8–8 with Figures 8–4(b) and 8–6[b].)

Observe that Figure 8–8 is symmetric about the point $k = N/2$. This is to be expected since $k = N/2$ corresponds to $\omega = \pi$ and $f = f_s/2$; the amplitude response function is symmetric about this point. Therefore, it is sufficient to plot the amplitude response function from $k = 0$ to $k = N/2$, as shown in Figure 8–9. (The frequency axis can be scaled for either ω or f, using Equations [8–9] and [8–11].)

Throughout this book there are many amplitude spectrum plots, $|X[\exp(j\omega)]|$ or $|H[\exp(j\omega)]|$, for finite duration sequences, $x(n)$ or $h(n)$. All of these plots were originally obtained using zero padding and the FFT, as described in this section.

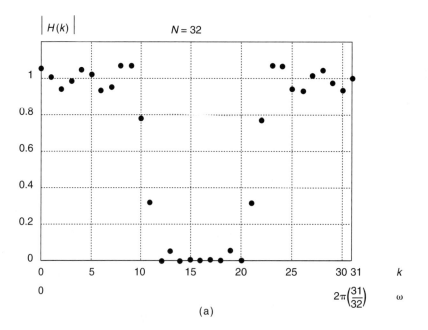

(a)

FIGURE 8–6

Amplitude response plot based on the 32-point FFT. Compare with Figure 8–8, in which a 512-point FFT is used instead.

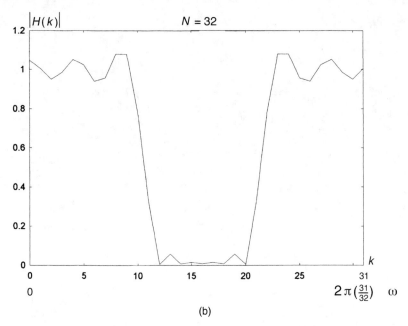

(b)

FIGURE 8–6
(continued)

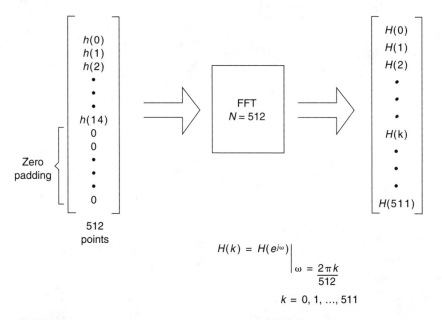

$$H(k) = H(e^{j\omega})\Big|_{\omega = \frac{2\pi k}{512}}$$

$$k = 0, 1, ..., 511$$

FIGURE 8–7
Zero padding a 15-point sequence $h(n)$ to a length of 512 points before taking the FFT. The FFT yields 512 samples of $H[\exp(j\omega)]$.

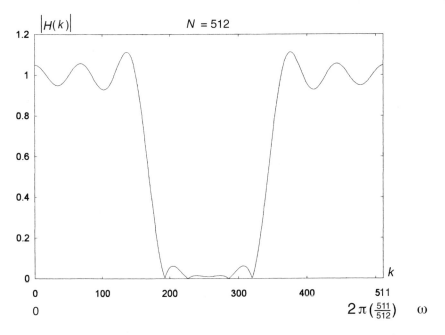

FIGURE 8–8
Amplitude response plot based on the 512-point FFT. There are enough sample points such that $|H(k)|$, when plotted, gives a good picture of $|H[\exp(j\omega)]|$.

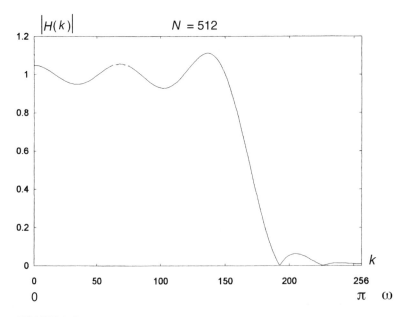

FIGURE 8–9
Since $|H[\exp(j\omega)]|$ is symmetric about $\omega = \pi$, it is sufficient to plot $|H(k)|$, $k = 0, 1, \ldots, N/2$.

8-4 THE DFT AND CONVOLUTION

Let $f(n)$ and $g(n)$ be N-point finite duration sequences ($n = 0, 1, \ldots, N - 1$). Let $F(k)$ and $G(k)$ denote the respective N-point DFTs

$$f(n) \xleftrightarrow[\text{N-point DFT}]{} F(k)$$

$$g(n) \xleftrightarrow[\text{N-point DFT}]{} G(k)$$

(8–21)

Let $Y(k)$ be formed by multiplying $F(k)$ and $G(k)$ together point by point:

$$Y(k) = F(k)G(k), k = 0, 1, ..., N - 1$$

(8–22)

Let $y(n)$ be the N-point sequence obtained by taking the inverse DFT of $Y(k)$:

$$y(n) \xleftrightarrow[\text{N-point DFT}]{} Y(k)$$

(8–23)

How is $y(n)$ related to $f(n)$ and $g(n)$? You might be tempted to say "convolution," since the inverse DTFT of $F[\exp(j\omega)]G[\exp(j\omega)]$ is in fact $f(n) * g(n)$. Note, however, that the sequence obtained by convolving $f(n)$ and $g(n)$ has a length of $2N - 1$ points. On the other hand, the sequence $y(n)$ obtained using the DFT operations described above has only N points; from this consideration alone, we know that $y(n)$ is *not* equal to $f(n) * g(n)$. At the same time, the way in which $f(n)$ and $g(n)$ are combined to form $y(n)$ is quite closely related to convolution. It can be shown that

$$y(n) = f(n) \otimes g(n) \xleftrightarrow[\text{N-point DFT}]{} Y(k) = F(k)G(k)$$

(8–24)

where $\otimes$ is used here to denote *circular convolution*:

$$y(n) = f(n) \otimes g(n) = \sum_{k=0}^{N-1} f(k)g\big[(n - k)\text{modulo } N\big], n = 0, 1, ..., N - 1$$

(8–25)

The modulo N notation requires some explanation. Let P and N be integers. To evaluate the expression "P modulo N," you must add to P or subtract from P enough integer multiples of N to bring the result within the range $[0, N - 1]$. Examples: 5 modulo 8 = 5; 10 modulo 8 = 2; –2 modulo 8 = 6; 25 modulo 8 = 1; –25 modulo 8 = 7.

The regular convolution of $f(n)$ and $g(n)$, as described in Chapter 5, is shown here again for the convenience of the reader:

$$f(n) * g(n) = \sum_{k=-\infty}^{\infty} f(k)g(n - k)$$

(8–26)

In general, regular convolution is *not* the same as circular convolution.

Fortunately, it turns out that the regular convolution of two finite duration sequences can be calculated by multiplying DFTs and taking the inverse *if the sequences are first zero-padded to sufficient length.* The circular convolution of two zero-padded sequences gives the same result as regular convolution (provided the zero padding is sufficient). Let's look at this relationship in more detail.

Let $x(n)$, $n = 0, 1, \ldots, L - 1$ and $q(n)$, $n = 0, 1, \ldots, P - 1$ be finite duration sequences of lengths L and P points, respectively. Let $y(n) = x(n) * q(n)$ (regular convolution). The sequence $y(n)$ has $L + P - 1$ points. The sequence $y(n)$ can be obtained as follows:

Step 1. Zero-pad sequences $x(n)$ and $q(n)$ such that each zero-padded sequence has an overall length of N points, where $N \geq L + P - 1$. In other words, both zero-padded sequences must have the same length (N), and N must be *at least* $L + P - 1$. (If the FFT is used, N must be a power of 2. For example, if $L + P - 1 = 53$, one would select $N = 64$.)

Step 2. Calculate the N-point DFTs:

$$\underset{\text{zero-padded}}{x(n)} \underset{N\text{-point DFT}}{\longleftrightarrow} X(k)$$

$$\underset{\text{zero-padded}}{q(n)} \underset{N\text{-point DFT}}{\longleftrightarrow} Q(k)$$

(8–27)

Step 3. Multiply the DFTs together point by point:

$$Y(k) = X(k)Q(k), k = 0, 1, \ldots, N - 1 \qquad (8\text{–}28)$$

Step 4. Take the N-point inverse DFT of $Y(k)$ to obtain $y(n)$:

$$y(n) \underset{N\text{-point DFT}}{\longleftrightarrow} Y(k) \qquad (8\text{–}29)$$

$y(n)$ will be equal to $x(n) * q(n)$, $n = 0, 1, \ldots, L + P\text{-}2$. If $N > L + P - 1$, then $y(n)$ will have zeros appended to it for $n = L + P - 1, \ldots, N - 1$.

One application of FFT-based convolution is the implementation of FIR filters. Instead of directly using the system difference equation as shown in Chapter 5 (see Figure 5–10 on page 121), two alternative methods are sometimes used: the *overlap-add* method and the *overlap-save* method. Both of these methods use the FFT to do convolution. It turns out that under certain conditions these methods offer computational savings relative to the direct form implementation (Figure 5–10). A detailed description of the overlap-add and overlap-save methods is beyond the scope of this book; the reader should consult the DSP literature for more information on this subject.

PROBLEMS

8–1. A continuous time signal $x_a(t)$ is sampled at $f_s = 1000$ Hz. A total of $N = 512$ points are taken to create a finite duration sequence $x(n)$, $n = 0, 1, \ldots, 511$.

    a. If a 512-point FFT is taken, what is the spacing (in Hertz) between the samples in the frequency domain?

    b. If $x(n)$ is zero-padded and a 2048-point FFT is taken, what is the spacing (in Hertz) between samples in the frequency domain?

    c. Once a sequence has been windowed (i.e., truncated), will any amount of zero padding undo the effect that the windowing has on the true spectrum?

    d. If the sampling frequency was too small, such that $X(\exp(j\omega))$ is corrupted by aliasing, will any amount of zero padding fix the problem (that is, undo the aliasing)?

8–2. The DTFT of an N-point rectangular window is described in Section 6–11. Suppose the DFT is taken of the rectangular window sequence itself, *without* any zero padding. Show that the resulting DFT will have the following values:

$$R(0) = N$$
$$R(1) = 0$$
$$R(2) = 0$$
$$\vdots$$
$$R(N - 1) = 0$$

HINT: Evaluate $R(e^{j\omega})$ (Equation [6–115]) at

$$\omega = \frac{2\pi}{N} k, k = 0, 1, \ldots, N - 1$$

NOTE: This is a good example of a case in which the DFT fails to give a good picture of the actual DTFT because of insufficient zero padding.

8–3. Calculate the number of complex multiplies required for a 4096-point DFT if the DFT is calculated directly using Equation (8-8). Then calculate the number of complex multiplications required if the FFT is used instead.

HINT: $4096 = 2^{12}$.

8–4. Given the following two $N = 2$-point sequences:

$$f(n) = \delta(n) + \delta(n - 1)$$
$$g(n) = \delta(n) + \delta(n - 1)$$

a. Compute the circular convolution $y(n) = f(n) \otimes g(n)$ using Equation (8–25).
b. Compute $y(n) = f(n) \otimes g(n)$ by using the DFT:
   i. Calculate $F(k)$ and $G(k)$, $k = 0,1$
   ii. Let $Y(k) = F(k)G(k)$, $k = 0,1$
   iii. Calculate $y(n)$, $n = 0$, 1 using the inverse DFT (Equation [8–12])
c. Calculate $y(n) = f(n) * g(n)$ (i.e., regular convolution) and compare this result with circular convolution. (The results should be different.)
d. Zero-pad $f(n)$ and $g(n)$ to a total length of $N = 3$ points. Then compute the circular convolution $y(n) = f(n) \otimes g(n)$ of the zero-padded sequences. (This time you should get the same result as with regular convolution).

8–5. Show that a sequence calculated by Equation (8–12) (the inverse DFT) is periodically extended if n is allowed to extend beyond the range $0 \le n \le N - 1$. In other words, $x(n) = x(n + N)$ for any value of n.

HINT:

$$x(n + N) = \frac{1}{N} \sum_{k=0}^{N-1} X(k) e^{j\frac{2\pi}{N} k(n+N)}$$

Use the fact that if k is any integer, $e^{j2\pi k} = 1$.

NOTE: The inverse DFT can be thought of as a discrete Fourier series (DFS) for a periodic sequence $x(n) = x(n + N)$. Compare Equations (8–12) and (8–8) with Equations (6–29) and (6–30).

Also show that if the DFT is calculated for values of k outside of the range $[0, N - 1]$, the sequence of numbers $X(k)$ is periodically extended. That is, $X(k) = X(k + N)$.

8–6. The purpose of this problem is to explore the relationship between the DFT and the Fourier series of a continuous time periodic signal.

Let $x_a(t)$ be periodic, with fundamental frequency f_o. Then $x_a(t)$ can be represented as a Fourier series:

$$x_a(t) = \sum_{k=-\infty}^{\infty} c_k e^{j2\pi k f_o t}$$

where c_k are the Fourier series coefficients. Now suppose $x_a(t)$ is sampled at f_s Hertz, where f_s is an *integer multiple* of the fundamental frequency f_o, that is,

$$f_s = N f_o$$

a. Show that the resulting sequence is

$$x_a(nT) = x(n) = \sum_{k=-\infty}^{\infty} c_k e^{j\frac{2\pi}{N}kn}$$

b. Show that $x(n)$ is a periodic sequence, with period $= N$.

c. Convince yourself that using the associative and commutative laws of addition allows us to write the equation in Problem 8–6(a)

$$x(n) = \sum_{k=0}^{N-1} \left[\sum_{l=-\infty}^{\infty} c_{(k-lN)} e^{j\frac{2\pi}{N}(k-lN)n} \right]$$

d. Show that since $\exp(-j2\pi m) = 1$ for any integer m, the equation in Problem 8–6(c) can be written as

$$x(n) = \sum_{k=0}^{N-1} \left[\sum_{l=-\infty}^{\infty} c_{(k-lN)} e^{j\frac{2\pi}{N}kn} \right]$$

e. The multiplier $\exp(j2\pi kn/N)$ is common to all terms in the inner sum of the equation in Problem 8–6(d). Also, we can multiply by N/N without changing anything. Convince yourself that the same equation can therefore be written as

$$x(n) = \frac{1}{N} \sum_{k=0}^{N-1} \left[N \sum_{l=-\infty}^{\infty} c_{(k-lN)} \right] e^{j\frac{2\pi}{N}kn}$$

f. If the equation in Problem 8–6(c) is evaluated for $n = 0, 1, \ldots, N - 1$, we obtain exactly one period of the sequence $x(n)$. This one period can be thought of as an N-point finite duration sequence $x(n)$, $n = 0, 1, \ldots, N - 1$; this sequence has a DFT, $X(k)$, given by Equation (8–8). With this in mind, compare the equation in Problem 8–6(e) with the inverse DFT formula (Equation [8–12]), and convince yourself that the relationship between $X(k)$ and the original Fourier series coefficients for the continuous time signal $x_a(t)$ is

$$X(k) = N \sum_{l=-\infty}^{\infty} c_{(k-lN)}$$

which means that $X(k)$ is in general an *aliased* version of the original Fourier series coefficients. (On the k axis, a copy of the sequence c_k is centered at all integer multiples of N. Aliasing occurs where these shifted copies overlap.)

g. Suppose $x_a(t)$ is bandlimited to Mf_o. Hertz. (This implies that the coefficients c_k are equal to zero outside of the range $[-M, M]$). Suppose the sampling frequency is chosen to avoid aliasing, that is, $f_s = Nf_o > 2Mf_o$. Show that in this case, the relationship between $X(k)$ and the original Fourier series coefficients is:

$$X(k) = \begin{cases} N c_k, & k = 0, 1, \ldots, M \\ N c_{(k-N)}, & k = N - M, N - M + 1, \ldots, N - 1 \end{cases}$$

9

DESIGN OF FIR FILTERS

9–1 INTRODUCTION

In this chapter, two methods for designing FIR lowpass, highpass, bandpass, and bandstop filters are presented: the Kaiser window method and the Parks-McClellan algorithm. Both methods are often provided with commercially available digital filter software design packages. Both methods create *linear phase* filters, which means that the resulting frequency response function has the form:

$$H\!\left(e^{j\omega}\right) = A(\omega)e^{-j\alpha\omega} \tag{9-1}$$

where $A(\omega)$ is a real function of ω, and $\alpha = (N-1)/2$ (N is the length of the impulse response). The significance and desirability of linear phase was discussed in Chapter 6. Furthermore, within the filter passband(s), $A(\omega)$ is both real and *non*negative. Therefore, within the passband(s), the following equations apply:

$$\left| H\!\left(e^{j\omega}\right) \right| = A(\omega) \tag{9-2}$$

and

$$\theta_H(\omega) = -\alpha\omega \tag{9-3}$$

The significance of Equation (9–3) is that within the passband(s), these filters have linear phase in the most exacting sense. In the stopband(s), $A(\omega)$ can be either positive or negative, which means that the phase response function in the stopbands has the jump discontinuities shown in Figure 6–34 (page 215). However, since the purpose of these filters is to greatly attenuate signals in the stopband(s) anyway, these jump discontinuities have little practical significance.

9–2 FIR FILTER DESIGN USING THE WINDOW METHOD

The design of a simple FIR lowpass filter was considered in Section 6–14; the results presented there will be referred to here as a way of introducing the window design method. The impulse response $h(n)$ of the causal, N-point FIR lowpass filter presented in Chapter 6 was obtained by shifting and truncating the impulse response of an ideal lowpass filter. Mathematically, the truncation operation can be represented as the result of multiplying by a rectangular window sequence:

$$h(n) = w(n)\left\{\frac{\sin\left[\omega_c\left(n - \frac{N-1}{2}\right)\right]}{\pi\left(n - \frac{N-1}{2}\right)}\right\} \tag{9–4}$$

where $w(n)$ is an N-point rectangular window:

$$w(n) = \begin{cases} 1, n = 0, 1, ..., N - 1 \\ 0, \text{otherwise} \end{cases} \tag{9–5}$$

The cutoff frequency is denoted ω_c. The filter length (N) can be even or odd. Equation (9–4) is the impulse response of an FIR filter designed using the window method, in which the window type is a rectangular window. In Chapter 6, the rectangular window was denoted $r(n)$. In this chapter, other window types will also be considered (in order to improve the characteristics of the resulting FIR filter), and all window types will be denoted $w(n)$.

Note that when length N is odd, Equation (9–4) will yield the indeterminate form 0/0 when $n = (N - 1)/2$. One must use L'Hopital's rule to establish the fact that when N is odd and $n = (N - 1)/2$, the result is $h(n) = w(n)(\omega_c/\pi)$. Note also that the impulse response of this FIR filter is symmetric about its midpoint $(N - 1)/2$; therefore, this filter has linear phase.

By appealing to the DTFT property that states that multiplication in the time domain is convolution in the frequency domain, we see that the frequency response function of this FIR filter can be expressed as

$$H\left(e^{j\omega}\right) = \frac{1}{2\pi} W\left(e^{j\omega}\right) * \left[e^{-j\left(\frac{N-1}{2}\right)\omega} H_i\left(e^{j\omega}\right)\right] \tag{9–6}$$

where $W[\exp(j\omega)]$ is the spectrum of the rectangular window (see Chapter 6, Section 6–11), and $H_i[\exp(j\omega)]$ is the frequency response function of an ideal lowpass filter (see Section 6–13). The symbol $*$ denotes convolution.

Recall that when two spectra are convolved, as in Equation (9–6), the resulting spectrum inherits characteristics from both of its parents. Thus the main lobe width and sidelobe amplitude of the rectangular window spectrum (see Figure 6–23, page 197) have a direct impact on the shape of the FIR lowpass filter frequency response function. Basically, $|H[\exp(j\omega)]|$ (see Figure 6–30, page 208) does indeed look like a lowpass filter, but the passband and stopband both suffer from excessive ripples. The peak ripple value (i.e., peak approximation error) in the stopband is $\delta = 0.09$. Since $20\log(0.09) \cong -21$, this filter is said to have 21 dB of attenuation in the stopband. *These relatively large ripples are a manifestation of the relatively large sidelobes in the rectangular window spectrum.*

The size of the sidelobes in the window spectrum can be reduced by modifying the shape of the window itself. Specifically, if we round off, or taper, the edges of the window, the sidelobes of the window spectrum will be reduced (as will be shown shortly with an example). In general, we will refer to a window with smoothed-off edges as a tapered window. Figure 9–1 illustrates the concept. Note that the tapering is symmetric; that is, a tapered window is symmetric about its midpoint $(N - 1)/2$. This means that if a tapered window

FIGURE 9–1

Rectangular window versus tapered window.

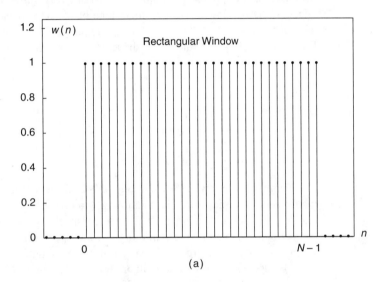

(a)

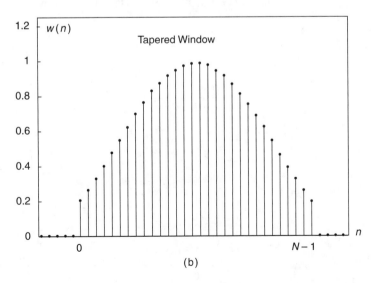

(b)

is used in Equation (9–4), the resulting FIR filter will still have an impulse response that is symmetric about its midpoint $(N-1)/2$, and thus will still have linear phase.

As a consequence of the fact that the sidelobes of the spectrum of a tapered window are reduced in amplitude, if a tapered window $w(n)$ is used in Equation (9–4), the ripples in the frequency response function of the FIR lowpass filter will be reduced in amplitude. This means that the passband will be more flat, and there will be more attenuation in the stopband. However, there is a price to be paid for this improvement: For a fixed length (N), the use of a tapered window will result in a wider transition bandwidth $\Delta\omega$ (which was first defined by Figure 6–30). In other words, *for a fixed length N, we can reduce the ripples at the cost of making the filter cutoff characteristic less sharp.* The reason is that tapering the edges of the window causes the main lobe of the window spectrum to increase in width. The wider this main lobe is, the more rounding off the convolution operation (shown in Equation [9–6]) will create in the shape of the frequency response function near the cutoff frequency.

To illustrate these effects, suppose we use a *Hamming window* (instead of a rectangular window) in Equation (9–4). The Hamming window is given by

$$w(n) - \begin{cases} 0.54 - 0.46\cos\left(\dfrac{2\pi n}{N-1}\right), n = 0, 1, ..., N-1 \\ 0, \text{otherwise} \end{cases} \tag{9–7}$$

The Hamming window is shown in Figure 9–2 (for $N = 16$); note that it is tapered and symmetric about its midpoint. The Hamming window spectrum is shown in Figure 9–3. The reader should compare Figure 9–3 with the spectrum of a 16-point rectangular window (Figure 6–23). Note that the Hamming window spectrum has much smaller sidelobes but a wider mainlobe, compared to the spectrum of a rectangular window.

Figure 9–4 shows the amplitude response function $|H[\exp(j\omega)]|$ of an $N = 29$ point FIR lowpass filter designed using a Hamming window in Equation (9–4). The reader should

FIGURE 9–2

Hamming window.

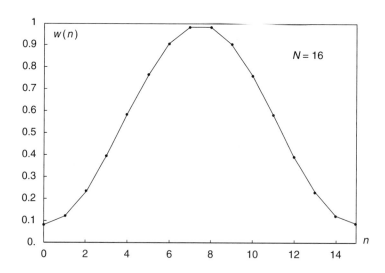

$N = 16$

FIGURE 9–3
Amplitude spectrum of a
Hamming window.

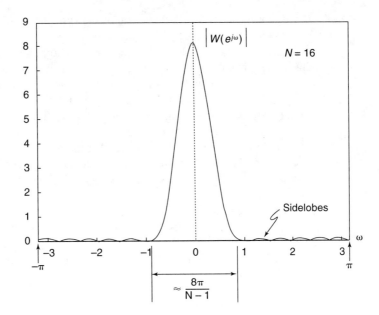

compare Figure 9–4 with the amplitude response function of a 29-point FIR lowpass filter designed with a rectangular window (Figure 6–30). Note that the ripples have been greatly reduced. In fact, the peak approximation error δ (Figure 9–4) is only (approximately) 0.0022, which is why it is so difficult to see on the plot. Since $20\log(0.0022) \cong -53$, this filter is said

FIGURE 9–4
Amplitude response function
of an $N = 29$-point FIR
lowpass filter designed by
shifting the impulse response
of an ideal lowpass filter and
then truncating it with a
Hamming window. Compare
Figure 9–4 with Figure 6–30
on page 208 (the rectangular
window case).

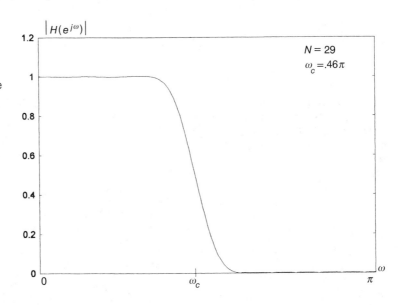

to have approximately 53 dB of attenuation in the stopband. (The filter designed with the rectangular window has only 21 dB of attenuation.) On the other hand, the transition bandwidth shown in Figure 9–4 has clearly increased relative to that shown in Figure 6–30. In fact, the transition bandwidth has more than tripled in width, making this filter less sharp.

As was the case with the filter designed using a rectangular window in Chapter 6, the transition band of the Hamming window-designed filter can be reduced by increasing length N, without having a significant impact on the amount of stopband attenuation. (This is because increasing N decreases the window spectrum main lobe width but has no significant impact on the sidelobe amplitude.) It should be clear that an FIR lowpass filter designed using a Hamming window will have to have a longer length (N) than a rectangular window-designed FIR lowpass filter having approximately the same transition bandwidth or sharpness.

The Hamming window is only one possible type of tapered window we could employ. There are other types of tapered windows; the differences between the various types have to do with the extent and exact shaping of the tapering function itself. Many types of tapered windows can be found in the literature: Bartlett window, Hanning window, Hamming window, Blackman window, Kaiser window, etc. The window type usually used in the design of FIR filters is the *Kaiser window*, which has an adjustable tapering parameter. But before considering the Kaiser window and the FIR filter design methodology associated with it, there are some general concepts that need to be firmly nailed down:

1. Figure 9–5a illustrates the concept of extent of tapering. The extent of tapering has a direct effect on the shape of the window spectrum. Specifically, for a given length N, the more tapered the window is, the smaller the sidelobes of the window spectrum will be, but the wider the main lobe will be (see Figure 9–5[b]). This will be illustrated in Section 9–3 for the Kaiser window (see Figures 9–6 and 9–7 for a sneak preview).

2. For a given length N, more tapering of the window results in an FIR filter frequency response function with a wider transition bandwidth (less sharp) but smaller ripples (flatter

FIGURE 9–5
(a) Extent of window tapering: "less tapering" vs. "more tapering." (b) Effect of extent of tapering on the spectrum of the window. (c) Effect of extent of tapering on the amplitude response function of the FIR lowpass filter.

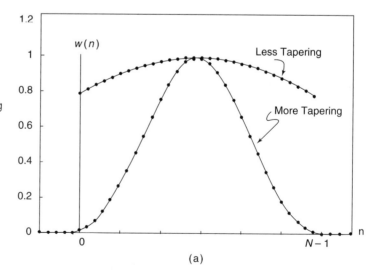

(a)

FIGURE 9–5
(continued)

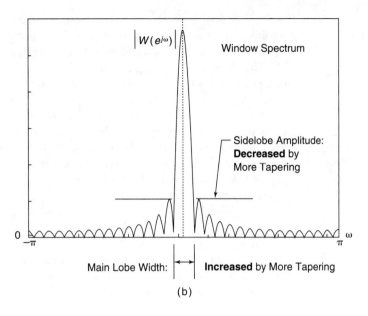

(b)

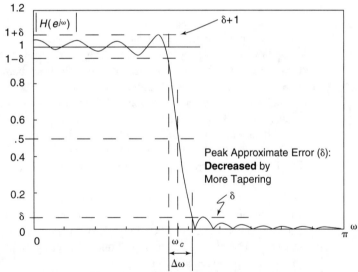

(c)

passband and greater stopband attenuation). (See Figure 9–5[c].) This will also be illustrated in the next section (see Figures 9–6 through 9–11 for a sneak preview).

3. For a given window tapering function (i.e., shape), the window spectrum main lobe width is inversely proportional to the length N, but the relative sidelobe amplitude of this spectrum is little affected by N. Thus, for an FIR filter designed using a given window tapering function, increasing N will cause the transition bandwidth to decrease, but will have little impact on the amount of stopband attenuation.

The specifications for a desired FIR lowpass filter usually include:

1. The desired cutoff frequency ω_c. (The cutoff frequency can also be specified in *hertz,* i.e., f_c instead of ω_c. If f_c is specified, the corresponding ω_c can be found using the conversion formula $\omega_c = (2\pi f_c)/f_s$, where f_s is the sampling frequency. The cutoff frequency in hertz will be a number between 0 and $f_s/2$.) Note that ω_c is halfway between the desired passband and stopband edges, and that the gain at this frequency is one-half of the passband gain, as shown, for example, in Figures 6–30, 9–4, and 9–5(c). In other words, the attenuation at the cutoff frequency is 6 dB, not 3 dB.

2. The maximum allowable transition bandwidth $\Delta\omega$. (We want the filter to be at least this sharp.) The transition bandwith can also be specified in hertz, that is, Δf instead of $\Delta\omega$. The conversion formula is $\Delta\omega = (2\pi\Delta f)/f_s$. For a definition of $\Delta\omega$, which depends on the location of the frequencies at which the filter gain is $(1 - \delta)$ and δ, refer to either Figure 6–30 or Figure 9–5(c).

3. The maximum allowable peak approximation error δ. This is the same thing as saying that we want at least A dB of attenuation in the stopband, where $A = -20\log(\delta)$. If the desired stopband attenuation (A dB) is specified, the corresponding value of δ is

$$\delta = 10^{\left(\frac{-A}{20}\right)} \qquad (9\text{--}8)$$

The peak approximation error δ is shown in Figures 6–30 and 9–5c.

Given these specifications, the design problem is to choose a length (N) and window tapering function that will result in the specifications being satisfied. Fortunately, this task is relatively simple if we use the Kaiser window. There are some well-known empirical formulas for selecting length N and Kaiser window tapering parameter β, given specifications (2) and (3) for FIR lowpass filters listed previously.

It should be pointed out that the frequency specifications (cutoff frequency and transition bandwidth) can be given in terms of a *normalized* sampling frequency of 1 Hz. To obtain normalized frequency specifications, divide all the unnormalized frequency specifications by the sampling frequency. For example, if the unnormalized specifications are $f_s = 1000$, $f_c = 250$, and $\Delta f = 50$, the normalized specifications are $f_s = 1$, $f_c = .25$, and $\Delta f = .05$. Either the normalized or unnormalized set of numbers will result in the same values of ω_c and $\Delta\omega$. (This is because the right side of the conversion formula $\omega = 2\pi f/f_s$ involves the ratio f/f_s.) The normalized and unnormalized set of frequency specifications will result in the *same* impulse response $h(n)$.

9–3 THE KAISER WINDOW

An N-point Kaiser window is given by

$$w(n) = \begin{cases} \dfrac{I_0\left(\beta\left\{1 - [(n-\alpha)/\alpha]^2\right\}^{1/2}\right)}{I_0(\beta)} & , n = 0, 1, ... N - 1 \\ 0, \text{otherwise} \end{cases} \qquad (9\text{--}9)$$

where $\alpha = (N-1)/2$, and $I_0(\ )$ is the zero order modified Bessel function of the first kind, which is defined by the following infinite series:

$$I_0(x) = 1 + \sum_{n=1}^{\infty} \left[\frac{(x/2)^n}{n!} \right]^2 \qquad (9\text{--}10)$$

The length N can be even or odd. β is the parameter that adjusts the degree of tapering. The larger β is, the more tapered the window is. (If $\beta = 0$, the Kaiser window becomes a rectangular window.) The value of β is determined using a set of formulas presented in Section 9–4.

Calculation of a Kaiser window function appears formidable because of the necessity of calculating the Bessel function $I_0(\ )$. Fortunately, there is a short recursive algorithm for accomplishing this task. A FORTRAN version of this algorithm is shown (it is a modified version of a program written by Kaiser):

```
        SUBROUTINE FI0   (X, Z)
C    Z = I0(X)  :  zero order modified Bessel  function of the first kind
        S=1.
        DS=1.
        D=0.0
    1   D=D + 2.0
        DS=DS * X * X / ( D * D)
        S= S + DS
        IF( DS .GE.   .2E-8*S ) GO TO 1
    2   Z=S
        RETURN
        END
```

Figure 9–6 shows $N = 30$-point Kaiser windows for $\beta = 3$ and $\beta = 6$.

NOTE: β does *not* have to be an integer. Observe that the $\beta = 6$ Kaiser window is more tapered than the $\beta = 3$ Kaiser window. Figure 9–7 shows the spectral amplitude function $|W[\exp(j\omega)]|$ for each of the windows of Figure 9–6. Observe that more tapering (larger β) reduces the amplitude of the sidelobes but increases the width of the main lobe.

Figure 9–8(a) shows the FIR filter impulse response $h(n)$ that results when an $N = 30$, $\beta = 3.0$ Kaiser window is used in Equation (9–4). (The cutoff frequency is $\omega_c = 0.46\pi$.) Figure 9–9(a) shows the amplitude response function $|H[\exp(j\omega)]|$ of this filter. Figure 9–10(a) shows the amplitude response in decibels (dB), that is, $20\log|H[\exp(j\omega)]|$. Figure 9–8(b) shows the FIR filter impulse response $h(n)$ that results when an $N = 30$, $\beta = 6.0$ Kaiser window is used in Equation (9–4). (The cutoff frequency is also $\omega_c = 0.46\pi$.) The difference between this figure and Figure 9–8(a) ($\beta = 3$ case) is that the tails of $h(n)$ die out more quickly. Figure 9–9(b) shows the amplitude response function $|H[\exp(j\omega)]|$ for the $\beta = 6$ case; Figure 9–10(b) shows the decibel amplitude response plot. If the frequency domain plots for the $\beta = 3$ and $\beta = 6$ cases are compared, it will be observed that increasing β (i.e., using more taper) results in less ripples (flatter passband and greater stopband attenuation) but also results in a wider transition band (the filter is less sharp). The fact that increasing β results in more stopband attenuation is especially evident when comparing the decibel plots (Figure 9–10[a] and [b]). (Note that $N = 30$ for both of these cases.) For convenience in comparing the frequency domain plots, they are superimposed in Figure 9–11.

FIGURE 9–6
N = 30-point Kaiser windows for β = 3 and β = 6. Note how β affects the extent of tapering.

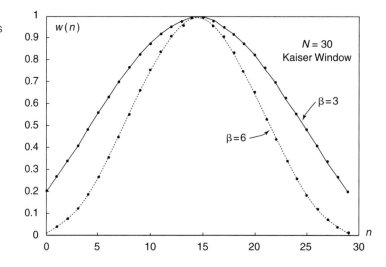

FIGURE 9–7
Amplitude spectra of N = 30-point Kaiser windows for β = 3 and β = 6. Note how extent of tapering (β) affects the main lobe width and sidelobe height.

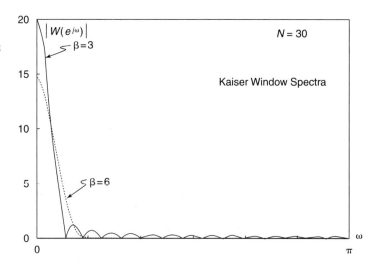

FIGURE 9–8
FIR lowpass filter impulse
responses for filters having
the same length (*N*) and
cutoff frequency but different
Kaiser window tapering
parameters (β = 3 and β = 6).

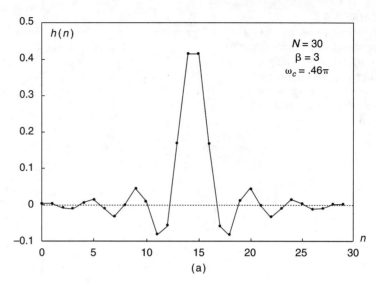

(a)

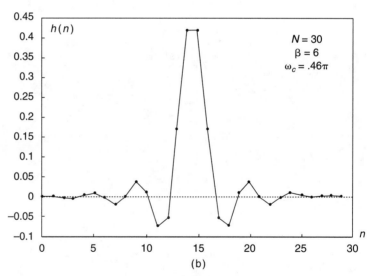

(b)

FIGURE 9–9
FIR lowpass filter amplitude response functions for filters having the same length (N) and cutoff frequency but different Kaiser window tapering parameters ($\beta = 3$ and $\beta = 6$).

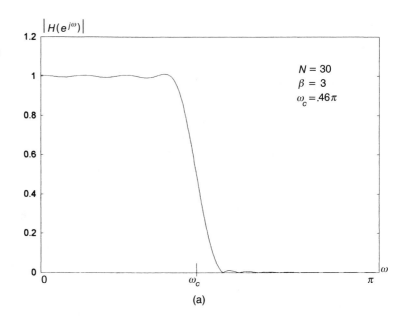

$N = 30$
$\beta = 3$
$\omega_c = .46\pi$

(a)

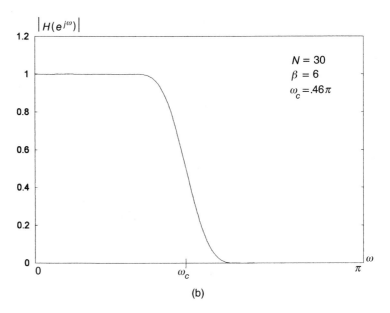

$N = 30$
$\beta = 6$
$\omega_c = .46\pi$

(b)

FIGURE 9–10
FIR lowpass filter decibel amplitude response functions for filters having the same length (N) and cutoff frequency but different Kaiser window tapering parameters (β = 3 and β = 6).

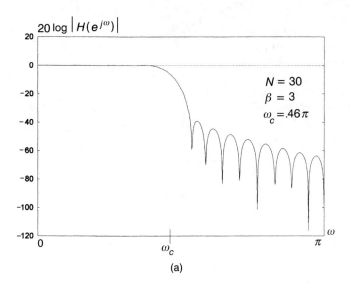

(a)

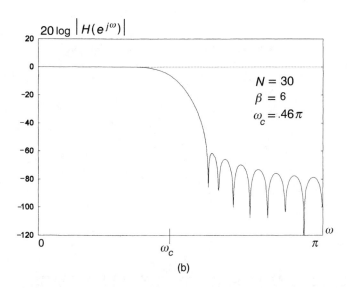

(b)

FIGURE 9–11
Direct comparison of amplitude and decibel amplitude response functions for FIR lowpass filters having the same length (N) and cutoff frequency but different Kaiser window tapering parameters ($\beta = 3$ and $\beta = 6$).

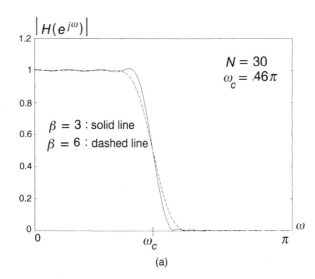

(a)

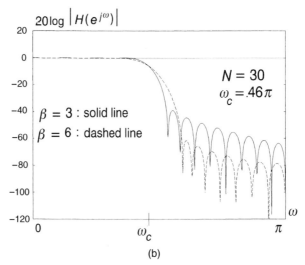

(b)

9–4 EMPIRICAL FORMULAS FOR FIR LOWPASS FILTER DESIGN USING THE KAISER WINDOW METHOD

Given the FIR lowpass filter specifications ω_c, $\Delta\omega$, and δ, one must first translate δ into the desired stopband attenuation in dB (A) using the formula $A = -20\log(\delta)$. The following empirical formulas can then be used to obtain appropriate values for Kaiser window tapering parameter β and FIR filter length N :

$$\beta = \begin{cases} 0.1102(A - 8.7), A > 50 \\ 0.5842(A - 21)^{0.4} + 0.07886(A - 21), 21 \leq A \leq 50 \\ 0.0, A < 21 \end{cases}$$

(9–11)

$$N = \frac{A - 8}{2.285\Delta\omega} + 1 \qquad (9\text{–}12)$$

The value of β is *not* usually an integer. However, length N must be an integer; therefore, the right side of Equation (9–12) must be rounded to the nearest integer. (These formulas were suggested by Dr. Kaiser in 1974.)

EXAMPLE 9–4.1

Design an FIR lowpass filter using the Kaiser window method. The specifications are

$$f_s = 1000 \text{ Hz}$$
$$f_c = 250 \text{ Hz}$$
$$\Delta f = 50 \text{ Hz}$$
$$\delta = .001 \quad (A = 60)$$

The *normalized* frequency specifications are $f_c = .25$ and $\Delta f = .05$. In terms of the frequency variable ω, the frequency specifications are $\omega_c = 0.5\pi$ and $\Delta\omega = 0.1\pi$. The empirical formulas for β and N suggest $\beta = 5.65326$ and $N = 73.438$. Rounding N to the nearest integer results in $N = 73$. A 73-point Kaiser window $w(n)$ was generated using Equation (9–9) (with the help of the recursive algorithm for evaluation of the Bessel function). The Kaiser window $w(n)$ was then used in Equation (9–4) to generate the $N = 73$-point FIR filter impulse response $h(n)$, which is shown in Figure 9–12. The amplitude response function $|H[\exp(j2\pi fT)]|$ is shown in Figure 9–13, which has been annotated to show that the transition bandwidth specification (Δf) has been approximately satisfied. (If N is increased, Δf will decrease.) The approximation error (δ) is too small to be seen in Figure 9–13. The decibel amplitude response plot, $20\log|H[\exp(j2\pi fT)]|$, is shown in Figure 9–14, which has been annotated to show that the specification $A = 60$ (i.e., $\delta = .001$) has been satisfied approximately.

FIGURE 9–12
Impulse response for
Example 9–4.1.

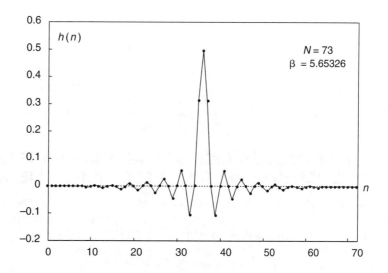

FIGURE 9–13
Amplitude response function
for Example 9–4.1.

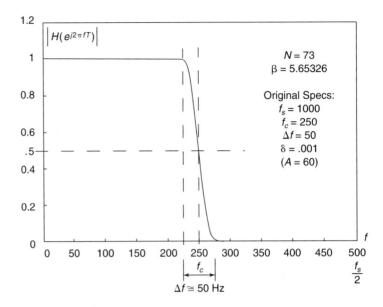

FIGURE 9–14
Decibel amplitude response
function for Example 9–4.1.

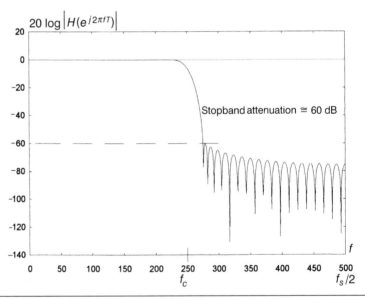

9–5   FIR BANDPASS, HIGHPASS, AND BANDSTOP FILTERS DESIGNED USING THE KAISER WINDOW METHOD

The frequency response function $H_i[\exp(j\omega)]$ of an *ideal bandpass filter* is shown on the interval $-\pi < \omega < \pi$ in Figure 9–15. The upper cutoff frequency is denoted ω_u; the lower cutoff frequency is denoted ω_l. Like any DTFT, this frequency response function is periodic (2π).

FIGURE 9–15

Ideal bandpass filter.

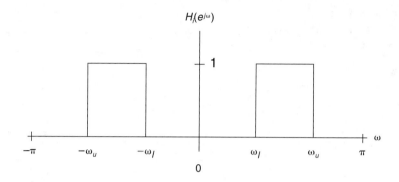

To find the impulse response $h_i(n)$ of an ideal bandpass filter, we can use the inverse DTFT integral:

$$h_i(n) = \frac{1}{2\pi} \int_{-\pi}^{\pi} H_i\left(e^{j\omega}\right) e^{j\omega n} d\omega \qquad (9\text{–}13)$$

For the problem at hand, this must be expressed as

$$h_i(n) = \frac{1}{2\pi} \int_{-\omega_u}^{-\omega_l} (1) e^{j\omega n} d\omega + \frac{1}{2\pi} \int_{\omega_l}^{\omega_u} (1) e^{j\omega n} d\omega \qquad (9\text{–}14)$$

After evaluating each integral, combining terms, and applying Euler's theorem (as in Equation [6–130] on page 206), we obtain

$$h_i(n) = \frac{\sin(\omega_u n)}{\pi n} - \frac{\sin(\omega_l n)}{\pi n} \qquad (9\text{–}15)$$

This impulse response is clearly the combination of the impulse responses of two ideal lowpass filters. (In fact, this result can be obtained directly, without resort to the inverse DTFT, by observing that an ideal bandpass filter frequency response function can be obtained by taking the frequency response function of an ideal lowpass filter with cutoff frequency ω_u and subtracting from it the frequency response function of an ideal lowpass filter with cutoff frequency ω_l.) The impulse response of an ideal bandpass filter is defined for all integer values of n ($-\infty < n < \infty$); it dies out as n gets larger in either the positive or negative direction. An ideal bandpass filter is therefore a noncausal system, and cannot be realized in the real world.

The impulse response $h(n)$ of a causal, N-point FIR bandpass filter can be obtained by shifting, truncating, and tapering the impulse response of an ideal bandpass filter:

$$h(n) = w(n)\left\{ \frac{\sin\left[\omega_u\left(n - \frac{N-1}{2}\right)\right] - \sin\left[\omega_l\left(n - \frac{N-1}{2}\right)\right]}{\pi\left(n - \frac{N-1}{2}\right)} \right\} \qquad (9\text{–}16)$$

where $w(n)$ is an N-point tapered window. As with the design of FIR lowpass filters, the Kaiser window is the "weapon of choice" because it offers great design flexibility.

FIGURE 9–16

FIR bandpass filter
parameters.

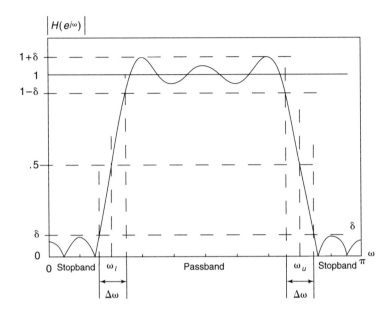

The specifications for a desired FIR bandpass filter include the upper and lower cutoff frequencies (ω_u and ω_l), the maximum transition bandwidth $\Delta\omega$, and the maximum approximation error δ or minimum stopband attenuation in dB, $A = -20\log(\delta)$. If the sampling frequency f_s is given, the frequency specifications f_u, f_l, and Δf in hertz may be specified instead. (As discussed earlier, the frequency specifications may also be normalized.) These specifications are illustrated in Figure 9–16. The empirical formulas shown earlier in conjunction with FIR lowpass filters for length N and Kaiser window parameter β (Equations [9–11] and [9–12]) can also be used for the design of FIR bandpass filters.

EXAMPLE 9–5.1

Design an FIR bandpass filter using the Kaiser window method. The specifications are

$$f_s = 1000\,Hz$$
$$f_l = 200\,Hz \quad (\omega_l = .4\pi)$$
$$f_u = 300\,Hz \quad (\omega_u = .6\pi)$$
$$\Delta f = 50\,Hz \quad (\Delta\omega = .1\pi)$$
$$\delta = .001 \quad (A = 60)$$

The empirical formulas for β and N suggest $\beta = 5.65326$ and $N = 73.438$. Rounding N to the nearest integer results in $N = 73$. A 73-point Kaiser window $w(n)$ was used in Equation (9–16) to generate the $N = 73$-point FIR filter impulse response $h(n)$, which is shown in

FIGURE 9–17
Impulse response for
Example 9–5.1.

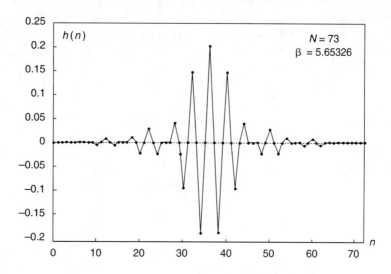

Figure 9–17. The amplitude response function $|H[\exp(j\omega)]|$ is shown in Figure 9–18, which has been annotated to show that the transition bandwidth specification has been approximately satisfied. (If N is increased, Δf will decrease.) The approximation error (δ) is too small to be seen in Figure 9–18. The decibel plot, $20\log|H[\exp(j\omega)]|$, is shown in Figure 9–19, which has been annotated to show that the specification $A = 60$ (i.e., $\delta = .001$) has been approximately satisfied.

FIGURE 9–18
Amplitude response function
for Example 9–5.1.

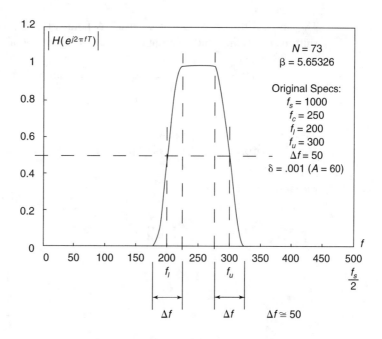

FIGURE 9–19
Decibel amplitude response
function for Example 9–5.1.

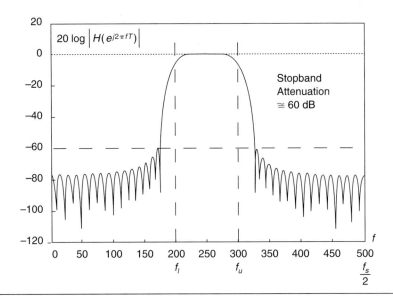

An *ideal highpass filter* with cutoff frequency ω_c, as shown in Figure 9–20, can be thought of as a bandpass filter with $\omega_l = \omega_c$, and $\omega_u = \pi$. Thus the impulse response of an ideal highpass filter is

$$h_i(n) = \frac{\sin(\pi n)}{\pi n} - \frac{\sin(\omega_c n)}{\pi n} \tag{9–17}$$

The impulse response $h(n)$ of a causal, N-point FIR highpass filter can be obtained by shifting, truncating, and tapering the impulse response of an ideal highpass filter:

$$h(n) = w(n) \left\{ \frac{\sin\left(\pi\left(n - \frac{N-1}{2}\right)\right)}{\pi\left(n - \frac{N-1}{2}\right)} - \frac{\sin\left(\omega_c\left(n - \frac{N-1}{2}\right)\right)}{\pi\left(n - \frac{N-1}{2}\right)} \right\} \tag{9–18}$$

where $w(n)$ is an N-point tapered window (usually a Kaiser window). For lowpass and bandpass filters, N can be even or odd; however, for highpass filters the best results are obtained when N is odd. (This is

FIGURE 9–20
Ideal highpass filter.

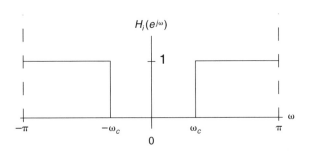

because the linear phase FIR filters discussed in this chapter all have the characteristic that if N is even, $H[\exp(j\omega)] = 0$ at $\omega = \pi$, which is not what we want in a highpass filter.) "It can be shown" that when N is odd, Equation (9–18) reduces to

$$h(n) = w(n)\left\{ \delta\left(n - \tfrac{N-1}{2}\right) - \frac{\sin\left(\omega_c\left(n - \tfrac{N-1}{2}\right)\right)}{\pi\left(n - \tfrac{N-1}{2}\right)} \right\} \tag{9–19}$$

where $\delta(n - (N-1)/2)$ is an impulse located at $n = (N-1)/2$.

The specifications for an FIR highpass filter include the cutoff frequency ω_c, the maximum transition bandwidth $\Delta\omega$, and the maximum approximation error δ or minimum stopband attenuation $A = -20\log(\delta)$. Specifications $\Delta\omega$ and δ are defined in a manner analogous to the lowpass filter case. The empirical formulas for length N and Kaiser window parameter β apply as before.

EXAMPLE 9–5.2

Design an FIR highpass filter using the Kaiser window method. The specifications are:

$$f_s = 1000\,Hz$$
$$f_c = 250\,Hz \quad (\omega_c = .5\pi)$$
$$\Delta f = 50\,Hz \quad (\Delta\omega = .1\pi)$$
$$\delta = .001 \quad (A = 60)$$

The empirical formulas for β and N suggest $\beta = 5.65326$ and $N = 73.438$. Rounding N to the nearest *odd* integer results in $N = 73$. A 73-point Kaiser window was used in Equation (9–19) to generate the $N = 73$-point FIR filter impulse response $h(n)$, which is shown in Figure 9–21.

FIGURE 9–21
Impulse response for
Example 9–5.2.

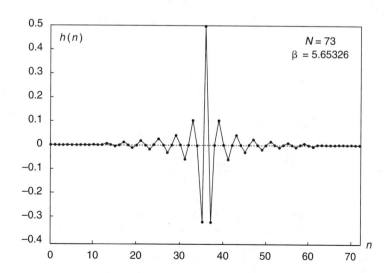

FIGURE 9–22
Amplitude response function
for Example 9–5.2.

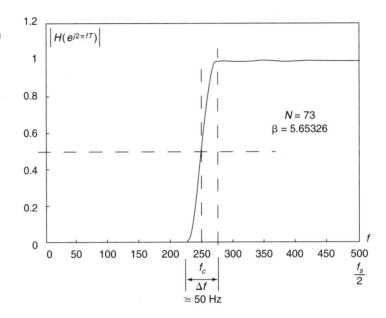

FIGURE 9–23
Decibel amplitude response
function for Example 9–5.2.

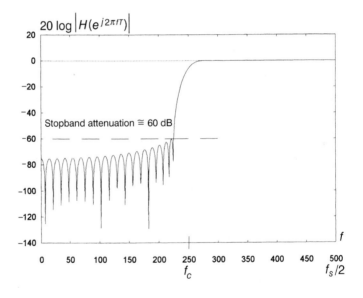

The amplitude response function $|H[\exp(j2\pi fT)]|$ is shown in Figure 9–22; the decibel amplitude response function $20\log|H[\exp(j2\pi fT)]|$ is shown in Figure 9–23. The frequency domain plots have been annotated to show that the specifications are approximately satisfied.

FIGURE 9–24

Ideal bandstop filter.

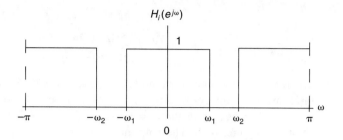

$H_i(e^{j\omega})$

The frequency response function of an *ideal bandstop filter* with lower cutoff frequency ω_1 and upper cutoff frequency ω_2 is shown in Figure 9–24. It is convenient to view this as a combination of a lowpass filter with cutoff frequency ω_1 and a highpass filter with a cutoff frequency of ω_2:

$$h_i(n) = \frac{\sin(\omega_1 n)}{\pi n} + \left[\frac{\sin(\pi n)}{\pi n} - \frac{\sin(\omega_2 n)}{\pi n}\right] \tag{9–20}$$

The impulse response $h(n)$ of a causal, N-point FIR bandstop filter can be obtained by shifting, truncating, and tapering the impulse response of an ideal bandstop filter:

$$h(n) = w(n)\left(\frac{\sin\left[\omega_1\left(n - \frac{N-1}{2}\right)\right]}{\pi\left(n - \frac{N-1}{2}\right)} + \left\{\frac{\sin\left[\pi\left(n - \frac{N-1}{2}\right)\right]}{\pi\left(n - \frac{N-1}{2}\right)} - \frac{\sin\left[\omega_2\left(n - \frac{N-1}{2}\right)\right]}{\pi\left(n - \frac{N-1}{2}\right)}\right\}\right) \tag{9–21}$$

where $w(n)$ is an N-point Kaiser window. Since an even value for N would force $H[\exp(j\omega)] = 0$ at $\omega = \pi$, an odd value for N is the best choice. If N is odd, Equation (9–21) reduces to

$$h(n) = w(n)\left\{\frac{\sin\left[\omega_1\left(n - \frac{N-1}{2}\right)\right]}{\pi\left(n - \frac{N-1}{2}\right)} + \delta\left(n - \frac{N-1}{2}\right) - \frac{\sin\left[\omega_2\left(n - \frac{N-1}{2}\right)\right]}{\pi\left(n - \frac{N-1}{2}\right)}\right\} \tag{9–22}$$

where $\delta(n - (N-1)/2)$ is an impulse located at $n = (N-1)/2$.

The specifications for a desired FIR bandstop filter include the two cutoff frequencies, maximum transition bandwidth $\Delta\omega$, and maximum approximation error δ or minimum stopband attenuation $A = -20\log(\delta)$. Specifications $\Delta\omega$ and δ are defined in a manner analogous to the bandpass filter case. The empirical formulas for length N and Kaiser window parameter β apply as before.

EXAMPLE 9–5.3

Design an FIR bandstop filter using the Kaiser window method. The specifications are:

$$f_s = 1000\text{Hz}$$
$$f_1 = 200\text{Hz} \quad (\omega_1 = .4\pi)$$

$$f_2 = 300\text{Hz} \quad (\omega_2 = .6\pi)$$

$$\Delta f = 25\text{Hz} \quad (\Delta\omega = .05\pi)$$

$$\delta = .01 \quad (A = 40)$$

The empirical formulas for β and N suggest $\beta = 3.3953$ and $N = 90.15$. Rounding N to the nearest odd integer results in $N = 91$. A 91-point Kaiser window was used in Equation (9–22) to generate the $N = 91$-point FIR filter impulse response $h(n)$, which is shown in Figure 9–25. The amplitude response function $\left|H[\exp(j2\pi fT)]\right|$ is shown in Figure 9–26. The decibel amplitude response function is shown in Figure 9–27. The frequency domain plots show that the specifications are approximately satisfied.

FIGURE 9–25

Impulse response for Example 9–5.3.

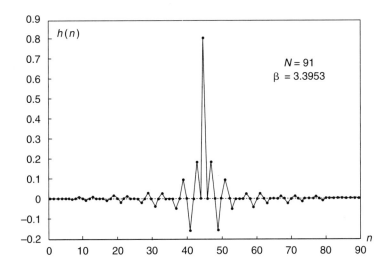

FIGURE 9–26

Amplitude response function for Example 9–5.3.

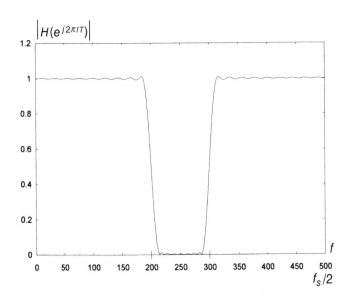

FIGURE 9–27
Decibel amplitude response
function for Example 9–5.3.

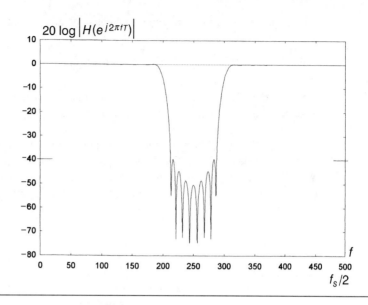

9–6 DESIGN OF FIR FILTERS USING THE PARKS-McCLELLAN ALGORITHM

In 1972, Parks and McClellan described an FIR filter design procedure which has become known as the Parks-McClellan algorithm. This is perhaps the most popular method of FIR filter design.

The Parks-McClellan algorithm is an *optimization* procedure. The impulse response $h(n)$ is selected such that the difference between the ideal filter frequency response function and the actual FIR filter frequency response function is minimized in some sense (this will be described later in the chapter). To introduce the basic idea, we will restrict our attention to odd length filters (i.e., N = odd); however, the Parks-McClellan algorithm can also handle the N = even case.

The N-point linear phase FIR filters we are interested in here have an impulse response $h(n)$ which is symmetric about its midpoint $(N - 1)/2$. If N is odd, $h(n)$ can be thought of as a shifted version of an even sequence $a(n)$, that is,

$$h(n) = a\left(n - \tfrac{N-1}{2}\right) \tag{9-23}$$

where $a(n)$ is symmetric about $n = 0$. The sequence $a(n)$ is the impulse response of a non-causal FIR filter with a frequency response function $A[\exp(j\omega)]$:

$$A\left(e^{j\omega}\right) = \sum_{n=-L}^{L} a(n)e^{-j\omega n} \tag{9-24}$$

where $L = (N - 1)/2$. Since $a(n)$ is an even sequence (that is, $a(n) = a(-n)$), $A[\exp(j\omega)]$ is a real function of ω. Using the fact that $a(n) = a(-n)$, along with Euler's theorem, allows us to express $A[\exp(j\omega)]$ as

$$A\left(e^{j\omega}\right) = a(0) + \sum_{n=1}^{L} 2a(n)\cos(\omega n) \tag{9-25}$$

The frequency response function of the causal version of this FIR filter is

$$H\left(e^{j\omega}\right) = A\left(e^{j\omega}\right)e^{-j\omega\left(\frac{N-1}{2}\right)} \qquad (9\text{--}26)$$

Since the only difference between $h(n)$ and $a(n)$ is a shift, as a practical matter we can consider designing frequency response function $A[\exp(j\omega)]$ instead of $H[\exp(j\omega)]$, and then taking care of the shift as a last step.

Let $H_i[\exp(j\omega)]$ be the frequency response function of an ideal filter (lowpass, bandpass, etc., as described earlier). $H_i[\exp(j\omega)]$ is a real function of ω and is equal to either 1 or 0. The design problem is to select coefficients $a(n)$ such that $A[\exp(j\omega)]$ is close to $H_i[\exp(j\omega)]$ in some sense. Let us define a difference or error function $E(\omega)$ as

$$E(\omega) = H_i\left(e^{j\omega}\right) - A\left(e^{j\omega}\right) \qquad (9\text{--}27)$$

and then pose the following problem: Choose the coefficients $a(n)$, $n = -L, \ldots, L$ to *minimize the maximum value of* $|E(\omega)|$ within the interval $0 \le \omega < \pi$. This is the basic idea behind the Parks-McClellan algorithm. (Minimizing the maximum value of $|E(\omega)|$ is essentially the same thing as minimizing the maximum value of the approximation error [ripple] in the FIR filter frequency response function.)

It turns out that instead of considering the entire interval $0 \le \omega < \pi$, it is necessary to carry out the minimization with respect to the passband and stopband only. This means that the width of the transition band must be specified in advance. Let us consider an ideal lowpass filter as an example. Figure 9–28(a) shows an ideal lowpass filter with cutoff frequency ω_c. Figure 9–28(b) shows the ideal lowpass filter that we specify with respect to the Parks-McClellan algorithm. The transition band characteristic is actually unspecified:

$$H_i\left(e^{j\omega}\right) = \begin{cases} 1, & 0 \le \omega \le \omega_p \\ \text{unspecified}, & \omega_p < \omega \le \omega_s \\ 0, & \omega_s < \omega < \pi \end{cases} \qquad (9\text{--}28)$$

Note the relationship between cutoff frequency ω_c and transition band edges ω_p and ω_s:

$$\omega_c = \frac{\omega_p + \omega_s}{2} \qquad (9\text{--}29)$$

In other words, the cutoff frequency is halfway between the transition band edges ω_p and ω_s. The transition bandwidth $\Delta\omega$ is

$$\Delta\omega = \omega_s - \omega_p \qquad (9\text{--}30)$$

If ω_c and $\Delta\omega$ are specified (as in the Kaiser window method), the transition band edges can be found as follows:

$$\omega_s = \omega_c + \frac{\Delta\omega}{2} \qquad (9\text{--}31)$$

FIGURE 9–28
(a) Ideal lowpass filter. (b) Ideal lowpass filter specified for the Parks-McClellan algorithm.

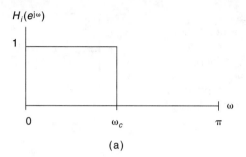

(a)

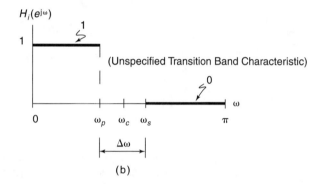

(b)

$$\omega_p = \omega_c - \frac{\Delta\omega}{2} \qquad\qquad (9\text{--}32)$$

A modification that is usually included allows the designer the option of having different amounts of ripple in the passband and stopband. (This is not possible when the Kaiser window design method is used.) In most FIR filter design programs that use the Parks-McClellan algorithm, the designer is asked to specify approximation error *weights* for each band. For example, if the designer specifies approximation error weights of 1.0 and 10.0 for the passband and stopband respectively, it means that the stopband ripple will be ten times bigger than the passband ripple. In other words, with respect to Figure 9–29, he is specifying that δ_2 will be 10 times bigger than δ_1. The specified error weights are incorporated into the design process by specifying a modified weighted error function:

$$E(\omega) = W(\omega)\left[H_i\left(e^{j\omega}\right) - A\left(e^{j\omega}\right)\right] \qquad\qquad (9\text{--}33)$$

where the weighting function $W(\omega)$ is defined such that the desired weighting will be achieved. The problem is to choose the coefficients $a(n)$ to minimize the maximum value of $|E(\omega)|$ in the passband and stopband. With filter length N, ω_p, and ω_s fixed, it turns out that the solution to the problem involves Chebyshev approximation theory (an advanced mathematical topic that is well beyond the scope of this book). The Parks-McClellan algorithm is an iterative method of finding the solution. Fortunately, the complicated computer program that implements this method is available to the public and has been incorporated into a number of FIR filter design packages.

FIGURE 9–29
Parks-McClellan algorithm
FIR lowpass filter specifications. Passband ripple (δ_1) and
stopband ripple (δ_2) can be
different values if the Parks-
McClellan algorithm is used.

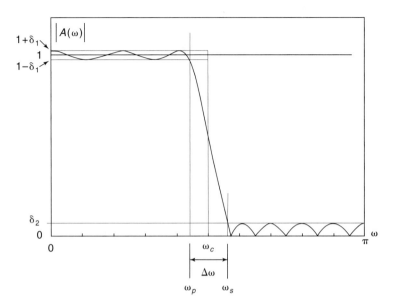

The *actual values* of δ_1 and δ_2 that are obtained after the Parks-McClellan algorithm has converged *depend on the filter length (N)*; the larger N is, the smaller the ripple will actually be. The weights only specify the *relative* size of the ripple in each band.

It is also important to note that *increasing N has no effect on the transition bandwidth* $\Delta\omega$, because the transition band edges ω_p and ω_s are fixed by the designer at the start of the procedure. (Note that with the Kaiser window method, the effect of changing N is quite different.)

An empirical formula for selecting filter length N for FIR filters designed using the Parks-McClellan algorithm has been suggested by Kaiser. Given desired ripple δ_1 and δ_2 (the actual values that are desired, not the relative values) and transition bandwidth $\Delta\omega$, (see Figure 9–29), a ballpark estimate of required length N is

$$N = \frac{-20 \log \sqrt{\delta_1 \delta_2} - 13}{2.324 \Delta\omega} + 1 \qquad (9\text{–}34)$$

Of course, N must be rounded to the nearest integer. If the designer specifies $\delta_1 = \delta_2 = \delta$ (i.e., same ripple in both bands), Equation (9–34) simplifies to

$$N = \frac{A - 13}{2.324 \Delta\omega} + 1 \qquad (9\text{–}35)$$

where $A = -20\log(\delta)$ is the desired stopband attenuation in dB. (These formulas will also give a ballpark estimate of N for bandpass filters, provided that the specified transition bandwidths are the same on both sides of the passband.)

Some adjustment in length N may prove to be necessary. In particular, if the resulting ripple is larger than originally specified, N must be increased. (It should also be noted again that for highpass and bandstop filters it is best to use $N =$ odd.)

FIR filters designed using the Parks-McClellan algorithm have the characteristic that all of the ripple peaks in the passband have equal height (the same is true in the stopband), as opposed to filters designed using the Kaiser window method, in which the ripple peaks decrease in amplitude as the frequency (ω) becomes more distant from the cutoff frequency (ω_c). For this reason, FIR filters designed using the Parks-McClellan algorithm are sometimes called *equiripple*-type filters. This characteristic will become clear in the examples that follow.

EXAMPLE 9–6.1

Design an FIR lowpass filter using the Parks-McClellan algorithm (using one of the available computer programs). The specifications are as follows:

$$f_s = 1000$$
$$f_c = 250 \quad (\omega_c = .5\pi)$$
$$\Delta f = 50 \quad (\Delta\omega = .1\pi)$$
$$\delta_1 = \delta_2 = .001$$

In order to compare the Parks-McClellan algorithm to the Kaiser window method, we have used the same specifications as used earlier in Example 9–4.1. We must use *normalized* frequency specifications in the computer program available to us. In particular, this program demands as input the normalized transition band edges f_p and f_{stop}, which are .225 and .275, respectively. (The normalized cutoff frequency is .25; the normalized transition bandwidth is .05.)

NOTE: We are naming the stopband edge f_{stop} instead of f_s so that the reader will not confuse this parameter with the sampling frequency.

The empirical formula for N (Equation [9.35]) suggests $N = 65.37$ (round off to $N = 65$) as the filter length. Therefore, the specifications entered into the program are $N = 65$, $f_p = .225$, and $f_{stop} = .275$. Since the desired ripple is the same in both bands, the specified approximation error weight is 1.0 for both bands.

Figure 9–30 shows the resulting impulse response $h(n)$. The amplitude response function $|H[\exp(j2\pi fT)]|$ is shown in Figure 9–31. In order to clearly illustrate the equiripple characteristic, the amplitude response function is shown in in greater detail in the passband and stopband in Figures 9–32 and 9.33, respectively. These figures also show that the specification $\delta_1 = \delta_2 \leq .001$ is not satisfied with a length of $N = 65$ (the ripple is slightly larger than desired). Figure 9–34 shows the decibel amplitude response function. Figure 9–35 shows the decibel plot in the stopband in greater detail; here again it is clear that the specifications have not been satisfied, since the stopband attenuation is roughly 58 dB and at least 60 dB is desired.

In order to satisfy the specifications with respect to passband and stopband ripple, the FIR filter length was increased slightly to $N = 67$ and the program was run again. Figure 9–36 shows the resulting impulse response $h(n)$. The amplitude response function is shown in Figure 9–37; the passband and stopband details are shown in Figures 9–38 and 9–39, respectively. From these figures it is clear that the specification $\delta_1 = \delta_2 \leq .001$ is satisfied.

FIGURE 9–30
Impulse response for
Example 9–6.1.

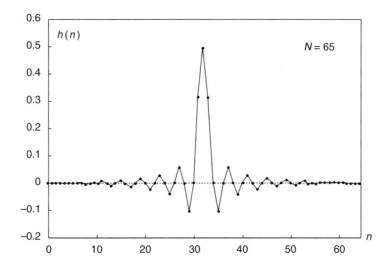

FIGURE 9–31
Amplitude response function
for Example 9–6.1.

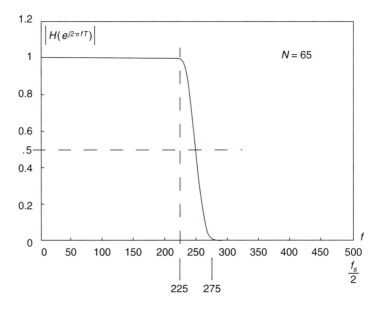

If you compare these results with Example 9–4.1, you will observe that for two FIR
lowpass filters with equivalent specifications with respect to transition bandwidth and
peak approximation error (δ), the minimum required length N is slightly less if the Parks-
McClellan method is used. However, note that the roll off characteristic in the stopband is
different for the two types of filters. In a filter designed with the Kaiser window method,
the stopband attenuation becomes greater (i.e., even better than desired) as the frequency
is increased away from the transition band; a filter designed with the Parks-McClellan algorithm does not have this roll off.

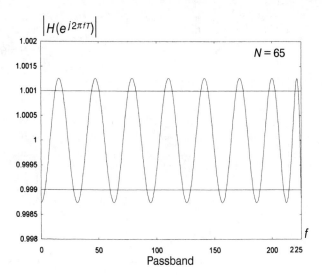

FIGURE 9–32
Passband detail for Example 9–6.1.

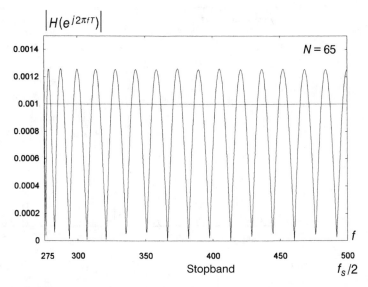

FIGURE 9–33
Stopband detail for Example 9–6.1.

The main advantage of the Parks-McClellan algorithm, as opposed to the Kaiser window method, is that the designer can specify different amounts of ripple in the passband and stopband, as illustrated by the next example.

FIGURE 9–34
Decibel amplitude response
function for Example 9–6.1.

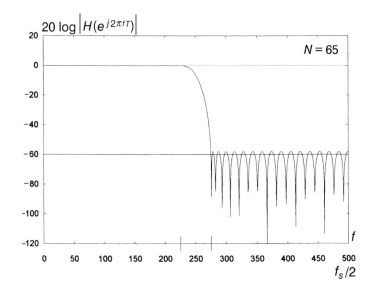

FIGURE 9–35
Stopband detail for
Example 9–6.1.

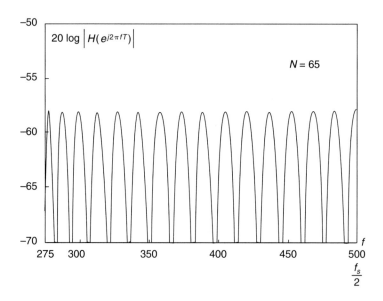

FIGURE 9–36
Impulse response for
Example 9–6.1
(*N* increased to 67).

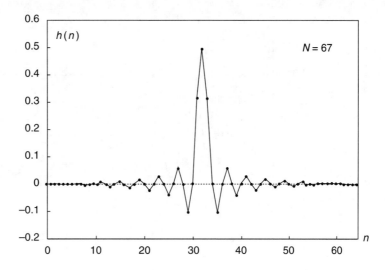

FIGURE 9–37
Amplitude response func-
tion for Example 9–6.1
(*N* increased to 67).

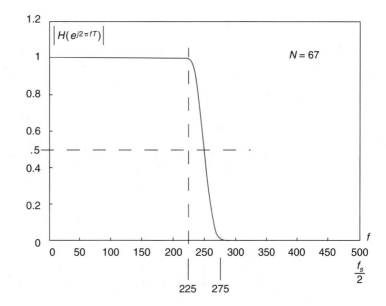

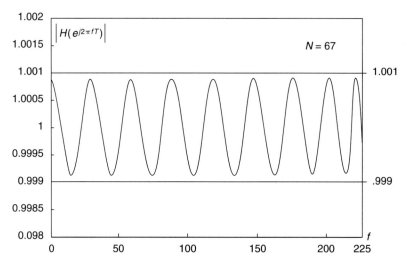

FIGURE 9–38
Passband detail for Example 9–6.1 (N increased to 67).

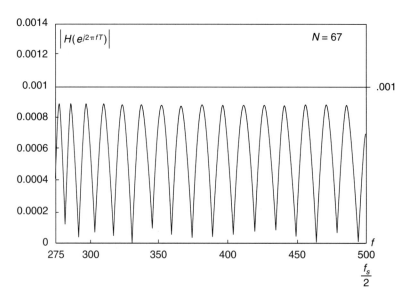

FIGURE 9–39
Stopband detail for Example 9–6.1 (N increased to 67).

EXAMPLE 9–6.2

Use the Parks-McClellan algorithm to design an FIR bandpass filter with the following specifications:

$$f_s = 1.0 \; (\text{normalized frequency specs})$$

$$\left.\begin{array}{l} f_1 = .05 \\ f_2 = .1 \\ f_3 = .25 \\ f_4 = .3 \end{array}\right\} \text{band edges}$$

$$\delta_1 = .001 \; (\text{passband ripple})$$

$$\delta_2 = .01 \; (\text{stopband ripple})$$

These specifications are illustrated in Figure 9–40. Observe that we have specified different amounts of ripple in the passband and stopbands. The approximation error weights are 1.0 for the passband and 10.0 for the stopbands. The transition bands are from .05 to 0.1 hertz and from 0.25 to 0.3 hertz. The passband is from 0.1 to 0.25 hertz. An *equivalent set of frequency specifications* is

$$f_s = 1.0$$
$$f_l = .075$$
$$f_u = .275$$
$$\Delta f = .05$$

FIGURE 9–40

Specifications for Example 9–6.2.

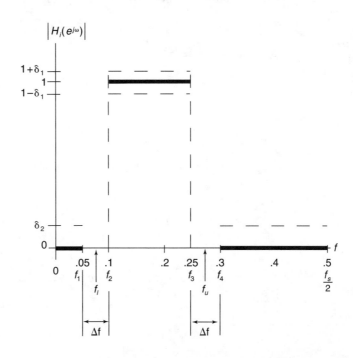

FIGURE 9–41

Impulse response for
Example 9–6.2.

FIGURE 9–41

Impulse response for
Example 9–6.2.

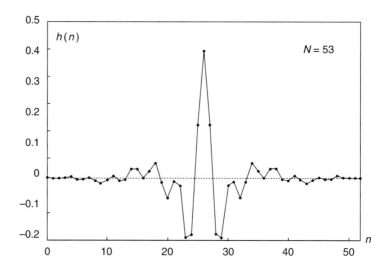

These specifications are also shown in Figure 9–40.

Since the stopband ripple has been specified as the same on both sides of the filter, and since the transition bandwidth Δf is also the same on both sides, we can still use Equation (9–34) to obtain a ballpark estimate of the required filter length N. The formula suggests $N = 52$ (after rounding to the nearest integer). However, it turns out that to satisfy the specifications for passband and stopband ripple, N must be increased to 53.

Figure 9–41 shows the resulting 53-point impulse response $h(n)$. Figure 9–42 shows the amplitude response function $|H[\exp(j2\pi fT)]|$. Examination of this plot shows that the

FIGURE 9–42

Amplitude response function
for Example 9–6.2.

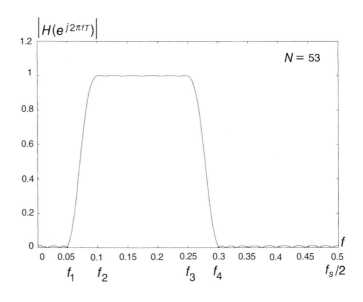

FIGURE 9–43
Passband detail for
Example 9–6.2.

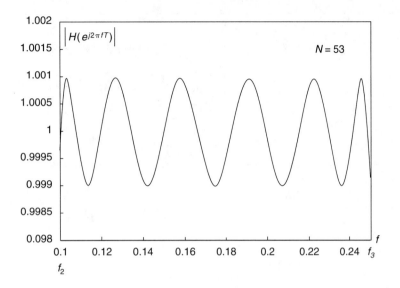

band edges are as specified. The amplitude response function in the passband is shown on
an expanded scale in Figure 9–43. Examination of this plot shows that the specification for
maximum passband ripple is satisfied. The decibel plot is shown in Figure 9–44. Since
$20\log(.01) = -40$ dB, it is clear from this plot that the specification for maximum stopband
ripple (i.e., minimum stopband attenuation) is satisfied.

FIGURE 9–44
Decibel amplitude response
function for Example 9–6.2.

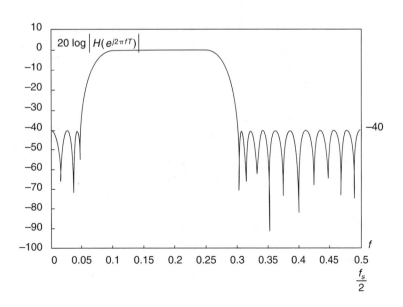

EXAMPLE 9–6.3

Use the Parks-McClellan algorithm to design an FIR bandstop filter with the following specifications:

$$f_s = 1.0 \text{ (normalized frequency specs)}$$

$$\left.\begin{aligned} f_1 &= .05 \\ f_2 &= .10 \\ f_3 &= .15 \\ f_4 &= .20 \end{aligned}\right\} \text{band edges}$$

$$\delta_1 = .01$$

$$\delta_2 = .001 \text{ (60 dB stopband attenuation)}$$

These specifications are illustrated in Figure 9–45. The approximation error weights are 10.0 for the passbands and 1.0 for the stopband. Since the passband ripple has been specified as the same on both sides of the filter, and since the transition bandwidth is $\Delta f = .05$ on both sides, we can still use Equation (9–34) to obtain an estimate for required filter length N. The formula suggests $N = 52$ (after rounding to the nearest integer), but since this is a bandstop filter, it is better to choose an odd length ($N = 53$).

Figure 9–46 shows the resulting 53-point impulse response $h(n)$. Figure 9–47 shows the amplitude response function $\left|H[\exp(j2\pi fT)]\right|$. Figure 9–48 shows the amplitude

FIGURE 9–45
Specifications for Example 9–6.3.

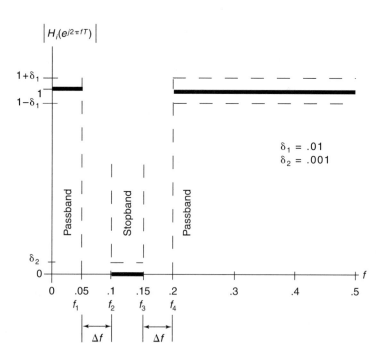

FIGURE 9–46
Impulse response for
Example 9–6.3.

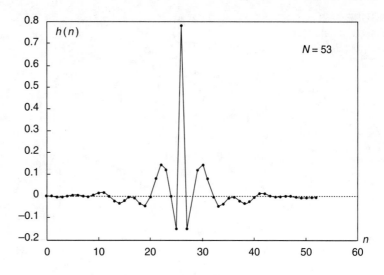

FIGURE 9–47
Amplitude response function
for Example 9–6.3.

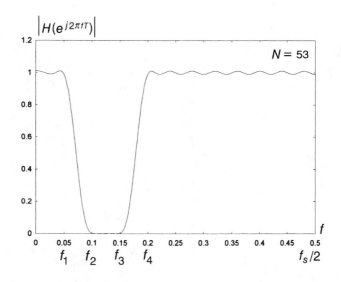

FIGURE 9–48
Passband detail for
Example 9–6.3.

FIGURE 9–48
Passband detail for
Example 9–6.3.

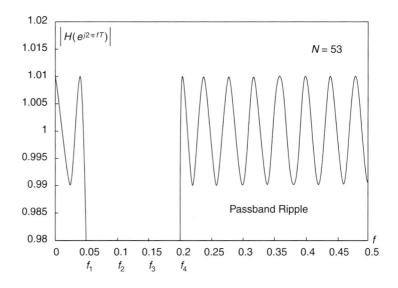

FIGURE 9–49
Decibel amplitude response
function for Example 9–6.3.

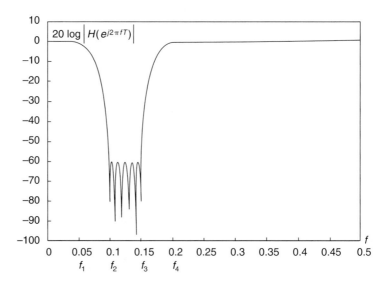

response function on a different vertical scale, which shows the detail of the passband ripple. Figure 9–49 shows the decibel plot $20\log|H[\exp(j2\pi fT)]|$. Examination of these plots shows that the specifications have been satisfied.

The Parks-McClellan algorithm is included in most digital filter design software packages. However, the details concerning how the filter specifications are presented to the software package vary; for this reason we have elected to avoid those details here.

9–7 EFFECTS OF COEFFICIENT QUANTIZATION IN FIR FILTERS

In this chapter we have discussed methods of designing FIR filters. The end result of the design process is an N-point finite-duration impulse response $h(n)$. As discussed in Chapter 5, the FIR filter is implemented with the difference equation:

$$y(n) = \sum_{k=0}^{N-1} h(k)x(n-k) \tag{9–36}$$

where $x(n)$ is the input and $y(n)$ is the output. Figure 5–10 (on page 121) shows the system block diagram for an FIR filter.

If the FIR filter difference equation is programmed on a microprocessor or DSP chip that uses fixed-point arithmetic (as opposed to floating point arithmetic), the values of the impulse response $h(n)$, that is, the coefficients $h(k)$ in Equation (9–36), must be *quantized*. For example, in an 8-bit fixed-point machine there are only 256 possible values that can be assigned to these coefficients; on a 16-bit machine there are 65,536 possible values. Let $\hat{h}(n)$ be the quantized values to be used in the fixed-point implementation. The quantized coefficients are related to the originally designed coefficients as follows:

$$\hat{h}(n) = h(n) + \varepsilon(n) \tag{9–37}$$

where $\varepsilon(n)$ is the quantization error. Let us denote the resulting frequency response function as $\hat{H}[\exp(j\omega)]$, which is related to the originally designed frequency response function $H[\exp(j\omega)]$ as follows:

$$\begin{aligned}
\hat{H}(e^{j\omega}) &= \sum_{n=0}^{N-1} \hat{h}(n)e^{-j\omega n} = \sum_{n=0}^{N-1} \left[h(n) + \varepsilon(n)\right]e^{-j\omega n} \\
&= \sum_{n=0}^{N-1} h(n)e^{-j\omega n} + \sum_{n=0}^{N-1} \varepsilon(n)e^{-j\omega n} \\
&= H(e^{j\omega}) + E(e^{j\omega})
\end{aligned} \tag{9–38}$$

In other words, as a result of coefficient quantization the originally designed frequency response function is distorted by the addition of an error function $E[\exp(j\omega)]$. However, since the quantization errors become smaller if the number of bits is increased, this error function becomes less significant if the number of bits is increased.

To illustrate the effect of coefficient quantization, consider once again the lowpass FIR filter of Example 9–4.1. In particular, consider Figure 9–14, which shows the decibel amplitude response plot obtained in the absence of coefficient quantization. Figure 9–50 shows how the decibel plot is altered if the coefficients are quantized to 16 bits (i.e., 65,536 possible values). There is a slight change in the shape of the stopband ripples, but the result is still reasonably close to what was originally specified. Figure 9–51 shows how the decibel plot is altered if the coefficients are quantized to 8 bits (i.e., 256 possible values). In this case, the distortion is quite serious. Although we still have a lowpass filter with the

FIGURE 9–50
Effect of quantizing the
filter coefficients to 16 bits.
Compare with Figure 9–14
(no quantization).

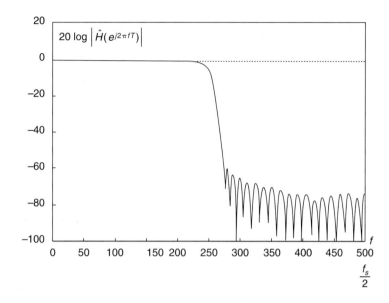

desired cutoff frequency, the amount of stopband attenuation is roughly 20 dB less than originally specified.

It is not intended to imply that 16-bit quantization never causes significant distortion in the shape of the frequency response function. Generally speaking, the larger the filter length (N), the more sensitive the frequency response function is to coefficient quantization. A more detailed consideration of this problem is beyond the scope of this book, but this section has at least served to make the reader aware of its existence.

FIGURE 9–51
Effect of quantizing the filter
coefficients to 8 bits.
Compare with Figure 9–14
(no quantization).

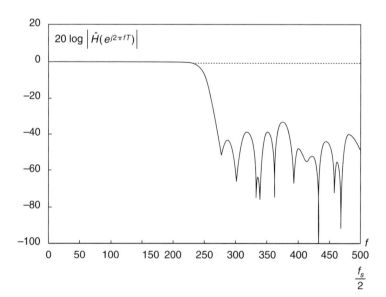

9–8 SCALING TO PREVENT OVERFLOW

Consider once again the FIR filter difference equation:

$$y(n) = \sum_{k=0}^{N-1} h(k)x(n-k) \tag{9–39}$$

where the coefficients $h(k)$ are selected using the design procedures described elsewhere in this chapter. The block diagram for this filter is shown in Figure 5–10. Recall that the input sequence $x(n)$ is created by an A/D converter and the output sequence $y(n)$ is presented to a D/A converter (see Figure 5–4 on page 112). The usual convention is that the fixed-point binary numbers presented to the processor by the A/D converter, and presented by the processor to the D/A converter represent real numbers in the range $[-1, 1)$. (The notation implies that -1 is included, but $+1$ is not.) If the processor is a fixed-point machine, only numbers in the range $[-1, 1)$ can be represented, and we must therefore ensure that no arithmetic overflow occurs at the summation node in Figure 5–10. If the processor is a floating-point machine, the output $y(n)$ must still be converted to binary numbers representing the range $[-1, 1)$ before being sent to the D/A converter; if the output of the summation node is outside of this range, an overflow condition still occurs.

Our anti-overflow strategy is to introduce a positive scaling constant c_1 into the system difference equation:

$$y(n) = \sum_{k=0}^{N-1} c_1 h(k)x(n-k) \tag{9–40}$$

The problem is to choose c_1 such that $|y(n) \le 1|$ is guaranteed. We begin by taking the absolute value on both sides of Equation (9–40):

$$|y(n)| = \left| \sum_{k=0}^{N-1} c_1 h(k)x(n-k) \right| \tag{9–41}$$

Next we appeal to the rule that states that the absolute value of a sum is less than or equal to the sum of the absolute values, that is,

$$\left| \sum_k a_k \right| \le \sum_k |a_k| \tag{9–42}$$

which allows us to write

$$|y(n)| \le \sum_{k=0}^{N-1} |c_1 h(k)x(n-k)| \tag{9–43}$$

that is,

$$|y(n)| \le c_1 \sum_{k=0}^{N-1} |h(k)||x(n-k)| \tag{9–44}$$

(Recall that $c_1 > 0$.) The next step is to use the fact that the input sequence $x(n)$ is bounded by ± 1. That is,

$$|x(n-k)| \leq 1 \qquad (9\text{–}45)$$

This allows us to write

$$|y(n)| \leq c_1 \sum_{k=0}^{N-1} |h(k)| \qquad (9\text{–}46)$$

At this point, observe that if c_1 is selected as follows:

$$c_1 = \frac{1}{\displaystyle\sum_{k=0}^{N-1} |h(k)|} \qquad (9\text{–}47)$$

then Equation (9–46) reads $|y(n)| \leq 1$, which is what we wanted to ensure.

Each filter coefficient $h(k)$ must be multiplied by scaling constant c_1. The result is the system difference equation:

$$y(n) = \sum_{k=0}^{N-1} \tilde{h}(k) x(n-k) \qquad (9\text{–}48)$$

where

$$\tilde{h}(k) = c_1 h(k) \qquad (9\text{–}49)$$

If this is to be implemented on an M-bit fixed-point machine, coefficients $h(k)$ must be quantized to 2^M possible bit patterns in the range $(-1, 1)$.

With respect to fixed-point (as opposed to floating-point) processors, there is an additional consideration. In an M-bit fixed-point processor, the result of every multiplication must be truncated to M bits. This has the effect of adding *noise* to the result of every multiplication, as shown in Figure 9–52(a). The overall result is that noise is added to the output signal, as shown in Figure 9–52(b). (The more bits there are, the smaller the noise power will be.) The noise power is independent of the scaling constant. On the other hand, the smaller the scaling constant is, the smaller the desired output signal power will be. That is, the smaller the scaling constant is, the smaller the output *signal-to-noise ratio* (SNR) will be. With this in mind, the strategy preferred by some designers is to choose a scaling constant larger than that given by Equation (9–47) but still small enough to ensure that overflow seldom occurs (although it is no longer guaranteed to never occur). In other words, the designer gives up the guarantee of no overflow ever in return for a better SNR. According to Parks and Burrus, the method often preferred is to choose a scaling constant (c_2) according to the formula

$$c_2 = \frac{1}{\max|H(e^{j\omega})|} \qquad (9\text{–}50)$$

In general, c_2 is greater than c_1. (If scaling constant c_2 is employed, replace c_1 with c_2 in Equation [9–49].)

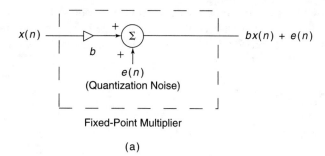

Fixed-Point Multiplier

(a)

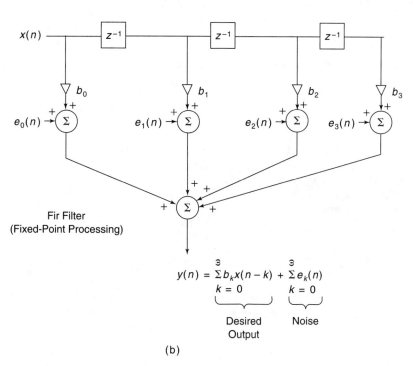

Fir Filter
(Fixed-Point Processing)

$$y(n) = \sum_{k=0}^{3} b_k x(n-k) + \sum_{k=0}^{3} e_k(n)$$

Desired   Noise
Output

(b)

FIGURE 9–52
FIR filter implemented with a fixed-point processor. Each fixed-point multipier injects
noise into the system.

It is interesting to note that for the FIR filters described in this chapter (where the
passband gain of the ideal filter is 1), Equation (9–50) reduces to

$$c_2 = \frac{1}{1+\delta} \tag{9-51}$$

where δ is the peak approximation error (ripple) discussed throughout this chapter. If δ is
relatively small (say 0.01 or less), then c_2 is approximately 1.

IIR filters must also be scaled to prevent overflow. The calculation of the required scaling constants for IIR filters is more complicated and is beyond the scope of this book. The reader is referred to the references in the Bibliography for more information.

PROBLEMS

9–1. Suppose

$$h(n) = \frac{\sin\left\{\omega_c\left[n - \left(\frac{N-1}{2}\right)\right]\right\}}{\pi\left[n - \left(\frac{N-1}{2}\right)\right]}$$

Use L'Hopital's rule to show that if N is odd,

$$h\left(\frac{N-1}{2}\right) = \frac{\omega_c}{\pi}$$

9–2. Suppose you have just designed an N-point FIR filter using one of the methods described in this chapter. You would like to obtain plots of $|H[\exp(j2\pi fT)]|$ and $20\log|H[\exp(j2\pi fT)]|$ on the interval $0 \le f \le f_s/2$ to check your design. Describe how you would use the FFT to do this. (Don't forget zero padding. Also remember that the FFT index k needs to be translated to frequency in hertz in order to make the plots clearer.)

9–3. Design an FIR lowpass filter using the Kaiser window method such that the following specifications are satisfied:

$$f_s = 11 \text{ kHz}$$
$$f_c = 3 \text{ kHz}$$
$$\Delta f = 300 \text{ Hz}$$
$$\text{Stopband attenuation} = 80 \text{ dB}$$

Specify the filter length, the Kaiser window tapering parameter (β), and the impulse response of the filter. Make a rough sketch of what you expect $20\log|H[\exp(j2\pi fT)]|$ to look like, showing f_c and Δf (in hertz), and also showing the stopband attenuation.

9–4. Find the *normalized* frequency specifications for the filter of Problem 9–3.

9–5. With respect to Problem 9–3, if the filter length (N) is made smaller than the length called for in the original design, what happens to the stopband attenuation and transition bandwidth?

9–6. Consider the ideal multiple bandpass filter shown in Figure P9–6. Find the impulse response of this filter.

HINT: Lowpass + Bandpass + Bandpass + Highpass.

9–7. Design an FIR bandpass filter using the Kaiser window method such that the following specifications are satisfied:

$$f_s = 10000 \text{ Hz}$$
$$f_l = 2000 \text{ Hz (lower cutoff frequency)}$$
$$f_u = 2500 \text{ Hz (upper cutoff frequency)}$$
$$\Delta f = 400 \text{ Hz}$$
$$\text{peak approximation error} = .0005$$

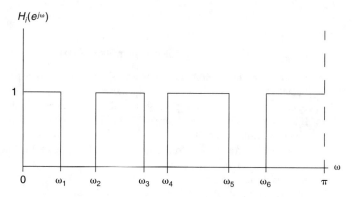

$H_i(e^{j\omega})$

FIGURE P9-6

Specify the filter length, the Kaiser window tapering parameter (β), and the impulse response of the filter. Make a rough sketch of what you expect $20\log|H[\exp(j2\pi fT)]|$ to look like, showing the cutoff frequencies, the transition bandwidths, and the attenuation in the stopbands.

9–8. Design an FIR highpass filter using the Kaiser window method such that the following specifications are satisfied:

$$f_s = 1 \text{ Hz}$$
$$f_c = 0.3 \text{ Hz}$$
$$\Delta f = 0.04 \text{ Hz}$$
$$\text{Stopband attenuation} = 40 \text{ dB}$$

a. Specify the filter length, the Kaiser window tapering parameter (β), and the impulse response of the filter. Make a rough sketch of what you expect $20\log|H[\exp(j2\pi fT)]|$ to look like, showing the cutoff frequency, the transition bandwidth, and the stopband attenuation.

b. If the sampling frequency is actually $f_s = 10$ kHz, what is the cutoff frequency and transition bandwidth in hertz?

9–9. Design an FIR bandstop filter using the Kaiser window method such that the following specifications are satisfied:

$$f_s = 48 \text{ kHz}$$
$$f_1 = 10 \text{ kHz (lower cutoff frequency)}$$
$$f_2 = 16 \text{ kHz (upper cutoff frequency)}$$
$$\Delta f = 2 \text{ kHz}$$
$$\delta = .0002$$

Specify the filter length, the Kaiser window tapering parameter (β), and the filter impulse response. Make a rough sketch of what you expect $20\log|H[\exp(j2\pi fT)]|$ to look like, showing $f_1, f_2, \Delta f$, and the stopband attenuation.

9–10. An FIR lowpass filter is designed using the Parks-McClellan algorithm. The specifications are

$$f_s = 1 \text{ Hz}$$
$$f_c = 0.2 \text{ Hz}$$

$$\Delta f = 0.04 \text{ Hz}$$
$$\text{Passband}: \delta = .0001$$
$$\text{Stopband attenuation} = 40 \text{ dB}$$

a. Find a filter length (N) that will come close to satisfying the specifications.
b. Specify the transition band edge frequencies.
c. Specify approximation error weights for the passband and stopband.
d. If the actual sampling frequency is $f_s = 25$ kHz, find the cutoff frequency and the transition bandwidth of the filter.
e. Describe the effect on the frequency response function of this filter if
  i. N is increased.
  ii. N is decreased.
f. Make a rough sketch of $20\log|H[\exp(j2\pi fT)]|$.

9–11. An FIR bandpass filter is designed using the Parks-McClellan algorithm. The specifications are

$$f_s = 1 \text{ Hz}$$
$$f_l = 0.2 \text{ Hz (lower cutoff frequency)}$$
$$f_u = 0.4 \text{ Hz (upper cutoff frequency)}$$
$$\Delta f = 0.02 \text{ Hz (on both sides of the passband)}$$
$$\text{Passband}: \delta = .001$$
$$\text{Stopbands: attenuation} = 80 \text{ dB}$$

a. Find a filter length (N) that will come close to satisfying the specifications.
b. Specify the edge frequencies for the transition bands.
c. Specify approximation error weights for the passband and stopbands.
d. If the actual sampling frequency is $f_s = 40$ kHz, find the cutoff frequencies and transition bandwidths of this filter.
e. Describe the effect on the frequency response function of this filter if:
  i. N is increased.
  ii. N is decreased.
f. Make a rough sketch of $20\log|H[\exp(j2\pi fT)]|$.

9–12. Suppose the following FIR filter has been designed:

$$h(n) = \frac{1}{4}\delta(n) - \frac{1}{2}\delta(n-1) + \delta(n-2) - \frac{1}{2}\delta(n-3) + \frac{1}{4}\delta(n-4)$$

The input to this filter is $x(n)$. The filter output is $y(n)$. Given that $|x(n)| \le 1$ is guaranteed, find the filter scaling constant that will guarantee $|y(n)| \le 1$. (The trivial solution $c = 0$ is excluded.)

9–13. Consider an FIR filter with impulse response $h(n)$ and length N. Show that if N is an *even* integer and $h(n)$ is symmetric about its midpoint (i.e., $h(n) = h(N-1-n)$), the frequency response function of this filter has the property that when $\omega = \pi$, $H[\exp(j\omega)] = 0$. In other words, $H[\exp(j\pi)] = 0$.

HINT: Try a simple example, such as $N = 4$.

10

DESIGN OF IIR FILTERS USING THE BILINEAR TRANSFORMATION

10–1   INTRODUCTION

As discussed in Chapter 5, an infinite impulse response (IIR) discrete time system is characterized by the inclusion of *feedback* terms in the system difference equation:

$$y(n) = \sum_{k=0}^{M} b_k x(n-k) - \sum_{k=1}^{N} a_k y(n-k) \qquad (10\text{--}1)$$

As also shown in Chapter 5, the transfer function for this system is

$$H(z) = \frac{\displaystyle\sum_{k=0}^{M} b_k z^{-k}}{1 + \displaystyle\sum_{k=1}^{N} a_k z^{-k}} \qquad (10\text{--}2)$$

It is desirable to express higher order IIR filter transfer functions as the product of first- and second-order transfer functions (or *sections*), that is, in *cascade* form. Let N be the filter order. If N is even, $H(z)$ can be expressed as

$$H(z) = \prod_{k=1}^{\frac{N}{2}} \frac{b_{0k} + b_{1k}z^{-1} + b_{2k}z^{-2}}{1 + a_{1k}z^{-1} + a_{2k}z^{-2}} \qquad (10\text{--}3)$$

If N is odd, $H(z)$ can be expressed as

$$H(z) = \frac{b_{00} + b_{10}z^{-1}}{1 + a_{10}z^{-1}} \prod_{k=1}^{\frac{N-1}{2}} \frac{b_{0k} + b_{1k}z^{-1} + b_{2k}z^{-2}}{1 + a_{1k}z^{-1} + a_{2k}z^{-2}} \qquad (10\text{--}4)$$

In cascade form, an IIR filter is realized by several difference equations *operating in sequence*. The first-order section in Equation (10–4) has the following difference equation:

$$y(n) = b_{00}x(n) + b_{10}x(n-1) - a_{10}y(n-1) \qquad \text{(10–5)}$$

where $x(n)$ is the input to this section and $y(n)$ is the output from this section. The kth second-order section in Equations (10–3) and (10–4) has the following difference equation:

$$y(n) = b_{0k}x(n) + b_{1k}x(n-1) + b_{2k}x(n-2) - a_{1k}y(n-1) - a_{2k}y(n-2) \qquad \text{(10–6)}$$

where $x(n)$ is the input to the kth section, and $y(n)$ is the output from the kth section. In cascade form, the output of the first difference equation becomes the input to the second difference equation, etc.

There are at least two good reasons why we want to express $H(z)$ in cascade form:

1. It can be shown that the cascade realization is less sensitive to coefficient roundoff than the direct realization of Equation (10–1).
2. The design method discussed in this chapter (bilinear transformation) is much easier to apply if the transfer function is in cascade form.

The design problem is to choose the coefficients (b_{ik} and a_{ik}) such that the frequency response function $H[\exp(j\omega)]$ has the desired characteristics (e.g., lowpass filter, bandpass filter, etc.).

For a given degree of sharpness with respect to the roll-off characteristics of the filter for frequencies outside of the passband, an IIR filter difference equation will require fewer terms than an FIR filter difference equation (there are $M + N + 1$ terms in Equation [10–1]). Thus IIR filters tend to be more efficient than FIR filters in a computational sense. On the other hand, FIR filters can have linear phase (the design methods described in Chapter 9 always result in linear phase filters), but IIR filters do not have this desirable characteristic. In fact, when using the IIR filter design method described in this chapter, it is only the qualities of the amplitude response function $|H[\exp(j\omega)]|$ that explicitly concern us, and we are tacitly assuming that *for the intended application* it is of no real concern that the phase response function might be highly nonlinear.

10–2 THE BILINEAR TRANSFORMATION

Consider a continuous time (i.e., analog) filter with transfer function $\hat{H}(s)$.

NOTE: We are placing the symbol $\wedge$ over the H to distinguish this transfer function from the discrete time transfer function $H(z)$. Information on $\hat{H}(s)$ for Butterworth, Chebychev, and elliptic filters is well-tabulated in the literature. Information for Butterworth and Chebyshev lowpass and highpass filters up to order $N = 8$ (in cascade form) is found in Chapter 4. (All of these continuous time filters are IIR types since the transfer functions have poles).

If faced with the task of designing a discrete time IIR filter, it is logical to ask if there is a way to take advantage of the vast store of available information on continuous time filters. For example, if the task at hand is to design a discrete time IIR lowpass filter, you might ask if there is a way to take a well-known continuous time lowpass filter transfer function $\hat{H}(s)$ and transform it to the desired $H(z)$. The answer to this question is yes, and there are several theoretical methods discussed in the DSP literature. The most popular method is known as the *bilinear transformation*.

Explaining the bilinear transformation requires mathematical manipulation of transfer functions and frequency response functions from both the continuous time world and the discrete time world. As an aid to the reader, the notation used in this chapter is summarized in Table 10–1. One thing that should be emphasized right at the start is that the discrete time filter and continuous time filter that are related through the bilinear transformation will have *different cutoff frequencies*. For lowpass and highpass filters, the cutoff frequency (in hertz) of the discrete time filter is denoted f_c, while the cutoff frequency of the corresponding continuous time filter is denoted $\hat{f}_c$.

The bilinear transformation is given by

$$H(z) = \hat{H}(s)\Big|_{s=\left(\frac{1-z^{-1}}{1+z^{-1}}\right)} \tag{10–7}$$

The notation implies that every place that s appears in the original continuous time transfer function, it is *replaced* with $(1 - z^{-1})/(1 + z^{-1})$. (If this concept is not clear, take a look ahead at Equations [10–19] and [10–26], both of which illustrate the idea.) The result in the frequency domain is that $|H[\exp(j\omega)]|$ retains the essential qualities of $|H_a(f)|$. For example, if $\hat{H}(s)$ is a Butterworth lowpass filter, $H(z)$ will also be a lowpass filter, and

TABLE 10–1
Notation Used in Chapter 10

Continuous Time (Analog) Systems

$\hat{H}(s)$: Transfer function

$H_a(f)$: Frequency response function $H_a(f) = \hat{H}(j2\pi f)$

   f : Frequency in hertz (Hz)

   $\hat{f}_c$: Cutoff frequency in Hz

   Ω : Frequency in radians per second $\Omega = 2\pi f$

   Ω_c : Cutoff frequency in radians per second $\Omega_c = 2\pi \hat{f}_c$

Discrete Time Systems

   $H(z)$: Transfer function

$H(e^{j\omega})$: Frequency response function

   f : Frequency in hertz (Hz)

   f_c : Cutoff frequency in Hz

   f_s : Sampling frequency in Hz

   ω : Frequency in radians per sample $\omega = \dfrac{2\pi f}{f_s}$

   ω_c : Cutoff frequency (on ω scale) $\omega_c = \dfrac{2\pi f_c}{f_s}$

$|H[\exp(j\omega)]|$ will have no ripples in the passband or stopband. However, the shape of the roll-off characteristic from passband to stopband will undergo some subtle changes, for reasons described below.

It can be shown (see Appendix C) that the frequency response function $H[\exp(j\omega)]$ of a discrete time IIR filter created using the bilinear transformation is related to the frequency response function $H_a(f)$ of the continuous time filter that was bilinearly transformed as follows:

$$H\left(e^{j\omega}\right) = H_a(f)\Big|_{f=\frac{1}{2\pi}\tan\left(\frac{\omega}{2}\right)} \tag{10–8}$$

Thus $H[\exp(j\omega)]$ is essentially a *frequency axis-warped* version of $H_a(f)$; the frequency axis-warping function is

$$f = \frac{1}{2\pi}\tan\left(\frac{\omega}{2}\right) \tag{10–9}$$

This warping function is plotted in Figure 10–1. The entire f axis, from $-\infty < f < \infty$, is mapped to the interval $-\pi < \omega < \pi$ on the ω axis. (Only positive frequencies are shown in Figure 10–1.) In effect, the entire shape of $H_a(f)$, defined on an infinite interval $-\infty < f < \infty$, must be squashed so that it fits into the finite interval $-\pi < \omega < \pi$. (The resulting $H[\exp(j\omega)]$ is periodically extended outside of this interval.) For positive frequencies, Figure 10–1 shows that this squashing is relatively linear for the frequency range $0 \le f < 0.159$, which is mapped to the interval $0 \le \omega < \pi/2$. For frequencies greater than $f = 0.159$, the squashing is very nonlinear.

Assuming we are talking about conventional lowpass filters, bandpass filters, etc., this squashing operation will preserve the essential qualities of $|H_a(f)|$. For instance, a squashed lowpass filter will still look like a lowpass filter. Consider a plot of a lowpass $|H_a(f)|$ drawn on a special surface that is deformable in the horizontal direction (i.e., in the direction of the frequency axis) but rigid in the vertical direction. If you squash the surface in the horizontal direction in the manner dictated by the frequency warping function, the plot will still look like a lowpass filter. If $|H_a(f)|$ is the shape of an ideal lowpass filter, it will still look like an ideal lowpass filter after being squashed because a squashed rectangle is still a rectangle. If $|H_a(f)|$ is a real-world lowpass filter, the passband-to-stopband roll-off characteristic will be altered (in general it will actually be sharper), but it will still look like a lowpass filter. If $|H_a(f)|$ originally had ripples in the passband and/or stopband, the squashed version will still have these ripples. If $|H_a(f)|$ has no ripples, the squashed version won't have ripples either.

Equation (10–9) shows that the relationship between the cutoff frequency of a discrete time filter (ω_c) and the cutoff frequency of the continuous time filter that was bilinearly transformed ($\hat{f}_c$) are related as follows:

$$\hat{f}_c = \frac{1}{2\pi}\tan\left(\frac{\omega_c}{2}\right) \tag{10–10}$$

Recall that in terms of frequency in hertz, the cutoff frequency (f_c) of a discrete time filter is related to ω_c as follows:

$$\omega_c = 2\pi\frac{f_c}{f_s} \tag{10–11}$$

FIGURE 10–1
Frequency warping function
associated with the bilinear
transformation.

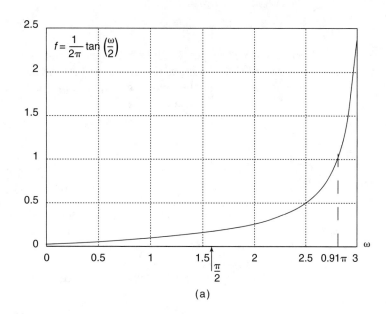

(a)

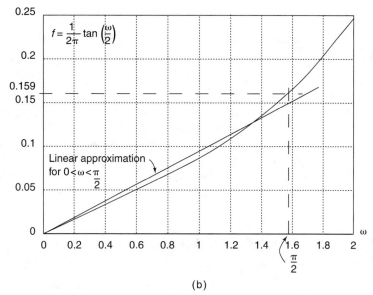

(b)

where f_s is the sampling frequency. By substituting Equation (10–11) into Equation (10–10), we find the relationship between the cutoff frequency (f_c) of the discrete time filter and the cutoff frequency $(\hat{f}_c)$ of the continuous time filter that was bilinearly transformed:

$$\hat{f}_c = \frac{1}{2\pi} \tan\left(\frac{\pi f_c}{f_s}\right) \qquad (10\text{–}12)$$

Note that $\hat{f}_c$ and f_c are *not* the same.

The cutoff frequency of a continuous time filter is often expressed in radians per second, that is, $\Omega_c = 2\pi\hat{f}_c$. (For example, see the discussion on Butterworth and Chebyshev lowpass and highpass filters in Chapter 4.) This can be substituted into Equations (10–10) and (10–12) to obtain the following relationships:

$$\Omega_c = \tan\left(\frac{\omega_c}{2}\right) \tag{10–13}$$

$$\Omega_c = \tan\left(\frac{\pi f_c}{f_s}\right) \tag{10–14}$$

If ω_c is specified in the design problem, Equation (10–13) is used to find the required value of Ω_c; if f_c and f_s are specified, Equation (10–14) is used instead. In either case, the process of determining Ω_c, given the desired cutoff frequency of the discrete time filter, is known as *prewarping* the frequency specifications.

For example, suppose the specifications for the discrete time filter are

$$\begin{array}{l} \text{Sampling frequency}: f_s = 1 \text{ KHz} \\ \text{Cutoff frequency}: f_c = 100 \text{ Hz} \end{array} \tag{10–15}$$

Then the continuous time filter to be bilinear transformed must have a cutoff frequency Ω_c of

$$\Omega_c = \tan\left(\frac{\pi \times 100}{1000}\right) = .3249 \tag{10–16}$$

NOTE: It is important to put your calculator in *radians* mode, not degrees, when evaluating the tangent function.

In this example the continuous time filter you have *in theory* must have a cutoff frequency of only .0517 hertz!. But you are not going to actually build this filter; you are only going to find the theoretical transfer function $\hat{H}(s)$ for this filter and then apply the bilinear transformation (Equation [10–7]).

10–3 THE DESIGN PROBLEM

The design problem considered here can be stated as follows: Given the transfer function $\hat{H}(s)$ of a particular type of continuous time filter (e.g., third-order Butterworth lowpass filter), use the bilinear transformation to design an IIR discrete time filter of the same type, with specified critical frequencies (e.g., for a lowpass filter the critical frequency is the cutoff frequency ω_c, which is a number between 0 and π). (The problem of determining the required filter order (N) will be considered in Section 10–5.)

The first step is always to *prewarp* the specified critical frequencies of the discrete time filter to determine the required critical frequencies of the continuous time filter to be bilinearly transformed. The prewarping formula is Equation (10–13) or Equation (10–14).

For lowpass and highpass filters there is only one critical frequency (ω_c) to prewarp. For bandpass or bandstop filters there are two critical frequencies: lower cutoff frequency ω_l and upper cutoff frequency ω_u, which become Ω_l and Ω_u after prewarping (these are the lower and upper cutoff frequencies of the continuous time bandpass or bandstop filter to be bilinearly transformed).

The second step is to take the given $\hat{H}(s)$ (in which the prewarped critical frequencies appear as parameters) and apply the bilinear transformation: every s is replaced with $(1 - z^{-1})/(1 + z^{-1})$. For filters of higher than second order, $\hat{H}(s)$ should be in cascade form (see Chapter 4) so that the bilinear transformation can be applied to each first- and second-order section separately. The result (after some algebra) will be a cascade form transfer function $H(z)$, as shown in Equations (10–3) or (10–4). The bilinear transformation preserves filter order: if $\hat{H}(s)$ is Nth order, the resulting $H(z)$ will also be Nth order.

The third step is to find the system difference equations. (When $H(z)$ is in cascade form, each section has its own difference equation.) The difference equations for first and second order sections are given in Equations (10–5) and (10–6). The coefficients b_{ik} and a_{ik} will depend on the prewarped critical frequencies and other fixed parameters appearing in the original $\hat{H}(s)$.

The general method outlined here is applicable to many filter types: lowpass, highpass, bandpass, and bandstop filters of the Butterworth, Chebychev, and elliptic varieties. However, in the examples that follow we will consider only Butterworth and Chebyshev lowpass filters. The transfer functions and frequency response characteristics of continuous time Butterworth and Chebyshev lowpass filters are covered in Chapter 4. (Recall from Chapter 4 that for Butterworth filters, $\Omega_k = \Omega_c$ in all filter sections. Keep this in mind when comparing the transfer functions described in the following examples with the transfer functions shown in Chapter 4.)

10–4 EXAMPLES

EXAMPLE 10–4.1

Design a first-order discrete time IIR lowpass filter based on the bilinear transformation of a first-order Butterworth lowpass filter. The specifications are

Passband gain = 1

Cutoff frequency = 2 KHz

Sampling frequency = 10 KHz

(Note that the cutoff frequency must be less than one-half of the sampling frequency.)
The prewarped cutoff frequency is obtained using Equation (10–14):

$$\Omega_c = \tan\left(\frac{\pi f_c}{f_s}\right) = \tan\left(\frac{\pi \times 2000}{10000}\right) = 0.72654 \qquad \textbf{(10–17)}$$

From Chapter 4, the transfer function of a first-order Butterworth lowpass filter with pass-band gain = 1 and cutoff frequency Ω_c is

$$\hat{H}(s) = \frac{\Omega_c}{s + \Omega_c} \tag{10–18}$$

To find $H(z)$, we use the bilinear transformation:

$$H(z) = \hat{H}(s)\Big|_{s=\frac{1-z^{-1}}{1+z^{-1}}} = \frac{\Omega_c}{\left(\dfrac{1-z^{-1}}{1+z^{-1}}\right) + \Omega_c} \tag{10–19}$$

After some algebraic manipulation, $H(z)$ can be expressed as

$$H(z) = \frac{\left(\dfrac{\Omega_c}{1+\Omega_c}\right) + \left(\dfrac{\Omega_c}{1+\Omega_c}\right)z^{-1}}{1 + \left(\dfrac{\Omega_c - 1}{\Omega_c + 1}\right)z^{-1}} \tag{10–20}$$

$H(z)$ is now in standard form, that is,

$$H(z) = \frac{b_0 + b_1 z^{-1}}{1 + a_1 z^{-1}} \tag{10–21}$$

where coefficients b_0, b_1, and a_1 are given by the formulas

$$\Omega_c = \tan\left(\frac{\pi f_c}{f_s}\right)$$

$$a_0 = \Omega_c + 1$$

$$a_1 = \frac{(\Omega_c - 1)}{a_0} \tag{10–22}$$

$$b_0 = \frac{\Omega_c}{a_0}$$

$$b_1 = \frac{\Omega_c}{a_0} = b_0$$

The difference equation for this filter is

$$y(n) = b_0 x(n) + b_1 x(n-1) - a_1 y(n-1) \tag{10–23}$$

where $x(n)$ is the input and $y(n)$ is the output.

The amplitude response function $|H[\exp(j2\pi fT)]|$ is shown in Figure 10–2.

FIGURE 10–2

Amplitude response function for Example 10–4.1.

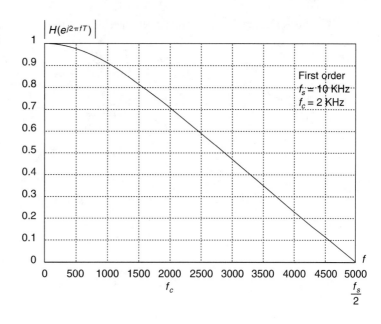

EXAMPLE 10–4.2

Design a second-order discrete time IIR lowpass filter based on the bilinear transformation of a second-order Butterworth lowpass filter. The specifications are

Passband gain = 1

Cutoff frequency = 2 KHz

Sampling frequency = 10 KHz

The prewarped cutoff frequency is obtained using Equation (10–14):

$$\Omega_c = \tan\left(\frac{\pi \times 2000}{10000}\right) = 0.72654 \tag{10–24}$$

From Chapter 4, the transfer function of a second order Butterworth lowpass filter with passband gain = 1 and cutoff frequency Ω_c is

$$\hat{H}(s) = \frac{\Omega_c^2}{s^2 + \Psi\Omega_c s + \Omega_c^2} \tag{10–25}$$

where $\Psi = \sqrt{2}$ (see Table 4–1, page 82). To find $H(z)$, we use the bilinear transformation:

$$H(z) = \hat{H}(s)\Big|_{s=\frac{1-z^{-1}}{1+z^{-1}}} = \frac{\Omega_c^2}{\left(\dfrac{1-z^{-1}}{1+z^{-1}}\right)^2 + \Psi\Omega_c\left(\dfrac{1-z^{-1}}{1+z^{-1}}\right) + \Omega_c^2} \tag{10–26}$$

After some algebraic manipulation, the standard form is obtained:

$$H(z) = \frac{b_0 + b_1 z^{-1} + b_2 z^{-2}}{1 + a_1 z^{-1} + a_2 z^{-2}} \tag{10–27}$$

where the formulas for the coefficients are

$$\Psi = \sqrt{2} \text{ (from Table 4 – 1)}$$

$$\Omega_c = \tan\left(\frac{\pi f_c}{f_s}\right)$$

$$a_0 = 1 + \Psi\Omega_c + \Omega_c^2$$

$$a_1 = \frac{\left(2\Omega_c^2 - 2\right)}{a_0}$$

$$a_2 = \frac{\left(1 - \Psi\Omega_c + \Omega_c^2\right)}{a_0} \tag{10–28}$$

$$b_0 = \frac{\Omega_c^2}{a_0}$$

$$b_1 = \frac{2\Omega_c^2}{a_0} = 2b_0$$

$$b_2 = \frac{\Omega_c^2}{a_0} = b_0$$

The difference equation for this filter is

$$y(n) = b_0 x(n) + b_1 x(n-1) + b_2 x(n-2) - a_1 y(n-1) - a_2 y(n-2) \tag{10–29}$$

The amplitude response function $|H[\exp(j2\pi fT)]|$ is shown in Figure 10–3.

It is interesting to consider this second-order example with different cutoff frequencies (same sampling frequency). Figure 10–4 shows $|H[\exp(j2\pi fT)]|$ for cutoff frequencies f_c = 1000 Hz, 2500 Hz, and 4000 Hz. The shape of the roll-off characteristic is somewhat different for each case, even though all three filters are based on a second-order Butterworth prototype. The difference in shape is due to the nonlinear character of the frequency axis-warping function.

FIGURE 10–3
Amplitude response function for Example 10–4.2.

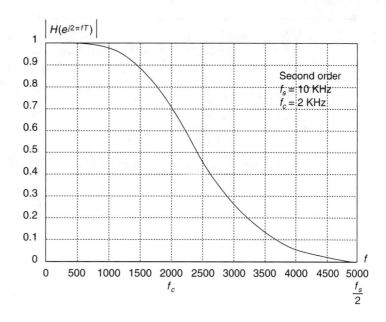

FIGURE 10–4
Amplitude response functions for second-order IIR lowpass filters designed by applying the bilinear transformation to analog Butterworth lowpass filters.

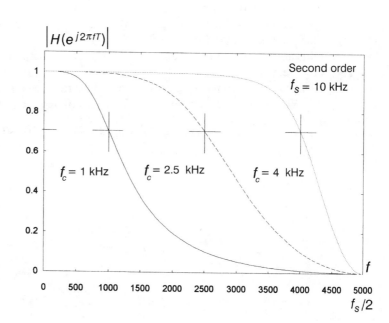

EXAMPLE 10–4.3

Design a fifth-order discrete time IIR lowpass filter based on the bilinear transformation of a fifth-order Butterworth lowpass filter. The specifications are

> Passband gain = 1
> Cutoff frequency = 2 KHz
> Sampling frequency = 10 KHz

The prewarped cutoff frequency is found using Equation (10–14):

$$\Omega_c = \tan\left(\frac{\pi \times 2000}{10000}\right) = .72654 \tag{10–30}$$

From Chapter 4, the transfer function of a fifth-order Butterworth lowpass filter with a passband gain of 1 is (in cascade form):

$$\hat{H}(s) = \left(\frac{\Omega_c}{s + \Omega_c}\right)\left(\frac{\Omega_c^2}{s^2 + \Psi_1\Omega_c s + \Omega_c^2}\right)\left(\frac{\Omega_c^2}{s^2 + \Psi_2\Omega_c s + \Omega_c^2}\right) \tag{10–31}$$

where $\Psi_1 = 0.618$ and $\Psi_2 = 1.618$ (from Table 4–1).

As discussed earlier, if the filter order is three or more we want to realize the discrete time filter in cascade form, as shown in Figures 10–5(a) and 10–5(b). Thus we apply the bilinear transformation to Equation (10–31) section by section. These first- and second-order sections have the same form as the previous two examples; therefore we can use the results obtained from these examples to find the required difference equations here.

The first-order section is realized by the difference equation:

$$p(n) = b_{00}x(n) + b_{10}x(n-1) - a_{10}p(n-1) \tag{10–32}$$

where

$$a_{00} = \Omega_c + 1$$
$$a_{10} = \frac{(\Omega_c - 1)}{a_{00}}$$
$$b_{00} = \frac{\Omega_c}{a_{00}} \tag{10–33}$$
$$b_{10} = \frac{\Omega_c}{a_{00}} = b_{00}$$

The difference equation for the first second-order section (i.e., the section with $\Psi_1 = 0.618$) is

$$r(n) = b_{01}p(n) + b_{11}p(n-1) + b_{21}p(n-2) - a_{11}r(n-1) - a_{21}r(n-2) \tag{10–34}$$

The difference equation for the second second-order section (where $\Psi_2 = 1.618$) is

$$y(n) = b_{02}r(n) + b_{12}r(n-1) + b_{22}r(n-2) - a_{12}y(n-1) - a_{22}y(n-2) \tag{10–35}$$

The coefficients b_{ik} and a_{ik} for the second-order sections ($k = 1, 2$) can be obtained with the following formulas:

$$a_{0k} = 1 + \Psi_k \Omega_c + \Omega_c^2$$

$$a_{1k} = \frac{2\Omega_c^2 - 2}{a_{0k}}$$

$$a_{2k} = \frac{1 - \Psi_k \Omega_c + \Omega_c^2}{a_{0k}}$$

$$b_{0k} = \frac{\Omega_c^2}{a_{0k}}$$

$$b_{1k} = 2b_{0k}$$

$$b_{2k} = b_{0k}$$

(10–36)

This filter is implemented by reading $x(n)$ and calculating $p(n)$, $r(n)$, and $y(n)$ in succession (Equations [10–32], [10–34], and [10–35]) during each iteration (n). The shift registers containing past values of $x(n)$, $p(n)$, $r(n)$, and $y(n)$ must be updated before incrementing n and going to the next iteration. The block diagram of this system is shown in Figure 10–5(b). (It is possible to find other sets of difference equations to implement a system such as the one described here. In other words, there are different *structures*, or block diagrams, that will achieve the same result. Alternate second-order structures will be considered in Section 10–6.)

The amplitude response function $|H[\exp(j2\pi fT)]|$ of this fifth-order filter is shown in Figure 10–6. The decibel plot $20\log|H[\exp(j2\pi fT)]|$ is shown in Figure 10–7. The impulse response $h(n)$ is shown in Figure 10–8.

Figure 10–9 shows $|H[\exp(j2\pi fT)]|$ for fifth-order lowpass filters designed using the method discussed previously with cutoff frequencies of $f_c = 1000$ Hz, 2500 Hz, and 4000 Hz.

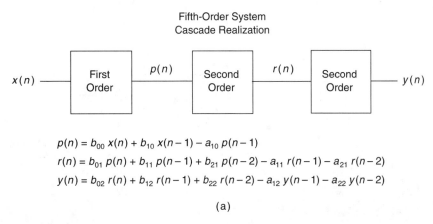

Fifth-Order System
Cascade Realization

$p(n) = b_{00}\, x(n) + b_{10}\, x(n-1) - a_{10}\, p(n-1)$

$r(n) = b_{01}\, p(n) + b_{11}\, p(n-1) + b_{21}\, p(n-2) - a_{11}\, r(n-1) - a_{21}\, r(n-2)$

$y(n) = b_{02}\, r(n) + b_{12}\, r(n-1) + b_{22}\, r(n-2) - a_{12}\, y(n-1) - a_{22}\, y(n-2)$

(a)

FIGURE 10–5
(a) Cascade realization for Examples 10–4.3 and 10–4.4. (b) Block diagrams for Examples 10–4.3 and 10–4.4. (See text for values of b_{ik} and a_{ik}.)

Fifth-Order System
Cascade Realization

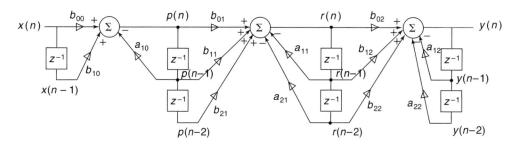

$$p(n) = b_{00}\,x(n) + b_{10}\,x(n-1) - a_{10}\,p(n-1)$$

$$r(n) = b_{01}\,p(n) + b_{11}\,p(n-1) + b_{21}\,p(n-2) - a_{11}\,r(n-1) - a_{21}\,r(n-2)$$

$$y(n) = b_{02}\,r(n) + b_{12}\,r(n-1) + b_{22}\,r(n-2) - a_{12}\,y(n-1) - a_{22}\,y(n-2)$$

(b)

FIGURE 10–5

(continued)

FIGURE 10–6
Amplitude response function
for Example 10–4.3.

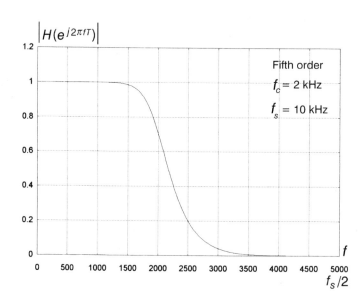

FIGURE 10–7
Decibel amplitude response
function for Example 10–4.3.

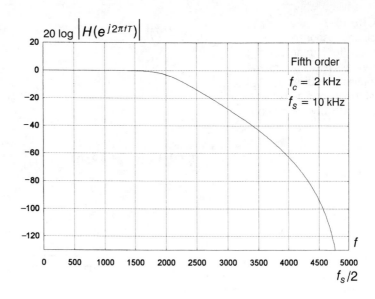

FIGURE 10–8
Impulse response for
Example 10–4.3.

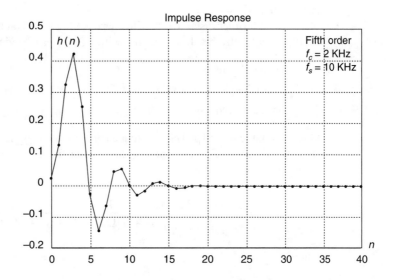

FIGURE 10–9

Amplitude response functions for fifth-order IIR lowpass filters designed by applying the bilinear transformation to analog Butterworth lowpass filters.

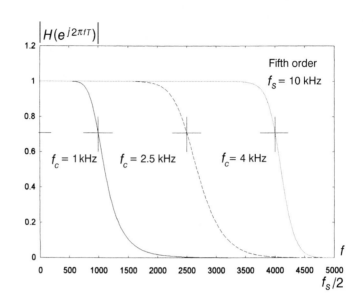

EXAMPLE 10–4.4

Design a fifth-order discrete time IIR lowpass filter based on the bilinear transformation of a fifth-order Chebyshev lowpass filter with 1 dB of passband ripple. The specifications are

Passband gain = 1

Cutoff frequency = 2 kHz.

Sampling frequency = 10 kHz.

Once again, the prewarped cutoff frequency is found using Equation (10–14) and turns out to be $\Omega_c = .72654$. From Chapter 4, the transfer function of a fifth-order continuous time Chebyshev lowpass filter is

$$\hat{H}(s) = \left(\frac{\Omega_0}{s + \Omega_0}\right)\left(\frac{\Omega_1^2}{s^2 + \Psi_1\Omega_1 s + \Omega_1^2}\right)\left(\frac{\Omega_2^2}{s^2 + \Psi_2\Omega_2 s + \Omega_2^2}\right) \qquad (10\text{–}37)$$

where

$$\Psi_1 = 0.7149026$$
$$\Psi_2 = 0.1799713$$
$$\Omega_0 = |p_0|\Omega_c = 0.2894934\Omega_c$$
$$\Omega_1 = |p_1|\Omega_c = 0.6552083\Omega_c$$
$$\Omega_2 = |p_2|\Omega_c = 0.9941403\Omega_c$$

(See Table 4–2[c] on page 85.)

FIGURE 10–10
Amplitude response function
for Example 10–4.4.

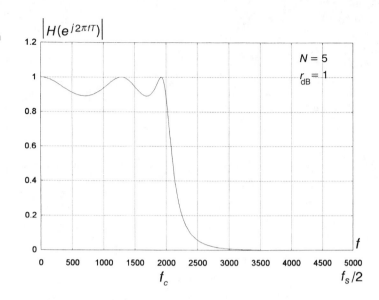

This fifth-order IIR Chebyshev lowpass filter is realized exactly as described in Example 10–4.3 (fifth-order IIR Butterworth lowpass filter), since the transfer functions have the same form. Therefore, Figures 10–5(a) and 10–5(b) also apply to the fifth-order Chebyshev lowpass filter. The only difference is that the formulas for determining the coefficients of the difference equations must be slightly modified to account for the fact that the value of critical frequency Ω is different in each section.

The first-order section is realized by the difference equation given in Equation (10–32). The formulas for the coefficients are given in Equation (10–33), except that Ω_c should be replaced by Ω_0.

The difference equations for the two second-order sections ($k = 1, 2$) are given in Equations (10–34) and (10–35). The formulas for the coefficients are given in Equation (10–36), except that Ω_c should be replaced by Ω_k.

The amplitude response function of this fifth-order IIR Chebyshev lowpass filter is shown in Figure 10–10.

10–5 REQUIRED FILTER ORDER

Figure 10–11 shows the amplitude response function (in dB) of a hypothetical discrete time IIR filter created by applying the bilinear transformation to a continuous time Butterworth lowpass filter transfer function. The desired cutoff frequency is f_c Hz. Also indicated is a higher frequency, denoted f_u, at which the desired attenuation is A dB. As usual, the sampling frequency is denoted f_s. To determine the required filter order, the first step is to prewarp both f_c and f_u:

$$\hat{f}_c = \frac{1}{2\pi}\tan\left(\frac{\pi f_c}{f_s}\right) \tag{10–38}$$

FIGURE 10–11
Determining required filter
order for IIR lowpass filters
designed by applying the
bilinear transformation to an
analog Butterworth lowpass
filter. (See text for equations.)

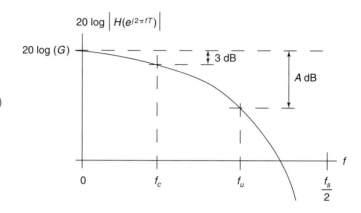

$$\hat{f}_u = \frac{1}{2\pi} \tan\left(\frac{\pi f_u}{f_s}\right) \tag{10–39}$$

The next step is to use the following equations (which are the same as Equations [4–67], [4–68], and [4–69] on page 102, except that f_c and f_u have been replaced by the prewarped frequencies $\hat{f}_c$ an $\hat{f}_u$, respectively):

$$a = 10^{\left(\frac{-A}{20}\right)} \tag{10–40}$$

$$x = \frac{\hat{f}_u}{\hat{f}_c}$$

$$N = \frac{\frac{1}{2} \log\left[\left(\frac{1}{a}\right)^2 - 1\right]}{\log[x]} \tag{10–41}$$

Of course, Equation (10–41) will not result in an integer value for N; therefore, the result must be rounded to the nearest integer.

Figure 10–12 shows the amplitude response function (in dB) of a hypothetical discrete time IIR lowpass filter created by applying the bilinear transformation to a continuous time Chebyshev lowpass filter transfer function. The passband ripple, in dB, is denoted r_{dB}. The desired cutoff frequency is f_c Hz. Once again, also shown is a higher frequency, denoted f_u, at which the desired attenuation is A dB. To determine the required filter order, the first step is to prewarp both f_c and f_u, as shown in Equations (10–38) and (10–39). The next step is to use the following equations (which are the same as Equations [4–79], [4–80], [4–81], [4–82], and [4–83] on pages 105 and 106, except that f_c and f_u have been replaced by $\hat{f}_c$ and $\hat{f}_u$, respectively):

$$a = 10^{\left(\frac{-A}{20}\right)} \tag{10–42}$$

$$\varepsilon^2 = 10^{\frac{r_{dB}}{10}} - 1 \tag{10–43}$$

FIGURE 10–12
Determining required filter order for IIR lowpass filters designed by applying the bilinear transformation to an analog Chebyshev lowpass filter. (See text for equations.)

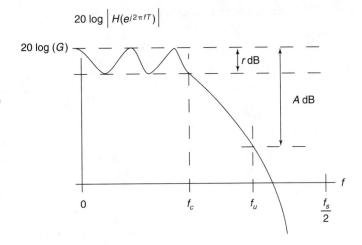

$$g = \sqrt{\frac{\left(\dfrac{1}{a}\right)^2 - 1}{\varepsilon^2}} \qquad\qquad (10\text{–}44)$$

$$x = \frac{\hat{f}_u}{\hat{f}_c} \qquad\qquad (10\text{–}45)$$

$$N = \frac{\log\left[g + \sqrt{g^2 - 1}\right]}{\log\left[x + \sqrt{x^2 - 1}\right]} \qquad\qquad (10\text{–}46)$$

(The result must be rounded to the nearest integer.)

The equations for determining the required filter order for a discrete time IIR highpass filter designed using the bilinear transformation can also be obtained by modifying the equations shown in Chapter 4. For highpass filters, there is a cutoff frequency f_c, along with a lower frequency f_r where the desired attenuation is A dB. These two frequencies must be prewarped to obtain frequencies $\hat{f}_c$ andy $\hat{f}_r$, which then replace f_c and f_r, in Equations (4–75) on page 104 and (4–89) on page 107.

EXAMPLE 10–5.1
The specifications for an IIR lowpass filter designed by applying the bilinear transformation to a Butterworth lowpass filter are as follows:

Sampling frequency = 10 kHz
Cutoff frequency = 3 kHz
20 dB of attenuation at f = 3.5 kHz

The required order is calculated as follows:

$$\hat{f}_c = \frac{1}{2\pi}\tan\left(\frac{\pi \times 3000}{10000}\right) = 0.219058$$

$$\hat{f}_u = \frac{1}{2\pi}\tan\left(\frac{\pi \times 3500}{10000}\right) = 0.312359$$

$$a = 10^{\left(\frac{-20}{20}\right)} = 0.1$$

$$x = \frac{\hat{f}_u}{\hat{f}_c} = \frac{0.312359}{0.219058} = 1.425919$$

$$N = \frac{\frac{1}{2}\log\left[\left(\frac{1}{.1}\right)^2 - 1\right]}{\log[1.425919]} = 6.48$$

If we choose $N = 7$, the specifications with respect to sharpness will be exceeded; on the other hand, if we choose $N = 6$, the specifications will not quite be satisfied.

EXAMPLE 10–5.2

The specifications for an IIR lowpass filter, to be designed by applying the bilinear trans-formation to a Chebyshev lowpass filter with 1 dB of passband ripple are as follows:

Sampling frequency = 10 kHz

Cutoff frequency = 3 kHz

20 dB of attenuation at $f = 3.5$ kHz

The required order is calculated as follows:

$$\hat{f}_c = \frac{1}{2\pi}\tan\left(\frac{\pi \times 3000}{10000}\right) = 0.219058$$

$$\hat{f}_u = \frac{1}{2\pi}\tan\left(\frac{\pi \times 3500}{10000}\right) = 0.312359$$

$$a = 10^{\left(\frac{-20}{20}\right)} = 0.1$$

$$\varepsilon^2 = 10^{\left(\frac{1}{10}\right)} - 1 = 0.258925$$

$$g = \sqrt{\frac{\left(\frac{1}{.1}\right)^2 - 1}{0.258925}} = 19.55377$$

$$x = \frac{\hat{f}_u}{\hat{f}_c} = \frac{0.312359}{0.219058} = 1.425919$$

$$N = \frac{\log\left[g + \sqrt{g^2 - 1} \right]}{\log\left[x + \sqrt{x^{2-1}} \right]} = 4.105$$

A fourth order filter will come very close to satisfying the specifications.

10–6 ALTERNATE SECOND-ORDER STRUCTURES

Consider the second-order transfer function, which is familiar by now:

$$H(z) = \frac{b_0 + b_1 z^{-1} + b_2 z^{-2}}{1 + a_1 z^{-1} + a_2 z^{-2}} \qquad (10\text{–}47)$$

Assume that $x(n)$ is the input signal and $y(n)$ is the output signal. In Chapter 5 it was shown that the following difference equation will implement this transfer function:

$$y(n) = b_0 x(n) + b_1 x(n-1) + b_2 x(n-2) - a_1 y(n-1) - a_2 y(n-2) \qquad (10\text{–}48)$$

This result has been used in the preceding sections of this chapter. The block diagram for this system is shown in Figure 10–13, labeled *Direct Form I*. In this section we want to take note of the fact that there are other sets of difference equations, or *structures,* that also realize the second-order transfer function of Equation (10–47). The two alternate structures to be considered here are called *Direct Form II* and *Transposed Direct Form II*. (There are other structures, such as Lattice structures, which will not be considered here.)

The Direct Form II structure realizes the second order transfer function with a set of two difference equations, used in the order shown here:

$$q(n) = x(n) - a_1 q(n-1) - a_2 q(n-2) \qquad (10\text{–}49)$$

$$y(n) = b_0 q(n) + b_1 q(n-1) + b_2 q(n-2) \qquad (10\text{–}50)$$

The block diagram for this system is shown in Figure 10–14, in which $q(n)$ appears as an intermediate result, so to speak. To verify that the transfer function for the Direct Form II

FIGURE 10–13
Direct Form I second-order section structure.

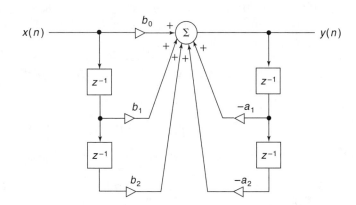

FIGURE 10–14
Direct Form II second-order
section structure.

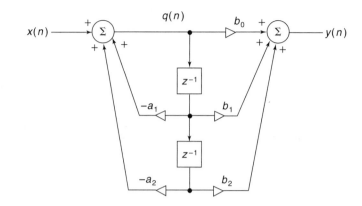

structure is indeed the same as Equation (10–47), we start by taking the Z transform on both sides of Equations (10–49) and (10–50):

$$Q(z) = X(z) - a_1 z^{-1} Q(z) - a_2 z^{-2} Q(z) \tag{10-51}$$

$$Y(z) = b_0 Q(z) + b_1 z^{-1} Q(z) + b_2 z^{-2} Q(z) \tag{10-52}$$

Solve Equation (10–51) for $Q(z)$:

$$Q(z) = X(z) \left(\frac{1}{1 + a_1 z^{-1} + a_2 z^{-2}} \right) \tag{10-53}$$

Next observe that Equation (10–52) can be written as

$$Y(z) = Q(z) \left(b_0 + b_1 z^{-1} + b_2 z^{-2} \right) \tag{10-54}$$

Substitute Equation (10–53) into Equation (10–54):

$$Y(z) = X(z) \left(\frac{1}{1 + a_1 z^{-1} + a_2 z^{-2}} \right) \left(b_0 + b_1 z^{-1} + b_2 z^{-2} \right) \tag{10-55}$$

If Equation (10–55) is solved for $H(z) = Y(z)/X(z)$, the result will obviously be the same as Equation (10–47). Therefore, the second-order Direct Form I and Direct Form II structures have the same transfer function.

The Transposed Direct Form II structure realizes the second-order transfer function with a set of three difference equations, used in the order shown here:

$$y(n) = b_0 x(n) + q(n-1) \tag{10-56}$$

$$q(n) = b_1 x(n) - a_1 y(n) + r(n-1) \tag{10-57}$$

$$r(n) = b_2 x(n) - a_2 y(n) \tag{10-58}$$

FIGURE 10–15

Transposed Direct Form II
second-order section
structure.

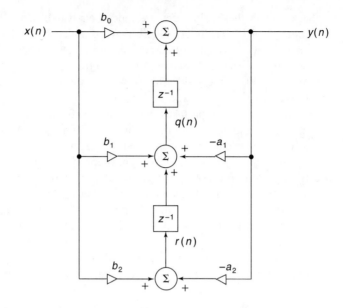

The block diagram for this system is shown in Figure 10–15. To verify that the transfer function for this structure is indeed the same as Equation (10–47), we start by taking the Z transform on both sides of Equations (10–56), (10–57), and (10–58):

$$Y(z) = b_0 X(z) + z^{-1} Q(z) \tag{10–59}$$

$$Q(z) = b_1 X(z) - a_1 Y(z) + z^{-1} R(z) \tag{10–60}$$

$$R(z) = b_2 X(z) - a_2 Y(z) \tag{10–61}$$

Substitute Equation (10–60) into Equation (10–59):

$$Y(z) = b_0 X(z) + b_1 z^{-1} X(z) - a_1 z^{-1} Y(z) + z^{-2} R(z) \tag{10–62}$$

Now substitute Equation (10–61) into Equation (10–62):

$$Y(z) = b_0 X(z) + b_1 z^{-1} X(z) - a_1 z^{-1} Y(z) + b_2 z^{-2} X(z) - a_2 z^{-2} Y(z) \tag{10–63}$$

Equation (10–63) can be rearranged as follows:

$$Y(z)\left(1 + a_1 z^{-1} + a_2 z^{-2}\right) = X(z)\left(b_0 + b_1 z^{-1} + b_2 z^{-2}\right) \tag{10–64}$$

If Equation (10–64) is solved for $H(z) = Y(z)/X(z)$, the result will obviously be the same as Equation (10–47). Therefore, the second-order Direct Form I, Direct Form II, and Transposed Direct Form II structures all have the same transfer function.

At this point the reader may be inclined to ask what difference it could possibly make which of these structures is used, since they have the same transfer function. The answer is that although these structures are indeed equivalent if the computations are done with infinite-precision arithmetic, there may be noticeable differences in performance if the computations are done with fixed-point arithmetic. When fixed-point arithmetic is used, the result of every multiplication must be truncated (or quantized) to accommodate the number of bits being used. This truncation introduces an error that can be modeled as noise. That is, noise is introduced into the system at each point in the structure at which a multiplication operation takes place. Figure 10–16 shows the basic idea. At the system output, the signal has two components: the *true* result (that is, the result that would be obtained if the calculations were done using infinite precision arithmetic) plus a noise signal. The strength of this noise signal depends on the structure as well as the actual coefficient values. It is not possible to state that in general one particular type of structure always produces the least amount of output noise. However, for a given set of coefficients, it is theoretically possible to determine which of these structures produces the least amount of noise. These calculations are beyond the scope of this book, but it is important to at least be aware that the problem exists. It should also be pointed out that there are also programming considerations that might dictate that one structure is better than another for some particular application. For example, the Direct Form II structure needs only two *past value* memory locations, for $q(n - 1)$ and $q(n - 2)$, but the Direct Form I structure needs four past value memory locations, for $x(n - 1)$, $x(n - 2)$, $y(n - 1)$, and $y(n - 2)$.

One final observation needs to be made before leaving this chapter. In Section 9–8, we considered the problem of scaling the coefficients of an FIR filter to prevent overflow when using a fixed-point machine to implement the filter. Naturally, the same problem also exists with IIR filters implemented on fixed-point machines, and it will come as no surprise to learn that the choice of structure has an impact on this problem. However, the coefficient scaling problem for IIR filters is not nearly as straightforward as is the case with FIR filters, and in any case is beyond the intended scope of this book. The reader will have to consult the more advanced DSP literature to learn about this problem in detail.

FIGURE 10–16

Fixed-point multiplication injects noise into the system. The structure type (for example, see Figures 10–13, 10–14, and 10–15) has an impact on the strength of the resulting noise appearing at the system output.

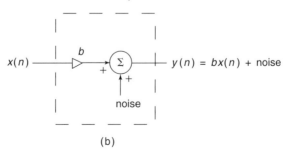

PROBLEMS

10–1. Verify that Equation (10–19) can be manipulated to result in Equation (10–21) (with the coefficient formulas of Equation [10–22]).

10–2. Verify that Equation (10–26) can be manipulated to result in Equation (10–27) (with the coefficient formulas of Equation [10–28]).

10–3. Given the transfer function of an analog *highpass* filter

$$\hat{H}(s) = \frac{s^2}{s^2 + \Psi\Omega_c s + \Omega_c^2}$$

where $\psi = \sqrt{2}$, $\Omega_c = 2\pi\hat{f}_c$, and $\hat{f}_c$ is the cutoff frequency of the analog highpass filter. Design a digital IIR highpass filter based on the bilinear transformation of $\hat{H}(s)$. The specifications for the digital highpass filter are

$$\text{Sampling frequency: } f_s = 10 \text{ kHz}$$
$$\text{Cutoff frequency: } f_c = 500 \text{ kHz}$$

Your answer should include the difference equation for this filter, with formulas provided for calculating the coefficients of this difference equation.

10–4. Suppose you want to design a digital IIR *bandpass* filter using the bilinear transformation. The specifications for this filter are

$$\text{Sampling frequency: } f_s = 10 \text{ kHz}$$
$$\text{Upper cutoff frequency: } 3 \text{ kHz}$$
$$\text{Lower cutoff frequency: } 2 \text{ kHz}$$

Find the upper and lower cutoff frequencies for the *analog* bandpass filter that will be bilinearly transformed to create the digital bandpass filter.

10–5. Consider the transfer function of an analog second-order *bandpass* filter:

$$\hat{H}(s) = \frac{(\Omega_u - \Omega_l)s}{s^2 + (\Omega_u - \Omega_l)s + \Omega_u\Omega_l}$$

where

$$\hat{f}_u \text{ is the upper cutoff frequency; } \Omega_u = 2\pi\hat{f}_u$$
$$\hat{f}_l \text{ is the lower cutoff frequency; } \Omega_l = 2\pi\hat{f}_l$$

Design a digital IIR bandpass filter based on the bilinear transformation of $\hat{H}(s)$. The specifications for the digital bandpass filter are

$$\text{Sampling frequency: } f_s = 10 \text{ kHz}$$
$$\text{Upper cutoff frequency: } f_u = 3 \text{ kHz}$$
$$\text{Lower cutoff frequency: } f_l = 2 \text{ kHz}$$

Your answer should include the difference equation for this filter, with formulas provided for calculating the coefficients of this difference equation.

10–6. Show that the bilinear transformation can be reversed, that is,

$$z = \frac{1+s}{1-s}$$

10–7. Let $s = a + jb$ be a point on the s plane. The bilinear transformation maps this point to a point on the z plane. Show that the following properties are valid:

    a. If $a = 0$, then $|z| = 1$. In other words, a point on the imaginary axis of the s plane is mapped to a point on the unit circle of the z plane.

    b. If $a < 0$, then $|z| < 1$. In other words, a point in the left half of the s plane is mapped to a point inside the unit circle of the z plane.

    c. If $a > 0$, then $|z| > 1$. In other words, a point in the right half of the s plane is mapped to a point outside the unit circle of the z plane.

    d. Problem 10–7(b) demonstrates that the bilinear transformation preserves BIBO stability. In other words, if $\hat{H}(s)$ represents a BIBO-stable system and $H(z)$ is created by applying the bilinear transformation to $\hat{H}(s)$, then $H(z)$ also represents a BIBO-stable system. Why?

    HINT: It has something to do with the location of the poles.

    HINT: If $s = a + jb$, the associated value of z can be found using the result of Problem 10–6:

$$z = \frac{1+s}{1-s} = \frac{1+a+jb}{1-a-jb}$$

Therefore,

$$|z|^2 = \frac{(1+a)^2 + b^2}{(1-a)^2 + b^2}$$

10–8. The specifications for an IIR lowpass filter designed by applying the bilinear transformation to a Butterworth analog lowpass filter are as follows:

$$\text{Sampling frequency: } f_s = 20 \text{ kHz}$$

$$\text{Cutoff frequency: } f_c = 4 \text{ kHz}$$

$$\text{At least 20 dB of attenuation at } f = 5 \text{ kHz}$$

Find the filter order required to satisfy these specifications.

10–9. The specifications for an IIR lowpass filter, designed by applying the bilinear transformation to a Chebyshev analog lowpass filter with 1.5 dB of ripple are as follows:

$$\text{Sampling frequency: } f_s = 20 \text{ kHz}$$

$$\text{Cutoff frequency: } f_c = 4 \text{ kHz}$$

$$\text{At least 20 dB of attenuation at } f = 5 \text{ kHz}$$

Find the filter order required to satisfy these specifications.

11

ADAPTIVE FIR FILTERS USING THE LMS ALGORITHM

11–1 GENERAL PROBLEM

The adaptive FIR filter problem in its most general form is shown in Figure 11–1. The FIR filter coefficients are automatically adjusted (adapted) by the LMS (least mean squared) algorithm, which is described in detail in this chapter. $x(n)$ is the input signal to be processed. $y(n)$ is the output of the FIR filter. $d(n)$ is another signal that is related to $x(n)$ in some sense. The nature of this relationship depends on the details of the problem at hand, but is presumably such than an FIR filter can be found that will make the relationship even stronger, that is, result in the relationship between $y(n)$ and $d(n)$ being stronger than the relationship between $x(n)$ and $d(n)$. In fact, finding an FIR filter that makes the relationship between $y(n)$ and $d(n)$ as close as possible is the desired result. Since $e(n)$ is related to $d(n)$ and $y(n)$ as

$$e(n) = d(n) - y(n) \qquad\qquad (11\text{–}1)$$

it follows that finding an FIR filter that makes $e(n)$ as small as possible in some average sense will achieve the same desired result. $e(n)$ is sometimes described as an *error signal.*

Another way of thinking of the problem is to find an FIR filter that will predict $d(n)$ at its output, given $x(n)$ at its input (to the extent that such prediction is possible). When viewed in this manner, $e(n)$ can be thought of as a *prediction error signal.*

The desired output from this system could be either $y(n)$ or $e(n)$, again depending on the specific problem at hand. For this reason, describing $e(n)$ as a type of error signal is perhaps misleading with respect to some applications. If this system were to work perfectly, $e(n)$ would be that component of $d(n)$ that is essentially unrelated to $x(n)$, and therefore not predictable from knowledge of $x(n)$. (More precisely, $e(n)$ would be that part of $d(n)$ that is essentially *uncorrelated* with $x(n)$.) On the other hand, $y(n)$ would be the component of $x(n)$

FIGURE 11–1
The adaptive FIR filter
problem in its most general
form.

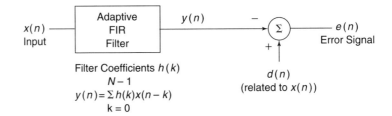

that is closely related to (i.e., *correlated* with) $d(n)$—that is, the component related closely enough so that the FIR filter can successfully predict it from knowledge of $x(n)$.

For specific adaptive filter problems, we must determine how to think of the problem in terms of the system in Figure 11–1. A specific example will be considered later in this chapter, in which $x(n)$ is a speech signal corrupted by an interfering tone (the frequency of which is not known in advance), and the desired result is to eliminate this tone.

The N-point FIR filter in Figure 11–1 has coefficients $h(k)$, $k = 0, 1, \ldots, N-1$. The output of the FIR filter is given by

$$y(n) = \sum_{k=0}^{N-1} h(k)x(n-k) \tag{11–2}$$

However, the filter coefficients $h(k)$ are not determined in advance and fixed in time; instead, they are *adapted* to the conditions at hand. The adaption algorithm is driven by both $x(n)$ and $e(n)$, as shown in Figure 11–2. With respect to the example suggested in the previous paragraph, if the goal is to eliminate an interfering tone from a speech signal without having advance information about the frequency of this tone, the filter coefficients must somehow be automatically adjusted to solve the problem. If the adaptive algorithm is successful, the system would become in effect a self-adjusting notch filter (i.e., a narrow bandstop filter).

FIGURE 11–2
The adaption algorithm (LMS
algorithm) is driven by the
input signal and the error
signal.

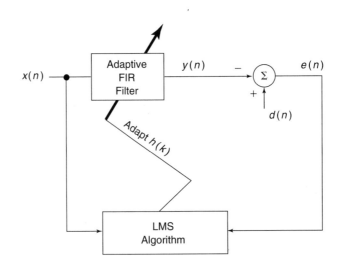

11–2 LMS ALGORITHM DERIVATION

It turns out that mathematically it is much more convenient to consider minimizing the *squared error*, that is, $e^2(n)$. Thus we begin this section by defining an *error strength* measure $J(n)$:

$$J(n) = e^2(n) \tag{11-3}$$

The problem is to adjust the FIR filter coefficients $h(0)$, $h(1)$, . . ., $h(N-1)$ to drive down $J(n)$. Note that $J(n)$ can be expressed as

$$J(n) = \left[d(n) - \sum_{k=0}^{N-1} h(k)x(n-k) \right]^2 \tag{11-4}$$

Thus at any particular time (n), $J(n)$ can be thought of as a function of N variables $h(0)$, $h(1)$, . . .,$h(N-1)$. If $d(n)$ and $x(n)$ were constant values, $J(n)$ would be a (nontime-varying) N-dimensional quadratic function; an N-dimensional parabola, so to speak:

$$J = \left[d - \sum_{k=0}^{N-1} h(k)x \right]^2 \tag{11-5}$$

This function has only one critical point (i.e., one point at which all of the partial derivatives with respect to $h(0)$, $h(1)$, . . ., $h(N-1)$ are equal to zero), and this critical point corresponds to the minimum value of J. Thus if x and d were constants, the values $h(0)$, $h(1)$, . . ., $h(N-1)$ that minimize J could be found by solving the following system of equations:

$$\frac{\partial J}{\partial h(0)} = 0$$

$$\frac{\partial J}{\partial h(1)} = 0 \tag{11-6}$$

$$\vdots$$

$$\frac{\partial J}{\partial h(N-1)} = 0$$

This system of equations can also be expressed as

$$\frac{\partial J}{\partial h(i)} = 0, i = 0,1,\ldots, N-1 \tag{11-7}$$

However, we do not have the luxury of using this solution, because d and x are in fact changing with time. $J(n)$ is actually a *time-varying N-dimensional quadratic function*: its exact shape and the location of its critical point (minimum value) are changing with time.

Suppose you are on the side of a mountain that is shrouded in clouds, and your goal is to get to the point of lowest elevation. How would you do it? One simple rule might govern here: *Go downhill*. We will use the same rule to try to work our way toward the minimum value of $J(n)$.

Let's begin with an initial guess, or initial position $[h(0), h(1), \ldots, h(i), \ldots, h(N-1)]$. Think of these as the coordinates describing an initial location on the side of this N-dimensional mountain. The problem is this: at each iteration we must adjust each of these coordinates such that we are moving downhill. Now consider the partial derivative of $J(n)$ with respect to $h(i)$: $\partial J(n)/\partial h(i)$. What does this derivative tell us (assuming we can calculate it)? If this derivative is *positive*, then we must *decrease* $h(i)$ to go downhill, and vice versa. Furthermore, the absolute value of this derivative tells us how steep this mountain is with respect to moving in the $h(i)$ direction. It can be shown that since $J(n)$ is a quadratic function, the larger $|\partial J(n)/\partial h(i)|$ is, the further away we are from the optimal value of $h(i)$. (That is, the steeper this mountain is, the further away we are from the bottom.) This suggests the following rule: The larger $|\partial J(n)/\partial h(i)|$ is, the larger the change in $h(i)$ should be. (The closer we get to the bottom, the smaller our changes should be, so that we can carefully zero in on the solution.)

Based on the above considerations, the following algorithm is suggested: At the nth iteration, we evaluate $\partial J(n)/\partial h(i)$ to determine how much to adjust $h(i)$ and whether the adjustment should be an increase or decrease. The new value of $h(i)$ at the $(n+1)$th iteration will be the current value plus a change that moves us in the downhill direction:

$$h(i)_{n+1} = h(i)_{n} - \Delta \frac{\partial J(n)}{\partial h(i)} \tag{11-8}$$

$$i = 0, 1, \ldots, N-1$$

where Δ is a parameter that governs the size of the change (step size). As Equation (11–8) suggests, at each iteration (n) this is done for all of the filter coefficients: $h(i)$, $i = 0, 1, \ldots, N-1$.

We must consider how to evaluate $\partial J(n)/\partial h(i)$, given that $x(n)$ and $e(n)$ are available. Observe that

$$\frac{\partial J(n)}{\partial h(i)} = \frac{\partial e^2(n)}{\partial h(i)} = 2e(n)\frac{\partial e(n)}{\partial h(i)} \tag{11-9}$$

By combining Equations (11–1) and (11–2), we obtain

$$e(n) = d(n) - h(0)x(n) - h(1)x(n-1) - \ldots$$
$$- h(i)x(n-i) - \ldots - h(N-1)x[n-(N-1)] \tag{11-10}$$

Therefore,

$$\frac{\partial e(n)}{\partial h(i)} = -x(n-i) \tag{11-11}$$

Using this result in Equation (11–9) results in

$$\frac{\partial J(n)}{\partial h(i)} = -2e(n)x(n-i) \tag{11-12}$$

The factor of 2 in Equation (11–12) can be gathered into parameter Δ, which must be adjusted for best results anyway. Thus the adaptive algorithm of Equation (11–8) can be expressed as follows:

$$h(i)_{n+1} = h(i)_n + \Delta e(n)x(n-i)$$
$$i = 0, 1, ..., N-1$$

(11–13)

Equation (11–13) is the LMS-adaptive FIR filter algorithm. As noted earlier, LMS stands for *least mean squared,* which refers to the more rigorous derivation of the algorithm found in more advanced textbooks. This algorithm is also called the Widrow algorithm, named after the researcher who first developed it.

Note that the LMS algorithm requires us to save past values of the input signal $x(n)$: $x(n-1), x(n-2), ..., x[n-(N-1)]$. Of course, we need to save these values anyway, because the FIR filter requires them (Equation [11–2]).

What about the size of parameter Δ? The trick is to find a value of Δ large enough to ensure that the adaptive filter will track changing conditions but small enough to insure that the adaptive system is stable. (If Δ is too large, then at each iteration the algorithm will overcompensate, which can cause an upward spiral away from the minimum point.) It can be shown that to ensure the stability of the LMS algorithm, the following constraint on the size of Δ must be obeyed:

$$\Delta < \frac{2}{N \cdot E[x^2(n)]}$$

(11–14)

where $E[x^2(n)]$ can be interpreted as the *average value* of $x^2(n)$. (To be more precise, it means the *expected value* of $x^2(n)$.) N is the number of FIR filter coefficients. In practice, it is probably a good idea to start with a conservative (i.e., small) value of Δ, and then tweak this value for best performance.

11–3 AN APPLICATION: SUPPRESSION OF NARROWBAND INTERFERENCE

In this section we will consider a particular application of an adaptive FIR filter controlled by the LMS algorithm. To begin, consider the system shown in Figure 11–3. *Note that the subsystem enclosed in the dashed line is the same as the general form of Figure 11–1.* In Figure 11–3, the input signal to be processed, $g(n)$, is the same as $d(n)$. The input to the adaptive filter, $x(n)$, is a delayed version of $g(n)$:

$$x(n) = d(n-D) = g(n-D)$$

(11–15)

where D is an integer delay factor. The output of the adaptive filter, $y(n)$, can be expressed as

$$y(n) = \sum_{k=0}^{N-1} h(k)x(n-k)$$
$$= \sum_{k=0}^{N-1} h(k)g[n-(D+k)]$$

(11–16)

FIGURE 11–3
Adaptive FIR filter system used to suppress narrowband interference. In this application, $e(n)$ is the system output. The subsystem within the dashed lines fits the general form shown in Figure 11–1.

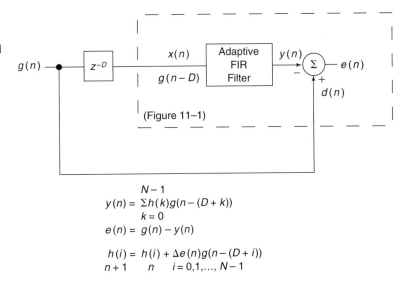

$$y(n) = \sum_{k=0}^{N-1} h(k)g(n-(D+k))$$

$$e(n) = g(n) - y(n)$$

$$\underset{n+1}{h(i)} = \underset{n}{h(i)} + \Delta e(n)g(n-(D+i)), \quad i = 0,1,\ldots, N-1$$

Observe that $y(n)$ is a linear combination of *past* values of $g(n)$. Thus $e(n)$ will be small if the adaptive FIR filter can create a good linear prediction of the current value of $g(n)$ from past values $g(n-D)$, $g(n-[D+1])$, $\ldots$, $g[n-(D+N-1)]$. If this system were to work perfectly, $e(n)$ would essentially be that part of $g(n)$ that *cannot* be predicted from past values $g(n-D)$, $\ldots$ $g[n-(D+N-1)]$. The output of the adaptive filter itself, $y(n)$, would essentially be that part of $g(n)$ that is highly predictable from those past values.

The LMS algorithm for this case is the same as described in the last section:

$$\underset{n+1}{h(i)} = \underset{n}{h(i)} + \Delta e(n)x(n-i), i = 0,1,\ldots, N-1 \qquad (11\text{–}17)$$

but since $x(n) = g(n-D)$, we can express the LMS algorithm in terms of input signal $g(n)$:

$$\underset{n+1}{h(i)} = \underset{n}{h(i)} + \Delta e(n)g[n-(D+i)], i = 0,1,\ldots, N-1 \qquad (11\text{–}18)$$

The system in Figure 11–3 is considered in the following context: Input signal $g(n)$ has two distinct components, that is, $g(n) = z(n) + w(n)$, where it is possible to select delay D such that one of these components is highly predictable from $g(n-D)$, $\ldots$ $g(n-(D+N-1))$ and the other component is essentially unpredictable from those past values. *The goal is to separate these two components.* D is chosen to enhance the unpredictability of one component without destroying the predictability of the other. (D is sometimes called a *decorrelation delay.*) If the unpredictable component is the desired signal, then $e(n)$ is the system output. On the other hand, if the predictable component is the desired signal, then $y(n)$ is the system output.

At this point, the reader may be asking, what good is this? To answer this question, we must relate signal predictability from past values to a concept that is more familiar:

signal *bandwidth*. The following principle, presented here in a nonrigorous and "it can be shown" basis, is crucial to this discussion:

> A narrowband *signal is much more predictable from its past values than a wideband signal.* The amount of delay (D) required to *decorrelate* a signal $x(n)$ from its past values $x(n - D)$, $x[n - (D + 1)]$, . . ., that is, the amount of delay required to make $x(n)$ essentially unpredictable from those past values, is inversely proportional to the bandwidth of $x(n)$.

This principle is best explained by considering two extreme cases. At one extreme, consider a signal consisting of a single sinusoidal tone. Since the spectrum of such a signal is an impulse in the frequency domain (see Section 6–5), the single sinewave is as narrowband as you can get. It's also extremely predictable: if the amplitude, frequency, and phase can be deduced, then the formula $x(n) = A\cos(\omega_o n + \phi)$ allows you to calculate its value at any point in time. Another way of looking at it is that a single sinewave has a regular pattern that, once identified, allows an observer to exactly predict its future values. (It doesn't get more predictable than this!) Thus if $x(n)$ is a single sinewave, it is highly predictable from past values $x(n - D)$, $x[n - (D + 1)]$, . . ., even if D is relatively large.

On the other extreme, consider a noise signal such as shown in Figure 11–4. This signal has no regular pattern, and it seems reasonable to suppose that any current value of this signal would be very difficult to predict from any of its past values. The amplitude spectrum of this signal, $|X[\exp(j\omega)]|$, is shown in Figure 11–5. This signal is clearly wideband, since it fills up the entire range of frequencies. If $x(n)$ is a signal like this, then it is very unpredictable from past values $x(n - D)$, $x[n - (D + 1)]$, . . ., even if D is small (say, $D = 1$).

For signals in between these extremes (a speech signal, for example), the minimum amount of delay (D) required to thoroughly decorrelate the signal from its past values

FIGURE 11–4

Wideband noise signal.

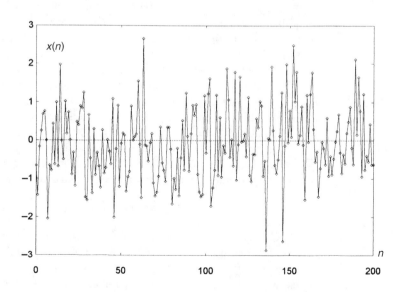

FIGURE 11–5
Amplitude spectrum of the
wideband noise signal of
Figure 11–4, obtained by
taking the FFT of the noise
signal.

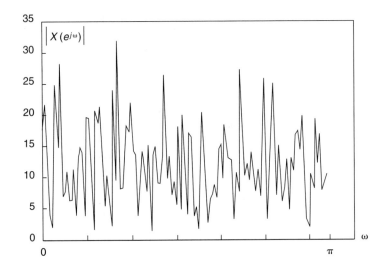

cannot be calculated exactly; some trial and error will be necessary. A rough but workable
rule of thumb that will put you in the right ballpark is

$$D \cong \frac{f_s}{f_{bw}} \tag{11-19}$$

where f_s is the sampling frequency and f_{bw} is the signal bandwidth. However, it may be necessary to make D larger than this formula would suggest.

Now that the relationship between predictability and bandwidth is understood—or
so we hope—we are in a position to state how the system of Figure 11–3 will work (ideally), in terms of its ability to separate signals with different bandwidths. The starting hypothesis is that input signal $g(n)$ is composed of two distinct components: a narrowband
signal added to a wideband signal. The goal is to seperate these components to the extent
possible. If the delay (D) has been chosen wisely, then the output of the adaptive FIR filter,
$y(n)$, will essentially be the narrowband component of $g(n)$. In other words, the adaptive
FIR filter becomes a self-adjusting bandpass filter; the output of this filter is subtracted
from $g(n)$ to produce $e(n)$. If we view the overall system of Figure 11–3 as having $g(n)$ as
the input and $e(n)$ as the output, then the system essentially acts as a self-adjusting bandstop (i.e., notch) filter.

This brings us to the application suggested by the title of this section (supression of
narrowband interference). Let's consider a specific case: Let $g(n)$ consist of a speech signal
$s(n)$ to which an interfering tone $w(n)$ (single sinusoid) has been added:

$$g(n) = s(n) + w(n) \tag{11-20}$$

The frequency of the interfering single sinusoid is unknown and possibly changing with
time. The speech signal $s(n)$ is wideband *relative to* the single sinusoid. (The speech signal
is not as wideband as the noise signal of Figure 11–4). The speech signal is the desired
signal; we need an adaptive notch filter to eliminate the interfering tone. This means, with
respect to Figure 11–3, that $e(n)$ will be the output signal in this application.

We must select a decorrelation delay (D) long enough to decorrelate $s(n)$ from $s(n - D)$, $s[n - (D + 1)]$, ... Let us suppose the sampling frequency is $f_s = 11$ kHz, and the band-width of $s(n)$ is 4 kHz. From Equation (11–19), we see that $D = 3$ might be sufficient. However, one can make a case for using a much longer delay (say, $D = 20$) on the grounds that this will more thoroughly decorrelate $s(n)$ from $s(n - D)$, . . . without decorrelating $w(n)$ from $w(n - D)$, . . . (since $w(n)$ is a sinusoid). (However, if $w(n)$ is another type of signal not quite as narrowband as a single sinusoid, one must be careful not to make D large enough to decorrelate both $s(n)$ and $w(n)$ from their past values.)

To see how well this system works, consider the following experimental results. Figure 11–6(a) shows a short speech signal, $s(n)$, obtained by speaking the phrase "testing one, two, three" into a microphone connected to a commercially available sound card. The sampling frequency was $f_s = 11$ kHz; there are approximately 28,000 points in $s(n)$. Figure 11–6(b) shows the first 1,500 points of $s(n)$; this segment consists entirely of negligible background noise and is shown here so that it can be compared with what follows. Figure 11–6(c) shows points 12,000 to 13,000 of $s(n)$, which is a short segment in the middle of the word "one." This figure shows some of the finer detail of the speech signal and is also provided so that it can be compared with what follows.

Speech signal $s(n)$ was then deliberately corrupted by adding to it a *sliding frequency* single tone (sometimes called a *chirp* signal). In other words, a signal $g(n)$ was created as

$$g(n) = s(n) + w(n) \tag{11–21}$$

$w(n)$ is the chirp signal:

$$w(n) = A\cos\left(2\pi[f(n)]Tn\right) \tag{11–22}$$

where $T = 1/f_s$ and

$$f(n) = 50 + \frac{2n}{30} \tag{11–23}$$

The amplitude of the chirp signal was selected to make the interference obvious and obnoxious to the listener. Note that the initial frequency (at $n = 0$) is 50 hertz; it slides up to approximately 1,900 hertz at the end of the speech signal.

Signal $g(n)$ (speech + chirp) is shown in Figure 11–7(a). The original speech is almost buried by the chirp signal. When you listen to it, the original speech is still understandable, but the chirp signal is loud and obnoxious. Figure 11–7(b) shows the first 1,500 points of $g(n)$. Since the first 1,500 points of $s(n)$ are negligible, Figure 11–7(b) essentially shows the chirp signal; the manner in which the frequency is slowly increasing is clearly illustrated. Figure 11–7(c) shows points 12,000 to 13,000 of $g(n)$, which should be compared to Figure 11–6(c), the corresponding segment of $s(n)$.

Signal $g(n)$ was then processed by the system of Figure 11–3, with $e(n)$ as the output. The parameters were selected as follows:

FIR filter length: $N = 25$

Decorrelation delay: $D = 20$

Step-size parameter: $\Delta = 1/(25 \times 50^2)$

Figure 11–8(a) shows the output signal $e(n)$; the original speech signal $s(n)$ has essentially been recovered. When you listen to it now, a very weak chirp signal can still be heard in the background, and there is a small change in the timbre of the speech, but intelligibility and speaker identity are well preserved. Figure 11–8(b) shows the first 1,500 points of $e(n)$, which clearly shows the system in the process of adapting at the beginning. This figure should be compared to Figures 11–6(b) and 11–7(b). Figure 11–8(c) shows points 12,000 to 13,000 of $e(n)$, which should be compared to Figures 11–6(c) and 11–7(c).

FIGURE 11–6

(a) Original speech signal, "testing one, two, three." (b) First 1,500 points of the original speech signal. (c) Short segment in the middle of the word "one" in the original speech signal.

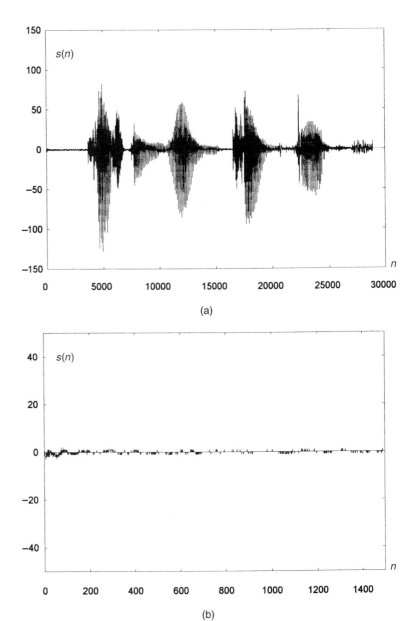

(a)

(b)

FIGURE 11–6
(continued)

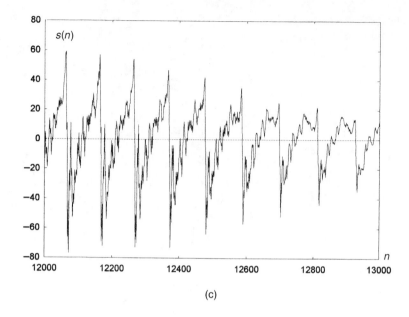

(c)

FIGURE 11–7
(a) "Testing one, two, three" speech signal corrupted by adding a chirp signal to it.
(b) First 1,500 points of the corrupted speech signal. (Compare with Figure 11–6[b].)
(c) Short segment of the corrupted speech signal in the middle of the word "one." (Compare with Figure 11–6[c].)

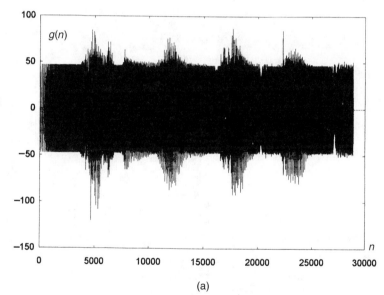

(a)

FIGURE 11–7
(continued)

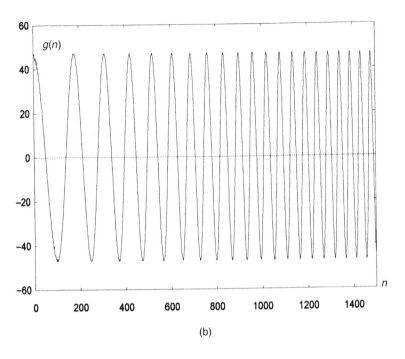

(b)

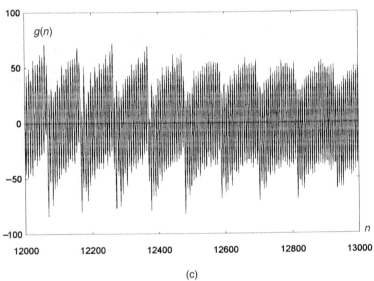

(c)

FIGURE 11–8
(a) Speech signal after being cleaned up by the adaptive system of Figure 11–3. With respect to the adaptive system, Figure 11–7(a) is the input signal, $g(n)$; Figure 11–8(a) is the output, $e(n)$. (Compare the output to the original speech signal shown in Figure 11–6[a].) (b) First 1,500 points of the cleaned-up speech signal. (c) Cleaned-up speech signal in the middle of the word "one." (Compare with Figures 11–6[c] and 11–7[c].)

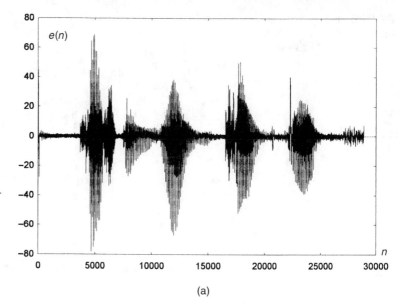

(a)

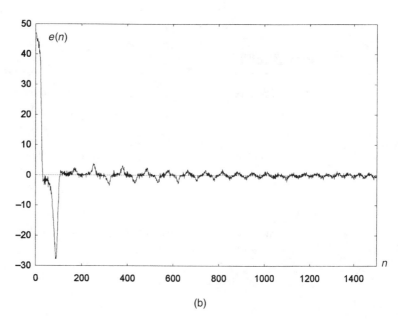

(b)

FIGURE 11–8
(continued)

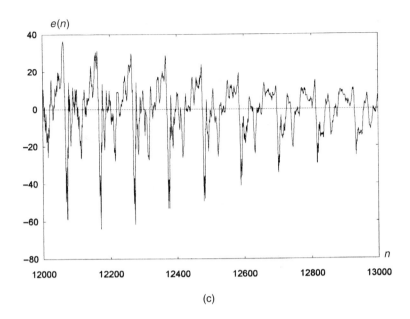

(c)

PROBLEMS

11–1. Consider the system shown in Figure 11–3. Suppose that the input signal is

$$g(n) = \begin{cases} 1, n \geq 0 \\ 0, n < 0 \end{cases}$$

Suppose also that the system parameters are as follows:

$$D = 2$$
$$N = 2$$
$$\Delta = 0.25$$

Suppose further that the starting time is $n = 0$ and that the filter coefficients have been initialized as follows:

$$h(0) \Big|_{n=0} = 0$$
$$h(1) \Big|_{n=0} = 0$$

Calculate $y(n)$, $e(n)$, and the new filter coefficients at each iteration, starting with $n = 0$ and continuing up to $n = 10$.

11–2. Repeat Problem 11–1 using $\Delta = 0.4$.

11–3. Repeat Problem 11–1 using

$$\Delta = 0.4, \ h(0)\Big|_{n=0} = 1, \ \text{and} \ h(1)\Big|_{n=0} = 1$$

12

RANDOM SIGNALS AND POWER SPECTRA, A/D CONVERSION NOISE, AND OVERSAMPLING

12–1 INTRODUCTION

As mentioned briefly in Chapter 5, a real-world analog-to-digital (A/D) converter must *quantize* each sample value since there are only a finite number of bits available to represent a signal value. Thus at any time n, the output of an A/D converter consists of the true signal value plus a *quantization error*. The quantization error can be modeled as an additional *noise* signal that is added to the true signal.

The quantization noise power can theoretically be reduced by increasing the number of bits employed by the A/D converter. However, it is also possible to reduce the quantization noise power with a technique called *oversampling*, which means operating the A/D converter at a sampling frequency significantly higher (as much as 64 or 128 times higher) than the theoretical minimum of $f_s = 2B$, where B is the signal bandwidth. At the output of the A/D converter, the true signal is still confined to its original bandwidth ($-B < f < B$), which is now much less than the entire usable bandwidth ($-f_s/2 < f < f_s/2$). However, it turns out that the spectrum of the quantization noise signal at the output of the A/D converter is evenly spread out over the entire usable bandwidth and does not depend on the sampling frequency. Therefore, by processing the output of the A/D converter with a discrete time lowpass filter with a cutoff frequency of B Hz, the part of the noise spectrum lying outside of the true signal bandwidth will be eliminated without changing the true signal itself, thus reducing the overall noise power with respect to the true signal power.

In order to explain the theory behind the oversampling scheme, it is necessary to introduce a new signal processing concept: the *random signal*. This will allow us to discuss the *power spectral density function* of the signals of interest, including the quantization noise signal. The necessary theory is covered in Sections 12–2, 12–3, and 12–4, followed by a detailed discussion of quantization noise and oversampling in Section 12–5.

On the face of it, oversampling appears to suffer from one serious drawback: If the sampling frequency is increased significantly, and we are dealing with a system that must operate in real time, then the processing rate (i.e., number-crunching rate) also must be increased significantly. Fortunately, it is possible to *decimate*, or *downsample*, the signal before sending it to the processor, which means that the number crunching can proceed at a slower pace. In addition, it is possible to *interpolate*, or *upsample*, a discrete time signal before sending it to a digital-to-analog (D/A) converter. The interpolation process in effect resamples the signal at a higher sampling frequency. (Recall from Chapter 7 that the distortion inherent in the D/A conversion process will be reduced if the sampling frequency is increased.) Decimation and interpolation are the subjects of Section 12–6. The overall oversampling-decimation-processing-interpolation-D/A conversion scheme is shown in Figure 12–17 in Section 12–6.

12–2 DISCRETE TIME RANDOM SIGNALS

We begin by noting that the term *random signal* is somewhat misleading because it is suggestive of meaningless garbage, or noise. (A somewhat fancier name for a random signal is *stochastic process*.) Actually, any signal that has an element of unpredictability with respect to some of its characteristics can be *modeled* as a random signal. If this view is adopted, it does not necessarily follow that the signal has no useful information. It is certainly true that noise is an example of a random signal, but not all random signals are noise. For example, communications signals are often modeled as random signals. If this seems strange, consider the fact that from the perspective of the receiver, such signals certainly have some unpredictable aspects, even if the signal is purely deterministic from the perspective of the transmitter.

Consider a signal $x(n)$. The lack of a formula allowing the precise value of this signal to be calculated for any time n could be viewed as the defining characteristic of a random signal. But even without such a formula, we could still consider asking questions about the *statistics* of the signal. For example, we might ask what value $x(n)$ is expected to have on the average. This is denoted $E[x(n)]$, that is, the *expected value* of $x(n)$. $E[x(n)]$ can be interpreted as the *DC value* of the signal. We might also ask what value $x^2(n)$ is expected to have, on the average. This is denoted $E[x^2(n)]$, which is interpreted to be the *average power* of the random signal. We can also ask questions about the *spectrum* or the *frequency content* of a random signal—we will have more to say about this later.

Readers who have taken a course in statistics or probability (which is not a prerequisite for this chapter!) should note that the expected value operator is being used in the same sense here as it is in those courses. In other words, $x(n)$ (for any particular time n) is being treated as a random variable x governed by some underlying probability density function. If $p(x)$ is the probability density function, then

$$E[x] = \int_{-\infty}^{\infty} xp(x)dx \qquad (12\text{–}1)$$

$$E[x^2] = \int_{-\infty}^{\infty} x^2 p(x)dx \qquad (12\text{–}2)$$

In a few special signal processing cases, the underlying probability density function $p(x)$ is actually known. (An example is the probability density function for the *quantization noise* created by an A/D converter.) In most cases, though, $p(x)$ is not really known, but is assumed to be Gaussian (i.e., normal) if an assumption about $p(x)$ must be made for some reason.

The interpretation of $E[x^2(n)]$ as the average power in the random signal should be justified. Let us suppose that the discrete time signal $x(n)$ is processed by a D/A converter to create a continuous time voltage waveform $x_a(t)$. If this voltage is placed across a resistive load R, the instantaneous power is $p(t) = x_a^2(t)/R$, and the average power is the average value (or expected value) of $x_a^2(t)/R$. Clearly this real-world average power (used to heat up the resistor) will be proportional to the average value, or expected value, of $x^2(n)$.

There is a lot of esoteric, mathematically intense theory under the heading of random signals or stochastic processes, but we want to avoid most of it here, since it is well beyond the intended scope of this textbook. Therefore, we will *assume* that our random signals have the kind of properties that make the theory practical to use and the mathematics bearable. To be specific, we will assume here that the random signals of interest are *wide sense stationary*, *zero mean*, and *ergodic*. The property wide sense stationary means, among other things, that $E[x(n)]$ and $E[x^2(n)]$ (that is, the mean value and average power respectively) do *not* depend on time (n). The property zero mean implies that $E[x(n)] = 0$ (that is, the DC value is zero). This is a reasonable assumption for communications signals and noise signals. The property ergodic means that expected values relating to signals can be determined using *time averaging*. For example, if a signal is ergodic, a reasonable estimate of $E[x(n)]$ can be found by taking an N-point segment of $x(n)$, adding up all the sample values in the segment, and dividing the sum by N. The same thing goes for finding an estimate of $E[x^2(n)]$: Take each value of $x(n)$ in an N-point-segment, square each value, add up the results, and divide the sum by N. Of course, the quality of these estimates depends on the size of N (the larger, the better).

One more comment is in order: For most real-world signals for which estimates of the mean value and average power must be obtained (the underlying probability density functions being unknown), it is impossible to determine whether or not the signal is really ergodic, and in any case time averaging is the only practical way to obtain these estimates. Therefore, as a practical matter, we have no choice but to assume that the signal is ergodic. As for the wide sense stationary property, it is usually reasonable to assume that this property at least holds true over some short-time interval of interest. The bottom line is that these assumptions make random signal theory a useful tool for *modeling* certain types of signal analysis problems.

The symbol usually employed for $E[x(n)]$ is $\mu_x(n)$. Since we are assuming $x(n)$ is wide sense stationary, the mean value does not depend on n; thus

$$E\big[x(n)\big] = \mu_x \qquad\qquad (12\text{–}3)$$

The symbol σ_x^2 is used to denote the *variance* of $x(n)$, that is, $E[(x(n) - \mu_x)^2]$. However, if we restrict ourselves to *zero mean* signals ($\mu_x = 0$), then it is proper to use σ_x^2 to denote the average power of $x(n)$:

$$E\big[x^2(n)\big] = \sigma_x^2 \qquad\qquad (12\text{–}4)$$

If $x(n)$ is zero mean, wide sense stationary, and ergodic, then σ_x^2 can be found (in theory) with an infinitely long time average, as follows:

$$\sigma_x^2 = \lim_{N \to \infty} \frac{1}{(2N+1)} \sum_{n=-N}^{N} x^2(n) \qquad (12\text{–}5)$$

Strictly speaking, the equality sign in Equation (12–5) must be read as *equal with probability 1*.

For any fixed segment length $(2N+1)$, we can form an *estimate* of the average power in the random signal, denoted $\hat{\sigma}_x^2$, as follows:

$$\hat{\sigma}_x^2 = \frac{1}{2N+1} \sum_{n=-N}^{N} x^2(n) \qquad (12\text{–}6)$$

Now denote $x_N(n)$ as the $2N+1$ point segment of $x(n)$, starting at $n = -N$ and ending at $n = N$. In other words,

$$x_N(n) = \begin{cases} x(n), -N \le n \le N \\ 0, \text{otherwise} \end{cases} \qquad (12\text{–}7)$$

Clearly, the average power estimate of Equation (12–6) can also be expressed in terms of $x_N(n)$:

$$\hat{\sigma}_x^2 = \frac{1}{(2N+1)} \sum_{n=-N}^{N} x_N^2(n) \qquad (12\text{–}8)$$

Let $X_N(e^{j\omega})$ be the DTFT of the sequence $x_N(n)$. According to *Parseval's theorem* (see Appendix D), the following relationship holds:

$$\sum_{n=-N}^{N} x_N^2(n) = \frac{1}{2\pi} \int_{-\pi}^{\pi} \left| X_N\left(e^{j\omega}\right)\right|^2 d\omega \qquad (12\text{–}9)$$

This implies that an estimate of the average power in the signal can be obtained in the following manner:

$$\hat{\sigma}_x^2 = \frac{1}{2\pi} \int_{-\pi}^{\pi} \frac{1}{(2N+1)} \left| X_N\left(e^{j\omega}\right)\right|^2 d\omega \qquad (12\text{–}10)$$

Note that $\hat{\sigma}_x^2$ is proportional to the *area* under the curve $[1/(2N+1)]|X_N(e^{j\omega})|^2$ between $\omega = -\pi$ and $\omega = \pi$. Note also that $|X_N(e^{j\omega})|^2$ is a real, nonnegative function. Therefore, it is reasonable to interpret $[1/(2N+1)]|X_N(e^{j\omega})|^2$ as a *frequency domain power density function* of some sort. In fact, $[1/(2N+1)]|X_N(e^{j\omega})|^2$ is an *estimator* for the true *power spectral density function*. The true power spectral density function is denoted $S_{xx}(\omega)$ and will be described shortly. Any *estimate* of $S_{xx}(\omega)$ is denoted $\hat{S}_{xx}(\omega)$; an *estimator* is a method for obtaining the estimate. Thus

$$\hat{S}_{xx}(\omega) = \frac{1}{(2N+1)} \left| X_N\left(e^{j\omega}\right)\right|^2 \qquad (12\text{–}11)$$

The true power spectral density function, $S_{xx}(\omega)$, tells us about the frequency content and bandwidth of a random signal $x(n)$. Like the above estimator, $S_{xx}(\omega)$ is a real, nonnegative function that has even symmetry about the point $\omega = 0$.

According to the *Wiener-Kinchine theorem*, the true power spectral density function is the DTFT of something called the *autocorrelation function*, denoted $r_{xx}(n)$, defined as

$$r_{xx}(n) = E\big[x(k)x(k+n)\big] \tag{12–12}$$

In other words, $S_{xx}(w) = \text{DTFT}[r_{xx}(n)]$. The reader may recall the following rule of thumb from an earlier chapter: Bandwidth is inversely proportional to pulse width. That is, if a time domain function $g(n)$ is narrow, then its DTFT $G(e^{j\omega})$ will be wide. The same rule of thumb applies here: If the autocorrelation function $r_{xx}(n)$ is narrow, then the power spectral density function $S_{xx}(\omega)$ will be wide. In other words, the narrower the autocorrelation function of a random signal is, the larger the bandwidth of that signal will be.

The reader may wonder what the autocorrelation function tells us about the characteristics of the random signal in the time domain. The absolute value of the autocorrelation function is, roughly speaking, a relative measure of the extent to which samples of the signal spaced n units apart are related to (correlated with) each other. That is, $|r_{xx}(n)|$ is a relative measure of how well the sample $x(k+n)$ can be predicted based on knowledge of $x(k)$. (It can be shown that the largest value of $|r_{xx}(n)|$ occurs when $n = 0$.) Thus if the autocorrelation function is wide, the implication is that samples of the signal spaced relatively far apart will still be related to each other fairly well; such a signal will have a relatively narrow bandwidth. On the other hand, if the autocorrelation function is narrow, the implication is that samples of the signal spaced relatively close together will not be related to each other very well; this signal will have a relatively wide bandwidth.

According to the Wiener-Kinchine theorem, the average power in a random signal can be found from the power spectral density function as follows:

$$\sigma_x^2 = r_{xx}(0) = E\big[x^2(k)\big] = \frac{1}{2\pi}\int_{-\pi}^{\pi} S_{xx}(\omega)\,d\omega \tag{12–13}$$

That is, the true average power in a random signal is proportional to the area under the actual power spectral density function curve. The reader should compare Equation (12–13) to Equation (12–10).

For the record, the precise relationship between $[1/(2N+1)]\,|X_N(e^{j\omega})|^2$ and $S_{xx}(\omega)$ is

$$S_{xx}(\omega) = E\left[\lim_{N\to\infty} \frac{1}{(2N+1)}\left|X_N\big(e^{j\omega}\big)\right|^2\right] \tag{12–14}$$

Equation (12–14) shows that $S_{xx}(\omega)$ is what we would expect $[1/(2N+1)]\,|X_N(e^{j\omega})|^2$ to be, on the average, if the sequence $x_N(n)$ is very long.

It is worth noting that except for a few special cases, the true power spectral density function $S_{xx}(\omega)$ of a random signal $x(n)$ cannot be found precisely. In general, we can only find an estimate of $S_{xx}(\omega)$. This is known in the signal processing community as the *spectral estimation* problem. It is true, as noted above, that $[1/(2N+1)]\,|X_N(e^{j\omega})|^2$ is an estimator for $S_{xx}(\omega)$. In fact, this particular estimator is sometimes called the *periodogram* in the spectral estimation literature. The periodogram has a long history (it was originally proposed in 1898 by Schuster). However, the periodogram is not an estimator of high quality. A better estimator can be found by partitioning $x(n)$ into contiguous finite-duration segments, finding the periodogram for each segment, and then finding the average of the periodograms. But this method also has drawbacks. There are no perfect spectral estimators, and an entire graduate course in electrical engineering can be built around the various methods proposed in the spectral estimation literature.

If a random discrete time signal $x(n)$ is obtained by sampling a random continuous time signal $x_a(t)$, then the usual considerations with respect to aliasing will still apply. A continuous time random signal will also have a power spectral density function associated with it. Let us denote this power spectral density function $S_{xxa}(f)$. (In this case, the power spectral density function is the Fourier transform of the autocorrelation function $r_{xxa}(t) = E[x_a(\alpha)x_a(\alpha + t)]$). The bandwidth of the signal is the highest frequency at which this power spectral density function is still significant. If the sampling frequency is not greater than two times this bandwidth, $S_{xx}(\omega)$ will be an aliased version of $S_{xxa}(f)$. The relationship between the two power spectral density functions turns out to be exactly what the theory discussed in Chapters 6 and 7 would suggest to the reader:

$$S_{xx}(2\pi fT) = \frac{1}{T} \sum_{n=-\infty}^{\infty} S_{xxa}(f - nf_s) \qquad (12\text{--}15)$$

where $f_s = 1/T$ is the sampling frequency and $\omega = 2\pi fT$ as usual. Assuming no aliasing, the relationship between $S_{xxa}(f)$ and $S_{xx}(\omega)$ is as shown in Figure 12–1. The original signal $x_a(t)$ is bandlimited to B Hz. $S_{xx}(\omega)$ is periodic (2π) like any other DTFT, but only the interval $-\pi \le \omega \le \pi$ is shown. Using the formula $\omega = 2\pi f/f_s$, we note that on the ω axis $S_{xx}(\omega)$ is band-limited to $\omega_B = 2\pi B/f_s$, as shown in Figure 12–1. The average power σ_x^2 in $x(n)$ is given by Equation (12–13), which is true in general. However, the actual limits of integration can be narrowed as follows:

$$\sigma_x^2 = \frac{1}{2\pi} \int_{\frac{-2\pi B}{f_s}}^{\frac{2\pi B}{f_s}} S_{xx}(\omega)d\omega \qquad (12\text{--}16)$$

FIGURE 12–1

Relationship between continuous time power spectral density function $S_{xxa}(f)$ and discrete time power spectral density function $S_{xx}(\omega)$ if the sampling frequency (f_s) is greater than $2B$.

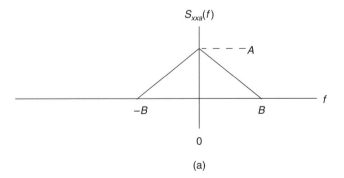

(a)

FIGURE 12–1
(continued)

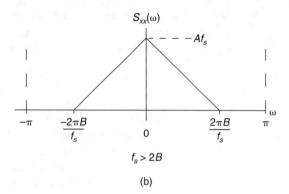

$f_s > 2B$

(b)

Note that the sampling frequency f_s does *not* change σ_x^2, even though an increase in f_s will cause $S_{xx}(\omega)$ to be narrower (relative to the ω axis). This is because an increase in f_s also increases the amplitude scale factor on $S_{xx}(\omega)$ (note the $1/T$ scale factor in Equation [12–15]).

12–3   RANDOM SIGNALS AND LSI SYSTEMS

Consider the discrete time, linear shift invariant system shown in Figure 12–2, where $x(n)$ is the input, $y(n)$ is the output, and $h(n)$ is the system impulse response. Note that although $x(n)$ and $y(n)$ might be modeled as random signals, the system itself is *not* random. The reader will recall from Chapter 6 that the input and output signals are related as follows:

$$y(n) = h(n) * x(n) \tag{12–17}$$

$$Y(e^{j\omega}) = H(e^{j\omega})X(e^{j\omega}) \tag{12–18}$$

If we square the magnitude on both sides of Equation (12–18), we obtain

$$\left|Y(e^{j\omega})\right|^2 = \left|H(e^{j\omega})\right|^2\left|X(e^{j\omega})\right|^2 \tag{12–19}$$

The reader has seen in the previous section how $|X(e^{j\omega})|^2$ is related to the estimator for the actual power spectral density function $S_{xx}(\omega)$. Therefore, with respect to Equation (12–19), the following relationship between the input and output power spectral density functions, stated here without proof, should not violate the readers' intuition:

$$S_{yy}(\omega) = \left|H(e^{j\omega})\right|^2 S_{xx}(\omega) \tag{12–20}$$

FIGURE 12–2
Discrete time LSI system
driven by a random input
signal.

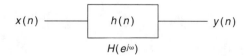

where $S_{yy}(\omega)$ is the power spectral density function of $y(n)$. We assume here that $x(n)$ is a zero mean signal; it can be shown that as a consequence, $y(n)$ will also be a zero mean signal. The average power in $y(n)$, denoted σ_y^2, is given by

$$\sigma_y^2 = \frac{1}{2\pi} \int_{-\pi}^{\pi} S_{yy}(\omega)d\omega = \frac{1}{2\pi} \int_{-\pi}^{\pi} \left| H\left(e^{j\omega}\right) \right|^2 S_{xx}(\omega)d\omega \qquad \textbf{(12–21)}$$

The implication of Equations (12–20) and (12–21) is that the power in a random signal can be reduced if parts of its power spectral density function are zeroed out (approximately) by a filter with frequency response function $H(e^{j\omega})$. This means, for instance, that the power in a wideband random signal will be reduced if this signal is processed by a lowpass filter. Figure 12–3 illustrates the concept: The input signal has a hypothetical power spectral density function $S_{xx}(\omega)$ as shown in Figure 12–3(a). The lowpass filter shown in Figure 12–3(b)

FIGURE 12–3

The relationship between the input and output power spectral density functions and the system frequency response function with respect to Figure 12–2.

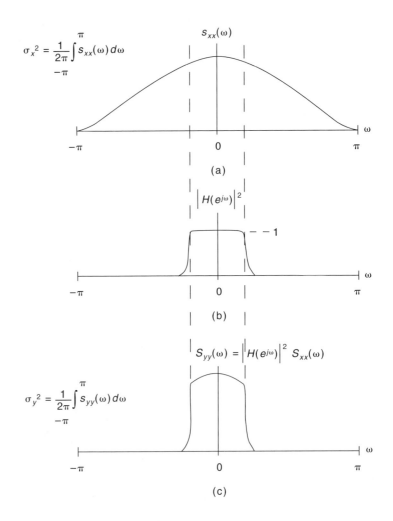

$$\sigma_x^2 = \frac{1}{2\pi} \int_{-\pi}^{\pi} s_{xx}(\omega)\,d\omega$$

$S_{xx}(\omega)$

(a)

$\left| H(e^{j\omega}) \right|^2$

$--1$

(b)

$S_{yy}(\omega) = \left| H(e^{j\omega}) \right|^2 S_{xx}(\omega)$

$$\sigma_y^2 = \frac{1}{2\pi} \int_{-\pi}^{\pi} s_{yy}(\omega)\,d\omega$$

(c)

has a cutoff frequency that is less than the bandwidth of the input signal, and a passband gain of 1. The resulting output signal has a power spectral density function $S_{yy}(\omega)$ as shown in Figure 12–3(c). The area under the $S_{yy}(\omega)$ curve is clearly less than the area under the $S_{xx}(\omega)$ curve; therefore, the output signal has less power than the input signal.

Consider also the discrete time, linear shift invariant system shown in Figure 12–4. The two random signals $x(n)$ and $g(n)$ are added together to create $z(n)$ before being processed by the system. The problem is to determine the relationship between the two input power spectral density functions, $S_{xx}(\omega)$ and $S_{gg}(\omega)$, and the output power spectral density function, $S_{yy}(\omega)$. From the earlier part of this section, we see that

$$S_{yy}(\omega) = \left| H\left(e^{j\omega}\right)\right| S_{zz}(\omega) \tag{12–22}$$

But how is $S_{zz}(\omega)$ related to $S_{xx}(\omega)$ and $S_{gg}(\omega)$? To answer this question, begin with the observation

$$S_{zz}(\omega) = \textbf{DTFT}\big[r_{zz}(n)\big] \tag{12–23}$$

where $r_{zz}(n) = E[z(k)z(k+n)]$. But $z(k) = x(k) + g(k)$ and $z(k+n) = x(k+n) + g(k+n)$. Therefore,

$$\begin{aligned} r_{zz}(n) &= E\big\{\big[x(k) + g(k)\big]\big[x(k+n) + g(k+n)\big]\big\} \\ &= E\big[x(k)x(k+n) + g(k)g(k+n) + g(k)x(k+n) + x(k)g(k+n)\big] \end{aligned} \tag{12–24}$$

It can be shown that the expected value operation can be distributed across the sum. That is,

$$\begin{aligned} r_{zz}(n) &= E\big[x(k)x(k+n)\big] + E\big[g(k)g(k+n)\big] + E\big[g(k)x(k+n)\big] \\ &\quad + E\big[x(k)g(k+n)\big] \end{aligned} \tag{12–25}$$

Using the definition of the autocorrelation function, we have

$$\begin{aligned} E\big[x(k)x(k+n)\big] &= r_{xx}(n) \\ E\big[g(k)g(k+n)\big] &= r_{gg}(n) \end{aligned} \tag{12–26}$$

Therefore,

$$r_{zz}(n) = r_{xx}(n) + r_{gg}(n) + E\big[g(k)x(k+n)\big] + E\big[x(k)g(k+n)\big] \tag{12–27}$$

FIGURE 12–4
If $x(n)$ and $g(n)$ are independent random signals and either (or both) are zero mean, then $S_{zz}(\omega) = S_{xx}(\omega) + S_{gg}(\omega)$.

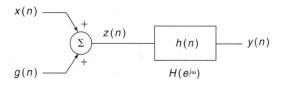

which is probably more complicated than the reader had hoped for, since $r_{zz}(n)$ is not in general just the sum of $r_{xx}(n)$ and $r_{gg}(n)$. This also implies that $S_{zz}(\omega)$ is not in general just the sum of $S_{xx}(\omega)$ and $S_{gg}(\omega)$. However, the situation is simplified if the two signals being added, $x(n)$ and $g(n)$, are *independent* (meaning, roughly, that one signal does not influence the other in any way). If two random variables are independent, the expected value of their product is equal to the product of their respective expected values. Therefore, if $x(n)$ and $g(n)$ are independent, Equation (12–27) becomes

$$r_{zz}(n) = r_{xx}(n) + r_{gg}(n) + E[g(k)]E[x(k+n)] + E[x(k)]E[g(k+n)] \qquad \textbf{(12–28)}$$

If in addition to being independent, either (or both) of the signals are zero mean (which is our usual assumption here), Equation (12–28) reduces to

$$r_{zz}(n) = r_{xx}(n) + r_{gg}(n) \qquad \textbf{(12–29)}$$

in which case it is also true that

$$S_{zz}(\omega) = S_{xx}(\omega) + S_{gg}(\omega) \qquad \textbf{(12–30)}$$

Therefore, for the system depicted in Figure 12–4, if $x(n)$ and $g(n)$ are independent, and either (or both) are zero mean, then

$$\begin{aligned} S_{yy}(\omega) &= \left|H\!\left(e^{j\omega}\right)\right|^2 \left\{ S_{xx}(\omega) + S_{gg}(\omega) \right\} \\ &= \left|H\!\left(e^{j\omega}\right)\right|^2 S_{xx}(\omega) + \left|H\!\left(e^{j\omega}\right)\right|^2 S_{gg}(\omega) \end{aligned} \qquad \textbf{(12–31)}$$

and the average power in the random signal $y(n)$ is given by

$$\begin{aligned} \sigma_y^2 &= \frac{1}{2\pi} \int_{-\pi}^{\pi} S_{yy}(\omega)\,d\omega \\ &= \frac{1}{2\pi} \int_{-\pi}^{\pi} \left|H\!\left(e^{j\omega}\right)\right|^2 S_{xx}(\omega)\,d\omega + \frac{1}{2\pi} \int_{-\pi}^{\pi} \left|H\!\left(e^{j\omega}\right)\right|^2 S_{gg}(\omega)\,d\omega \end{aligned} \qquad \textbf{(12–32)}$$

12–4 ZERO MEAN WHITE NOISE RANDOM SIGNAL

For the purposes of this chapter, we can define a *zero mean white noise random signal* $x(n)$ as a signal with the three following properties:

1. $E[x(n)] = 0$.
2. Two samples of the signal taken at different times are *independent*. That is, $x(k)$ and $x(k+n)$ are independent (unless $n = 0$).
3. White noise is independent with respect to any other signal.

As mentioned in the last section, if two random variables are independent, the expected value of their product is equal to the product of their expected values. Therefore, for the white noise signal $x(n)$:

$$E[x(k)x(k+n)] = E[x(k)]E[x(k+n)], n \neq 0 \qquad \textbf{(12–33)}$$

But since $E[x(n)] = 0$, it follows that

$$E[x(k)x(k+n)] = 0, n \neq 0 \tag{12-34}$$

When $n = 0$, we have

$$E[x(k)x(k+0)] = E[x^2(k)] = \sigma_x^2 \tag{12-35}$$

As noted earlier in this chapter, $E[x(k)x(k+n)]$ is the *autocorrelation function* for the signal $x(n)$, denoted $r_{xx}(n)$. Using the results from Equations (12–34) and (12–35), it can be seen that for a zero mean white noise signal,

$$r_{xx}(n) = \sigma_x^2 \delta(n) \tag{12-36}$$

where $\delta(n)$ is the discrete time unit impulse. Since the power spectral density function is the DTFT of the autocorrelation function, the power spectral density function $S_{xx}(\omega)$ for a zero mean white noise signal is constant across the entire spectrum:

$$S_{xx}(\omega) = \sigma_x^2 \tag{12-37}$$

This is illustrated in Figure 12–5. Note that the Wiener-Kinchine theorem is easy to verify in this case:

$$\frac{1}{2\pi} \int_{-\pi}^{\pi} S_{xx}(\omega)d\omega = \frac{1}{2\pi} \int_{-\pi}^{\pi} \sigma_x^2 d\omega = \sigma_x^2 \tag{12-38}$$

FIGURE 12–5
Autocorrelation function (a) and power spectral density function (b) for a zero mean white noise random signal $x(n)$ having average power $\sigma_x^2 = E[x^2(n)]$

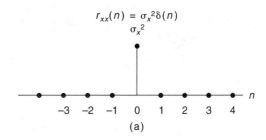

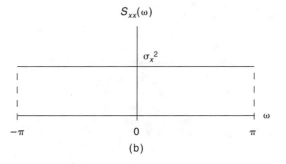

An example of a zero mean white noise signal is one created by the following hypothetical process: For each n, select a real number at random from the range $-\Delta/2 \le x \le \Delta/2$. Each selection must be made without bias and without reference to any previous selections. The random signal created in this manner is a uniformly distributed zero mean white noise signal. Under many realistic contitions, a uniformly distributed zero mean white noise signal is a good model for the quantization error associated with an analog-to-digital converter, if Δ is the quantization interval. This is the subject of Section 12–5.

12–5 A/D CONVERSION, QUANTIZATION NOISE, AND OVERSAMPLING

Consider an N-bit A/D converter with an input voltage range bounded by $\pm V$ volts. The analog input signal to the A/D converter is $x_a(t)$. The A/D converter samples $x_a(t)$ once every T seconds and outputs an N-bit binary number that is the *quantized* value of the sample $x_a(nT)$. N-bits can represent $Q = 2^N$ quantized voltage levels. Each level corresponds to a change in the least significant bit (LSB) of the converter output. The distance between levels, in volts, is denoted Δ:

$$\Delta = \frac{2V}{Q - 1} = \frac{2V}{2^N - 1} \tag{12–39}$$

This is illustrated for a simple case ($N = 3$) in Figure 12–6.

The A/D converter *in effect* takes an input sample $x_a(nT)$ and assigns it to the nearest level. This nearest level is the quantized value of $x_a(nT)$, which will virtually always be different than the true value of $x_a(nT)$. The difference between the true value and the quantized value is called the *quantization error*. It should be clear that the quantization error is bounded by $\pm$ half of Δ:

$$-\frac{\Delta}{2} \le \text{quantization error} \le \frac{\Delta}{2} \tag{12–40}$$

FIGURE 12–6
$N = 3$-bit A/D converter with input voltage range $\pm V$. An input voltage is assigned to one of the 2^N quantized voltage levels. Δ is the spacing between levels. The difference between the true input value and the quantized value is the quantization error, which can be modeled as zero mean white noise with power $\sigma^2_e = \Delta^2/12$.

$N = 3$ bits
$Q = 2^N = 8$ levels

$\Delta = \frac{2v}{7}$

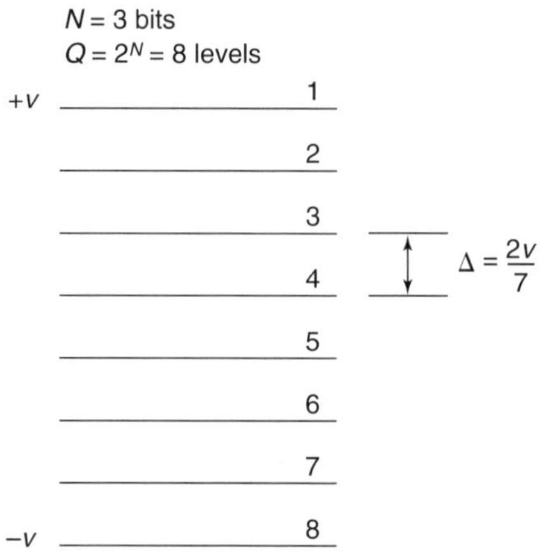

Let us determine what this implies about the discrete time signal that constitutes the output of the A/D converter. First, define

$$x(n) = x_a(nT) \qquad (12–41)$$

Note that $x(n)$ is not the exact output of the A/D converter because, as we have seen, there is a quantization error associated with each sample. Let $e(n)$ be the quantization error at time n. Let $\hat{x}(n)$ be the actual A/D converter output. The relationship between these signals is

$$\hat{x}(n) = x(n) + e(n) \qquad (12–42)$$

In other words, the signal appearing at the output of the A/D converter is the sum of two components: the true signal $x(n)$ (which an *ideal* A/D converter would produce) plus a *quantization noise* signal $e(n)$. The true signal $x(n)$ has an average power $\sigma_x^2 = E[x^2(n)]$. The quantization noise signal has an average power $\sigma_e^2 = E[e^2(n)]$. Obviously we would like the true signal power to dominate the noise power. The A/D converter output signal $\hat{x}(n)$ can be characterized by an A/D conversion *signal-to-noise ratio (SNR)*, defined as follows:

$$\text{SNR} = \frac{\sigma_x^2}{\sigma_e^2} \qquad (12–43)$$

Signal-to-noise ratio is often expressed in terms of decibels, as follows:

$$\text{SNR}_{dB} = 10\log(\text{SNR}) = 10\log\sigma_x^2 - 10\log\sigma_e^2 \qquad (12–44)$$

Let us now attempt to characterize $e(n)$ statistically. First of all, it is reasonable to assume that for any time n, $e(n)$ is in effect a random variable with possible values in the range $-\Delta/2 \leq e(n) \leq \Delta/2$. It is also reasonable to assume that any one value in this range is as likely as any other value. In other words, we can think of $e(n)$ as being created by the "pick a number between $-\Delta/2$ and $\Delta/2$" process described in the last section. This means that $e(n)$ is a uniformly distributed random variable. Since positive and negative values are equally likely, it turns out that $e(n)$ is a zero mean random signal. (A formal demonstration of this fact requires the use of Equation [12–1].) Using Equation (12–2), it can also be shown that the average power σ_e^2 in $e(n)$ is given by

$$\sigma_e^2 = E\left[e^2(n)\right] = \frac{\Delta^2}{12} \qquad (12–45)$$

The relationship between the SNR and the number of bits (N) in the A/D converter can be developed as follows. From Equations (12–39), (12–43), and (12–45), we have

$$\text{SNR} = \frac{\sigma_x^2}{\sigma_e^2} = \frac{12\sigma_x^2}{\Delta^2} = \frac{12\sigma_x^2\left(2^N - 1\right)^2}{(2V)^2} = 3\frac{\sigma_x^2}{V^2}\left(2^N - 1\right)^2 \qquad (12–46)$$

Therefore, SNR_{dB} can be calculated as follows:

$$\text{SNR}_{dB} = 10\log(3) + 10\log\left(\frac{\sigma_x^2}{V^2}\right) + 10\log\left(2^N - 1\right)^2 \qquad (12–47)$$

For real-world A/D converters, where $N = 8$, 10, or 12, the approximation $2^N - 1 \cong 2^N$ is valid. Therefore,

$$\text{SNR}_{dB} \cong 10\log(3) + 10\log\left(\frac{\sigma_x^2}{V^2}\right) + 10\log\left(2^{2N}\right)$$

$$\cong 4.77 + 10\log\left(\frac{\sigma_x^2}{V^2}\right) + 6.02N$$

(12–48)

Equation (12–48) shows that for each bit added to the A/D converter, the SNR_{dB} is improved by approximately 6 dB. (This is the well-known "6 dB per bit" rule of thumb.) It is intuitive that increasing the number of bits will decrease the average squared quantization error, since the quantization levels will be closer together.

Under many realistic conditions, it is at least plausible to argue that for all practical purposes, quantization errors occurring at different times are independent. This is not strictly true, but the assumption is often used because it results in a model that is mathematically tractable and is true enough to lead to good engineering results. This assumption of independence implies that $e(n)$ can be modeled as a uniformly distributed white noise signal, as described in Section 12–4. Therefore, the power spectral density function of this noise, denoted $S_{ee}(\omega)$, is flat across the entire spectrum:

$$S_{ee}(\omega) = \sigma_e^2 = E\left[e^2(n)\right]$$

(12–49)

This is shown in Figure 12–7. It is important to note that $S_{ee}(\omega)$ covers the entire range $-\pi \leq \omega \leq \pi$ regardless of the sampling frequency of the A/D converter. In other words, the quantization noise power at the output of the A/D converter is always evenly distributed across the frequency (Hz) range $-f_s/2 \leq f \leq f_s/2$. On the other hand, the total quantization noise power σ_e^2 depends on Δ, not on the sampling frequency. Therefore, increasing the sampling frequency does not change σ_e^2, but it does serve to spread out the noise power over a wider part of the spectrum.

FIGURE 12–7
Power spectral density function of quantization noise, which is modeled as zero mean white noise with power $\sigma_e^2 = \Delta^2/12$.

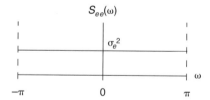

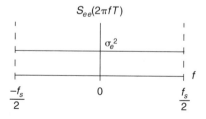

Since $e(n)$ is being modeled as zero mean white noise, we assume it is independent of the true signal $x(n)$. Here again, this is an assumption that is not strictly true, but is true enough for practical purposes. Let $S_{\hat{x}\hat{x}}(\omega)$, $S_{ee}(\omega)$, and $S_{\hat{x}\hat{x}}(\omega)$ be the power spectral density functions for signals $x(n)$, $e(n)$, and $\hat{x}(n)$, respectively. Since $x(n)$ and $e(n)$ are independent, it follows (from Section 12–3) that

$$S_{\hat{x}\hat{x}}(\omega) = S_{xx}(\omega) + S_{ee}(\omega) \qquad \text{(12–50)}$$

Figure 12–8(a) shows the power spectral density function $S_{xxa}(f)$ of a hypothetical analog signal $x_a(t)$ bandlimited to B Hz. Figures 12–8(b), 12–8(c), and 12–8(d) show both $S_{xx}(\omega)$ and $S_{ee}(\omega)$ for three different sampling frequencies: $f_s = 2B$, $4B$, and $6B$, respectively. Refer to Figure 12–1 to see how the bandwidths (on the ω axis) and amplitude scale factors are determined. The true signal power and quantization noise power (and thus the SNR) is the same in Figures 12–8(b), 12–8(c), and 12–8(d). This result is generalized in Figure 12–9, in which the sampling frequency is $f_s = 2MB$; M is the oversampling factor (we must have $M \geq 1$ to avoid aliasing). The reader is invited to use the Wiener-Kinchine theorem (Equation [12–13]) to verify that in Figure 12–9(b), the average power in $x(n)$ is equal to BA and therefore does not depend on M (i.e., does not depend on the sampling frequency).

Observe that as the sampling frequency is increased above the theoretical minimum of $2B$, some of the quantization noise power spectral density function lies in the frequency range beyond that covered by the signal itself. This out-of-signal bandwidth noise can be eliminated using a discrete time lowpass filter with a cutoff frequency on the ω scale of $\omega_c = \pi/M$ without altering the desired signal itself, as shown in Figures 12–10 and 12–11. This will improve the SNR at the output of the filter. The scheme depicted in Figure 12–10 (A/D converter with a discrete time lowpass filter in cascade) can be thought of as an oversampling A/D conversion *system*.

The *resolution* (in bits) of an oversampling A/D conversion system is the *effective* number of bits with respect to SNR. As noted earlier, each bit added to the A/D converter results in a 6 dB improvement in SNR_{dB}. Therefore, for every 6 dB of improvement obtained using the oversampling/filtering scheme, the *resolution* is said to have been improved by 1 bit, even though the actual number of bits (N) has not changed.

To simplify the analysis, assume that the lowpass filter depicted in Figures 12–10 and 12–11 is an ideal lowpass filter with a passband gain of unity. The output of the filter, denoted $g(n)$, has two components, signal plus noise:

$$g(n) = x(n) + \hat{e}(n) \qquad \text{(12–51)}$$

Again note that the filter does not change the desired signal, which is $x(n)$ both before and after the filter. On the other hand, the filter does alter the noise, which is $e(n)$ at the filter input and $\hat{e}(n)$ at the output. Since the two inputs are assumed to be independent, the power spectral density function at the filter output is

$$S_{gg}(\omega) = \left| H\left(e^{j\omega}\right) \right|^2 S_{xx}(\omega) + \left| H\left(e^{j\omega}\right) \right|^2 S_{ee}(\omega)$$
$$= S_{xx}(\omega) + S_{\hat{e}\hat{e}}(\omega) \qquad \text{(12–52)}$$

FIGURE 12–8
A continuous time random signal with bandwidth B hertz having the power spectral density function depicted in (a) is sampled at (b) $f_s = 2B$; (c) $f_s = 4B$; (d) $f_s = 6B$. $S_{xx}(\omega)$ is the power spectral density function of the sequence at the A/D converter output that would result if there was no quantization error. $S_{ee}(\omega)$ is the power spectral density function of the quantization noise, which spans the entire range $-\pi < \omega < \pi$ regardless of the sampling frequency. The power spectral density function at the output of the A/D converter is the sum of $S_{xx}(\omega)$ and $S_{ee}(\omega)$.

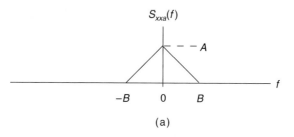

(a)

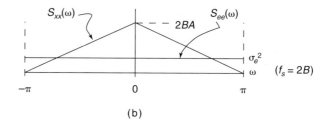

(b)

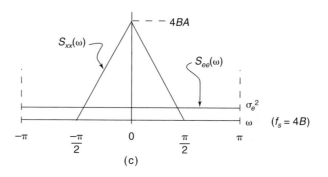

(c)

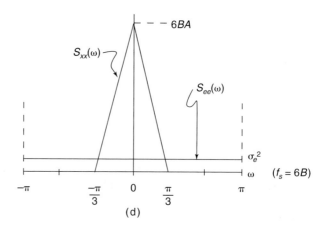

(d)

FIGURE 12–9
Generalization of Figure 12–8. If $f_s = 2MB$, $S_{xx}(\omega)$ is band-limited to $\omega = \pi/M$, but $S_{ee}(\omega)$ spans the entire range $-\pi < \omega < \pi$ regardless of the sampling frequency.

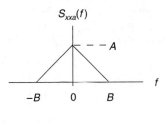

(a)

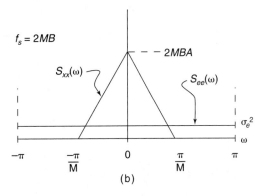

(b)

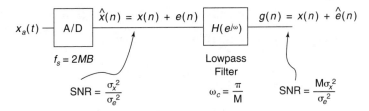

FIGURE 12–10
A lowpass filter with cutoff frequency $\omega_c = \pi/M$ will eliminate the part of $S_{ee}(\omega)$ outside of the frequency range covered by $S_{xx}(\omega)$, thus reducing the quantization noise power without reducing the signal power.

This result is depicted in Figure 12–11. The desired signal power at the filter output is the same as it is at the filter input: $\sigma_x^2 = E[x^2(n)]$. On the other hand, the noise power at the filter output, denoted $\sigma_{\hat{e}}^2$, has been reduced as follows:

$$\sigma_{\hat{e}}^2 = \frac{1}{2\pi} \int_{-\pi/M}^{\pi/M} S_{\hat{e}\hat{e}}(\omega)d\omega = \frac{1}{2\pi} \int_{-\pi/M}^{\pi/M} \sigma_e^2 d\omega$$

$$\sigma_{\hat{e}}^2 = \frac{\sigma_e^2}{M}$$

(12–53)

FIGURE 12–11
Frequency domain plots that
go along with Figure 12–10.

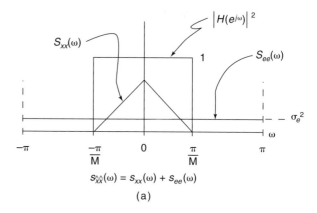

$$S_{\hat{x}\hat{x}}(\omega) = S_{xx}(\omega) + S_{ee}(\omega)$$

(a)

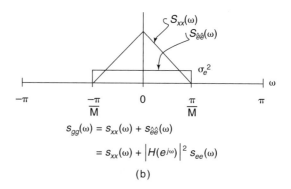

$$S_{gg}(\omega) = S_{xx}(\omega) + S_{\hat{e}\hat{e}}(\omega)$$

$$= S_{xx}(\omega) + \left|H(e^{j\omega})\right|^2 S_{ee}(\omega)$$

(b)

This means that the signal-to-noise ratio at the filter output is

$$\underset{\text{filter output}}{\text{SNR}} = \frac{\sigma_x^2}{\sigma_{\hat{e}}^2} = M \frac{\sigma_x^2}{\sigma_e^2} = M \times \underset{\text{filter input}}{\text{SNR}} \tag{12–54}$$

In other words, the oversampling/filtering scheme has multiplied the SNR by factor M. In terms of decibels, we have

$$\underset{\text{filter output}}{\text{SNR}_{\text{dB}}} = \underset{\text{filter input}}{\text{SNR}_{\text{dB}}} + 10 \log M \tag{12–55}$$

This important result shows that the SNR_{dB} at the filter output is increased by $10 \log M$ dB. For example, if $M = 2$, the improvement is 3 dB; if $M = 4$, the improvement is 6 dB; etc. Observe that for every *quadrupling* of M, the SNR_{dB} is improved by 6 dB, which is exactly the same improvement that would be obtained by increasing the actual number of A/D converter bits (N) by 1 bit. Thus the *resolution* of the oversampling A/D conversion *system* is increased by 1 bit for every quadrupling of the oversampling factor M.

Let R be the resolution (in bits) of the oversampling A/D conversion system. The relationship between R, N, and M is as follows:

$$R = N + \frac{1}{2} \log_2 M \tag{12–56}$$

If Equation (12–56) is solved for M, we obtain a formula that tells us what the oversampling factor must be to obtain a desired R bits of resolution, given an A/D converter with N bits:

$$M = 2^{2(R-N)} \qquad\qquad (12\text{–}57)$$

For example, to obtain $R = 10$ bits of resolution from a system using an $N = 8$ bit A/D converter, the oversampling factor must be $M = 2^4 = 16$. To put this in perspective, suppose the signal being sampled, $x_a(t)$, has a bandwidth of $B = 20$ kHz (a music signal, for example). To obtain 10 bits of resolution with an 8-bit A/D converter, the sampling frequency would have to be $f_s = 2MB = 2 \times 16 \times 20000 = 640$ kHz. On the other hand, to obtain $R = 16$ bits of resolution with an $N = 8$ bit A/D converter, the oversampling factor must be $M = 65536$, which translates to an unrealistic sampling frequency of 2.62 GHz.

The *dynamic range* of an A/D converter is defined as the SNR obtained when the input signal is a full range sinusoid; that is, $x_a(t) = V\cos(2\pi f_o t + \theta)$. The RMS or effective value of this sinusoid is $V/\sqrt{2}$. But the RMS value is actually the square root of the average squared value, which is $V^2/2$. Therefore, it should not violate the reader's intuition to learn ("it can be shown") that the average power of the sampled version of this sinusoid, $\sigma_x^2 = E[x^2(n)]$, is also equal to $V^2/2$. (Formal proof of this fact is beyond the scope of this text.) Let D_{dB} denote the dynamic range in dB. By substituting $\sigma_x^2 = V^2/2$ into Equation (12–48), we obtain the following:

$$D_{\text{dB}} \cong 10\log(3) + 10\log\left(\frac{1}{2}\right) + 10\log\left(2^{2N}\right) \qquad\qquad (12\text{–}58)$$

$$D_{\text{dB}} \cong 1.76 + 6.02N$$

Thus the dynamic range is also improved by 6 dB per bit. An 8-bit A/D converter has a dynamic range of approximately 50 dB, a 10-bit A/D converter has a dynamic range of approximately 62 dB, and so on.

The dynamic range (in dB) of an oversampling/filtering A/D conversion system can be found by using Equations (12–58) and (12–55):

$$D_{\text{dB}\atop\text{oversampling}} \cong 1.76 + 6.02N + 10\log M \qquad\qquad (12\text{–}59)$$

For example, the dynamic range of an A/D system using an $N = 8$-bit A/D converter with an oversampling factor of $M = 64$ is approximately 68 dB, which is the equivalent of 11 bits of resolution.

12–6 DECIMATION AND INTERPOLATION

In the previous section we observed that oversampling, that is, using a sampling frequency $f_s = 2MB$ significantly greater than the theoretical minimum $(2B)$, improves the signal to noise ratio with respect to quantization error when combined with an appropriate digital lowpass filter. In Chapter 7 it was also observed that using a larger sampling frequency also reduces the distortion created by the real-world digital-to-analog conversion process. Note also that when the sampling frequency is significantly larger than the theoretical minimum, the design requirements for the analog anti-aliasing lowpass filter placed before the A/D converter, as well as the analog reconstruction filter placed after the D/A converter, are simplified because the cutoff characteristic of these filters need not be as sharp.

Of course, if the digital signal processing system in question must operate in real time, increasing the sampling frequency will increase the number of mathematical operations the processor must carry out per unit time. If the sampling frequency is too large, the processor might not be able to keep up the pace. Note also that if the sampled signal is to be stored in memory for later retrieval, increasing the sampling frequency will increase the amount of memory required (since for a given length of time, more samples will be taken). Clearly, a high sampling frequency can potentially create serious problems with respect to both cost (dollars) and implementation.

Fortunately, there is a way around this problem. We begin with the observation that the signal at the output of the lowpass filter in Figure 12–10, $g(n)$, is bandlimited to $\omega_B = \pi/M$ on the ω scale, as shown in Figure 12–11. On the f (hertz) scale, this means that $g(n)$ is bandlimited to B Hz. For the purposes of this discussion, it is convenient to think of $g(n)$ as a discrete time signal obtained by sampling (at $f_s = 2MB$) a hypothetical continuous time signal $g_a(t)$ that is bandlimited to B Hz. Although the actual sampling frequency is $f_s = 2MB$, $g_a(t)$ *could have been, in theory,* sampled at $f_s = 2B$ without any aliasing. Figure 12–12(a) shows the hypothetical Fourier transform of $g_a(t)$, denoted $G_a(f)$; Figures 12–12(b) and 12–12(c) show the DTFT of $g(n)$, denoted $G(e^{j\omega})$, resulting from sampling frequencies $f_s = 2MB$ and $f_s = 2B$, respectively. Figure 12–12 clearly shows that if the sampling frequency at the output of the lowpass filter could somehow be reduced down to $f_s = 2B$, the resulting sequence would still represent the same continuous time signal without aliasing. If this could be done, the processor could operate at a lower rate. A moment's reflection will reveal that this effect can be obtained in a very simple manner, assuming that oversampling factor M is an integer: If we save only every Mth sample of $g(n)$ and discard the $M - 1$ samples in between, the resulting sequence will be the same as the one obtained by (theoretically) resampling $g_a(t)$ at $f_s = 2B$. This process is called *decimation* or *downsampling*.

The decimation process is indicated schematically as shown in Figure 12–13. The notation M↓ means *decimate by factor M*. To distinguish between the original sequence $g(n)$ and its decimated version, the decimated version is denoted $g_D(n)$. Mathematically, the relationship between the two sequences can be expressed as

$$g_D(n) = g(Mn) \tag{12–60}$$

The oversampling/filtering/decimation scheme is summarized by Figure 12–14. However, if the system is implemented exactly as suggested by this figure, the lowpass filter must still operate at the high sampling frequency ($2MB$). This lowpass filter must also be implemented by the processor, so it would be nice if this filter could also be operated at the lower sampling frequency ($2B$). This can be done by embedding the decimation operation within the filter itself.

Assume that the lowpass filter in Figure 12–14 is an N-point FIR filter, as shown in Figure 12–15. In this FIR filter, $g(n)$ is calculated as follows:

$$g(n) = \sum_{k=0}^{N-1} h(k)\hat{x}(n - k) \tag{12–61}$$

FIGURE 12–12

If a signal having spectrum $G_a(f)$, shown in (a), is sampled at $f_s = 2MB$, the resulting discrete time signal will have the spectrum depicted in (b). If the discrete time signal is then decimated by factor M, the spectrum is changed to that depicted in (c).

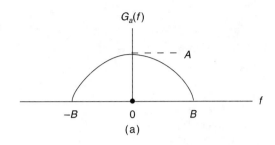

(a)

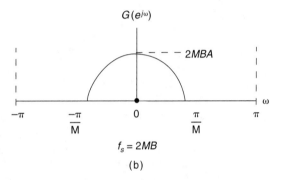

$f_s = 2MB$

(b)

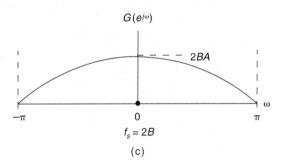

$f_s = 2B$

(c)

Equations (12–60) and (12–61) can be combined as follows:

$$g_D(n) = g(Mn) = \sum_{k=0}^{N-1} h(k)\hat{x}(Mn - k) \qquad (12\text{–}62)$$

The sequence $g_D(n)$ can be viewed as a weighted sum of sequences; that is,

$$g_D(n) = h(0)\hat{x}(Mn) + h(1)\hat{x}(Mn - 1) + h(2)\hat{x}(Mn - 2) + h(3)\hat{x}(Mn - 3) + \cdots \qquad (12\text{–}63)$$

FIGURE 12–13

The decimation (i.e., downsampling) operation is represented by the symbol $M\downarrow$.

$$g(n) \longrightarrow \boxed{M\downarrow} \longrightarrow g_D(n) = g(Mn)$$

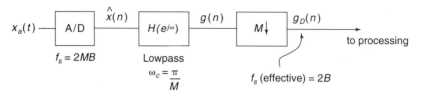

FIGURE 12–14
Oversampling/lowpass filtering/decimation scheme designed to reduce the signal-to-noise ratio by initially oversampling by factor M, but allowing the main processing to function at the rate $f_s = 2B$.

The kth sequence, $\hat{x}(Mn - k)$, is created by first delaying $\hat{x}(n)$ by k units of time and *then* decimating it by factor M. In other words, each of the delayed versions of $\hat{x}(n)$ must be separately decimated. Equation (12–62), which is the difference equation for the combined lowpass/decimation filter, can therefore be realized using the structure shown in Figure 12–16. (Sometimes this is just called a *decimation filter.*) Observe that the multiplication and addition operations are performed at the reduced sampling rate (2B).

Let us now turn our attention to the big picture shown in Figure 12–17. The discrete time signal $g_D(n)$, obtained as described above, will be processed at the effective sampling rate $f_s = 2B$. (Exactly what the processing software consists of depends on what the system designer is trying to accomplish; this is not an issue here.) The output of the processing block is a discrete time signal $y(n)$, which must be converted to a continuous time signal $y_a(t)$ by a D/A converter. For reasons covered in Chapter 7, it is desirable to operate the D/A converter at a rate much higher than $f_s = 2B$. Without sacrificing much generality, let's say that we want the D/A converter to operate at $f_s = 2MB$, where M is the oversampling factor used at the system input.

It is convenient to think of $y(n)$ as a discrete time signal obtained by sampling (at $f_s = 2B$) a hypothetical continuous time signal $y_a(t)$, bandlimited to B hertz. Although the actual sampling frequency at this point is $f_s = 2B$, $y_a(t)$ *could have been, in theory,* sampled at the higher rate $f_s = 2MB$. Figure 12–18(a) shows the hypothetical Fourier transform of $y_a(t)$,

FIGURE 12–15
FIR lowpass filter followed by sequence decimation by factor M, as depicted in Figure 12–14. If the decimation operation follows the filtering operation, then the filter must operate at the higher rate $f_s = 2MB$.

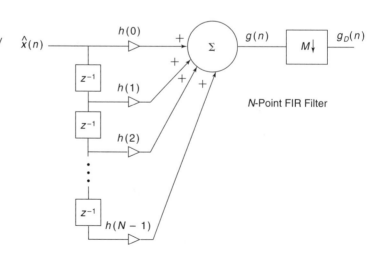

FIGURE 12–16
FIR decimation filter that combines the operations of lowpass filtering and decimation such that the multiply-and-sum operation functions at the reduced rate $f_s = 2B$.

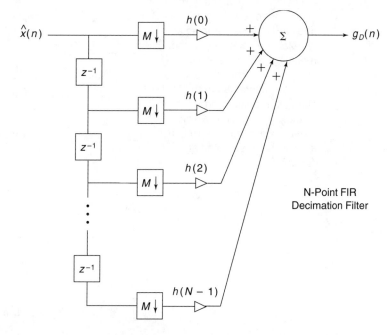

denoted $Y_a(f)$. Figure 12–18(b) shows the DTFT of the signal $y(n)$, denoted $Y(e^{j\omega})$, that we actually have at the output of the processing block. Figure 12–18(c) shows the DTFT of the signal $y_I(n)$, denoted $Y_I(e^{j\omega})$, that would be obtained if $y_a(t)$ were actually sampled at the higher rate $f_s = 2MB$. The signal $y_I(n)$ *is what we actually want to present to the D/A con-verter.* Obviously we can't physically resample $y_a(t)$ at the higher rate because $y_a(t)$ doesn't actually exist at this point in the process. However, there is an algorithm known as *interpo-lation* or *upsampling* that will convert $y(n)$ to $y_I(n)$. (The conversion is approximate but is close enough for practical purposes.) This is the interpolation filter indicated in Figure 12–17. Basically, the interpolation algorithm changes $y(n)$ such that its DTFT is trans-formed from Figure 12–18(b) to Figure 12–18(c).

To explain the interpolation process, let's begin with the following observation. Sup-pose we take the signal $y(n)$ and insert $(M-1)$ zero-valued samples between each sample, as shown in Figure 12–19. (Figure 12–19 illustrates the specific case $M = 4$). Call the resulting

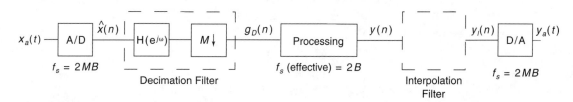

FIGURE 12–17
The interpolation filter (see text) in effect *upsamples* the sequence $y(n)$ back to the oversampling rate $f_s = 2MB$ so that the D/A converter can operate at the higher sampling frequency. This will improve the performance with respect to D/A conversion distortion.

FIGURE 12–18
If a hypothetical signal $y_a(t)$ having spectrum $Y_a(f)$ as depicted in (a) is sampled at $f_s = 2B$, the spectrum of the resulting sequence is as shown in (b). The interpolation filter described in the text changes the spectrum of (b) to that shown in (c), which is what would result if $y_a(t)$ had actually been sampled at $f_s = 2MB$.

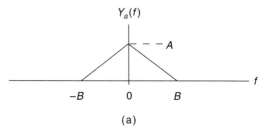

$Y_a(f)$

$-B$ 0 B

A

(a)

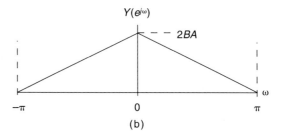

$Y(e^{j\omega})$

$2BA$

$-\pi$ 0 π

(b)

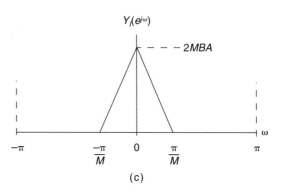

$Y_i(e^{j\omega})$

$2MBA$

$-\pi$ $\dfrac{-\pi}{M}$ 0 $\dfrac{\pi}{M}$ π

(c)

FIGURE 12–19
The zero insertion operation, shown here for $M = 4$, is the first step in the interpolation process.

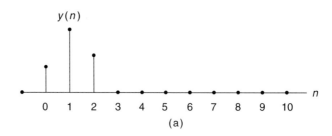

$y(n)$

0 1 2 3 4 5 6 7 8 9 10 n

(a)

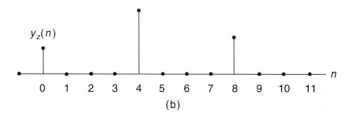

$y_z(n)$

0 1 2 3 4 5 6 7 8 9 10 11 n

(b)

FIGURE 12–20
The zero insertion operation is
represented by the symbol $M\!\uparrow$.

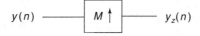

$$y(n) \longrightarrow \boxed{M\!\uparrow} \longrightarrow y_z(n)$$

signal $y_z(n)$. This *zero insertion* process is indicated schematically by the symbol $M\!\uparrow$ as shown in Figure 12–20. The relationship between $y(n)$ and $y_z(n)$ can be expressed as follows:

$$y_Z(n) = \begin{cases} y\!\left(\dfrac{n}{M}\right), & \text{if } n = \text{integer multiple of } M \\ 0 & , \text{ otherwise} \end{cases} \tag{12–64}$$

The DTFT of $y_z(n)$, denoted $y_Z(e^{j\omega})$, is calculated as follows:

$$Y_Z\!\left(e^{j\omega}\right) = \sum_{n=-\infty}^{\infty} y_Z(n)e^{-j\omega n} \tag{12–65}$$

Considering Equation (12–64), this DTFT can also be expressed as

$$Y_Z\!\left(e^{j\omega}\right) = \sum_{\substack{n=-\infty \\ n=\text{integer multiple of } M}}^{\infty} y\!\left(\frac{n}{M}\right)e^{-j\omega n} \tag{12–66}$$

which in turn can be expressed as

$$Y_Z\!\left(e^{j\omega}\right) = \sum_{n=-\infty}^{\infty} y(n)e^{-jM\omega n} \tag{12–67}$$

The DTFT of $y(n)$ itself can be expressed as

$$Y\!\left(e^{j\omega}\right) = \sum_{n=-\infty}^{\infty} y(n)e^{-j\omega n} \tag{12–68}$$

When Equations (12–67) and (12–68) are compared, the relationship between the DTFTs of $y(n)$ and $y_z(n)$ is seen to be

$$Y_Z\!\left(e^{j\omega}\right) = Y\!\left(e^{j(M\omega)}\right) \tag{12–69}$$

In other words, the only difference between these two DTFTs is that the ω axis has been rescaled: $Y_Z(e^{j\omega})$ looks like $Y(e^{j\omega})$ except that the *period* has been changed from 2π to $2\pi/M$. This relationship is illustrated in Figure 12–21(a) and 12–21(b) for the specific case $M = 4$. At this point the reader should ask the following question: What could be done to $y_Z(n)$ to transform its DTFT (Figure 12–21[b]) to that which is desired, namely $Y_I(e^{j\omega})$ as shown in Figure 12–18(c)? Observe that the part of $Y_Z(e^{j\omega})$ in the interval $-\pi/M \leq \omega \leq \pi/M$ looks exactly like the desired $Y_I(e^{j\omega})$, to within an amplitude scale factor. The rest of $Y_Z(e^{j\omega})$ needs to be eliminated; this can be accomplished with a very sharp lowpass filter with a cutoff frequency of $\omega_c = \pi/M$, as shown also in Figure 12–21. The part of the spectrum eliminated by the filter is sometimes called the *image spectrum*. The lowpass filter used here is sometimes called an *anti-image filter*. (Observe that if the passband gain of the lowpass filter is M instead of 1, the amplitude scaling factor will be corrected.) In effect, the

FIGURE 12–21
Zero insertion changes the spectrum from (a) to (b). (The $M = 4$ case is shown here.) When the signal with the inserted zeros is processed by an ideal lowpass filter (c) with cutoff frequency $\omega_c = \pi/M$, the resulting signal has the spectrum shown in (d), which is exactly what we would get if the hypothetical underlying continuous time signal was actually resampled at the higher rate $f_s = 2MB$.

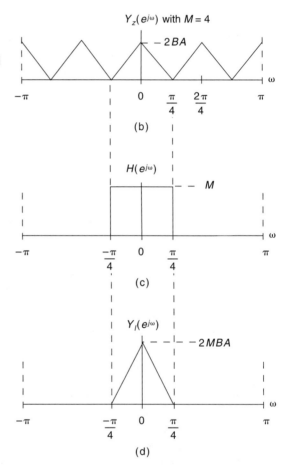

lowpass filter fills in the zero sample values appearing in $y_z(n)$ to create the interpolated sequence $y_I(n)$. (In order for this to work perfectly, the anti-image filter would have to be an ideal lowpass filter.)

The interpolation (upsampling) process described here is summarized in Figure 12–22, which represents the interpolation filter shown in the overall system of Figure 12–17. If the

FIGURE 12–22
The interpolation filter combines the zero insertion
operation with a lowpass filter.

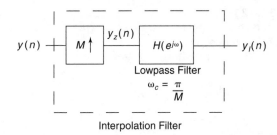

Interpolation Filter

interpolation filter is directly implemented, as shown in Figure 12–22, the structure shown in Figure 12–23 will result (assuming an FIR filter is used). Note that this filter will have to operate at the higher sampling rate, however. Fortunately, the standard FIR filter structure shown in Figure 12–23 can be transposed, resulting in the structure shown in Figure 12–24. We leave it to the reader to show that both FIR structures have the same transfer function, that is,

$$H(z) = \sum_{k=0}^{N-1} h(k)z^{-k} \tag{12–70}$$

The transposed structure is advantageous in this case because the operations *insert zero values* and *multiply by a constant* will commute. If the zero insertion operation is done *after* the multiply by a constant operation instead of before, we will avoid wasting valuable processing time multiplying by zero. This observation leads to the more efficient structure of Figure 12–25, in which the multiplications are carried out at the lower rate $f_s = 2B$.

FIGURE 12–23
Direct implementation of
Figure 12–22 using an FIR
lowpass filter. The filter must
operate at the higher rate
$f_s = 2MB$.

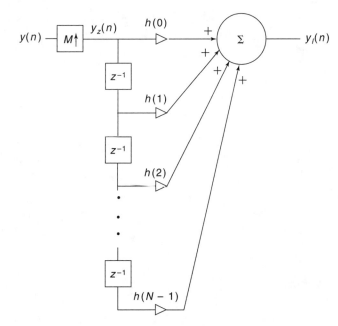

FIGURE 12–24
Transposed version of the
FIR filter structure shown in
Figure 12–23.

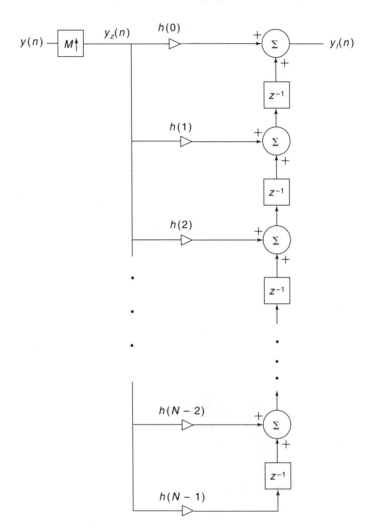

FIGURE 12–25
Efficient implementation
of the interpolation filter,
based on the fact that the
zero insertion operation
can be carried out *after* the
multiplications shown in
the transposed structure of
Figure 12–24. The multiplications can be carried out at
the lower rate ($f_s = 2B$).

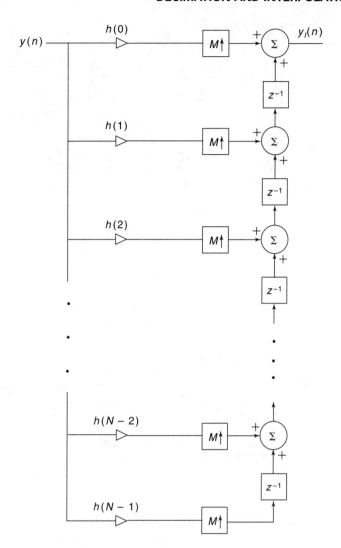

PROBLEMS

12–1. The power spectral density function, $S_{xx}(\omega)$, is a real function of ω. Does it follow that the autocorrelation function must be an even function of n, that is, $r_{xx}(n) = r_{xx}(-n)$? Why or why not?

12–2. Given that $S_{xx}(\omega)$ is the DTFT of $r_{xx}(n)$, use the inverse DTFT integral to show that

$$r_{xx}(0) = \tfrac{1}{2\pi} \int_{-\pi}^{\pi} S_{xx}(\omega)d\omega$$

12–3. Suppose $x(n)$ is a zero mean, wide sense stationary random signal having the power spectral density function shown in Figure P12–3.
  a. Find the autocorrelation function $r_{xx}(n)$. Verify that if the bandwidth is reduced, the autocorrelation function becomes wider.
  b. Find σ_x^2.

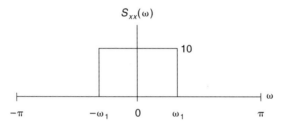

FIGURE P12–3

12–4. Suppose $x(n)$ is zero mean white noise with power $\sigma_x^2 = 10$. Suppose $x(n)$ is the input to an ideal lowpass filter with a cutoff frequency of $\omega_c = \pi/4$ and a passband gain of 1. The filter output signal is $y(n)$.
  a. Sketch $S_{yy}(\omega)$.
  b. Find the power in the output signal (σ_y^2).

12–5. Suppose $x(n)$ is zero mean white noise with power $\sigma_x^2 = 10$. Suppose $x(n)$ is the input to an ideal *bandpass* filter with lower cutoff frequency $\omega_l = \pi/4$, upper cutoff frequency $\omega_u = \pi/2$, and a passband gain of 1. The filter output signal is $y(n)$.
  a. Sketch $S_{yy}(\omega)$.
  b. Find the power in the output signal (σ_y^2).

12–6. Consider the system shown in Figure P12–6(a). $e(n)$ is zero mean white noise with power $\sigma_e^2 = 0.25$. $x(n)$ is the random signal having the power spectral density function shown in Figure P12–6(b). The ideal lowpass filter has cutoff frequency ω_c. The output signal $y(n)$ has two components: a signal component, the part caused by $x(n)$, and a noise component, the part caused by $e(n)$. Find the output SNR_{dB} if
  a. $\omega_c = \pi/8$
  b. $\omega_c = \pi/4$
  c. $\omega_c = \pi/2$
  d. $\omega_c = \pi$ (i.e., no filter)

12–7. Consider an A/D converter with an input voltage range of $\pm V$ volts (see Figure 12–6). Suppose the input signal is $x_a(t) = A\cos(2\pi f_o t + \phi)$, where ϕ is a random phase (meaning, in effect, that the value of the phase shift was selected at random when the signal generator was turned on). Suppose this signal is sampled to create a random discrete time signal $x(n)$. It can be shown that $E[x^2(n)] = \sigma_x^2 = A^2/2$. Find the SNR_{dB} at the A/D converter output if
  a. $A = V$
  b. $A = V/2$
  c. $A = V/4$

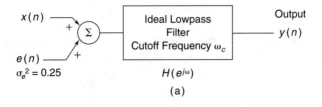

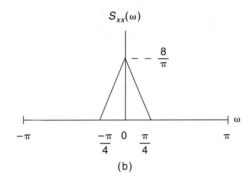

(b)

FIGURE P12–6

How does reducing the peak value of the input sinusoid change the effective resolution (in bits) of the A/D converter?

12–8. Consider the oversampling A/D conversion system shown in Figure 12–10. If the A/D converter has 10 bits, find the oversampling factor (M) such that the effective resolution of the oversampling A/D conversion system is 12 bits.

12–9. Using the sequence $x(n)$ shown in Figure P12–9, show that the sequence $x(Mn - k)$ is created by first delaying $x(n)$ by k samples and then decimating the delayed sequence by factor M. Use $M = 3$ and $K = 2$.

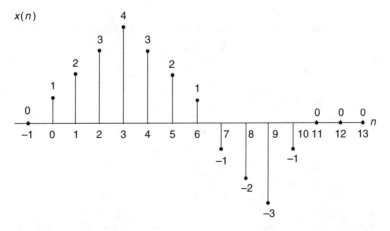

FIGURE P12–9

12–10. Show that with respect to input signal $y_z(n)$ and output signal $y_I(n)$, the structures shown in Figures 12–23 and 12–24 have the same transfer function.

12–11. A/D conversion quantization error (e) is a random variable uniformly distributed on the interval $-\Delta/2 < e < \Delta/2$. Therefore, the probability density function for random variable e is

$$p(e) = \begin{cases} \dfrac{1}{\Delta}, & \dfrac{-\Delta}{2} < e < \dfrac{\Delta}{2} \\ 0, \text{otherwise} \end{cases}$$

Note that $\int_{-\infty}^{\infty} p(e)de = 1$. Use Equation (12–2) to show that $\sigma_e^2 = \Delta^2/12$, as claimed by Equation (12–45). In other words, show that

$$\int_{-\infty}^{\infty} e^2 p(e)de = \frac{\Delta^2}{12}$$

12–12. Consider the system shown in Figure P12–12. Note that $Sxx(\omega)$ is as shown in Figure 12–1, with $A = 1/\pi B$; note also that $\sigma_e^2 = 0.25$. The ideal lowpass filter has a cutoff frequency of $\omega c = \pi/M$. Calculate the SNR_{dB} at the output if

   a. $M = 1$
   b. $M = 2$
   c. $M = 4$

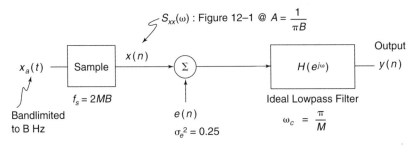

FIGURE P12–12

12–13. Suppose $y(n) \leftrightarrow Y(e^{j\omega})$ as in Figure 12–21(a). Suppose $y_z(n)$ is created as shown by Equation (12–64). Insert $M - 1$ zero-valued samples between each sample of $y(n)$. Sketch $Y_z(e^{j\omega})$ if

   a. $M = 2$
   b. $M = 3$
   c. $M = 8$

A

CONVOLUTION

Consider an LSI system with input $x(t)$, impulse response $h(t)$, and output $y(t)$. In the Laplace transform domain, we calculate $Y(s) = H(s)X(s)$. Substituting the Laplace transform integrals for $H(s)$ and $X(s)$, we have

$$Y(s) = \left[\int_{-\infty}^{\infty} h(\lambda)e^{-s\lambda}d\lambda\right] \cdot \left[\int_{-\infty}^{\infty} x(t)e^{-st}dt\right] \tag{A–1}$$

We have used λ as the dummy variable in the first integral because the next step is to express Equation (A–1) as a double integral:

$$Y(s) = \int_{-\infty}^{\infty} \int_{-\infty}^{\infty} h(\lambda)x(t)e^{-s(t+\lambda)}dtd\lambda \tag{A–2}$$

Equation (A–2) can be expressed as

$$Y(s) = \int_{-\infty}^{\infty} h(\lambda)\left[\int_{-\infty}^{\infty} x(t)e^{-s(t+\lambda)}dt\right]d\lambda \tag{A–3}$$

Change variables on the inner integral of Equation (A–3):

$$t + \lambda = \alpha$$
$$\therefore t = \alpha - \lambda \tag{A–4}$$
$$dt = d\alpha$$

Thus, Equation (A–3) can be written as

$$Y(s) = \int_{-\infty}^{\infty} h(\lambda)\left[\int_{-\infty}^{\infty} x(\alpha - \lambda)e^{-s\alpha}d\alpha\right]d\lambda \tag{A–5}$$

Let's rearrange Equation (A–5) and change the names of the dummy variables from λ and α to τ and t, respectively. (The change in names is merely cosmetic but serves to make the result look like a Laplace transform.)

$$Y(s) = \int_{-\infty}^{\infty}\left[\int_{-\infty}^{\infty} h(\tau)x(t-\tau)d\tau\right]e^{-st}dt \qquad \text{(A–6)}$$

Equation (A–6) says that $Y(s)$ can be found by taking the Laplace transform of the function of t enclosed by brackets, that is,

$$Y(s) = \text{Laplace}\left\{\int_{-\infty}^{\infty} h(\tau)x(t-\tau)d\tau\right\} \qquad \text{(A–7)}$$

But since $Y(s)$ is also the Laplace transform of output signal $y(t)$, we conclude that

$$y(t) = \int_{-\infty}^{\infty} h(\tau)x(t-\tau)d\tau \qquad \text{(A–8)}$$

Equation (A–8) is a *convolution* integral. It says that $y(t)$ is equal to the convolution of $h(t)$ with $x(t)$. Convolution is often denoted by a * operator, that is,

$$y(t) = h(t) * x(t) = \int_{-\infty}^{\infty} h(\tau)x(t-\tau)d\tau \qquad \text{(A–9)}$$

It can be shown that convolution commutes. In other terms,

$$y(t) = h(t) * x(t) = x(t) * h(t) = \int_{-\infty}^{\infty} x(\tau)h(t-\tau)d\tau \qquad \text{(A–10)}$$

Let's take a closer look at the convolution integral (we will use Equation [A–10]). Since the integral is with respect to the variable τ, the functions inside the integral must be thought of as functions of τ. There are two functions of τ, $x(\tau)$ and $h(t-\tau)$, that are multiplied together to create some new function of τ. $x(\tau)$ is fixed in position on the τ axis, but the position of $h(t-\tau)$ depends on the particular value of t; thus the product $x(\tau)h(t-\tau)$ forms a different function of τ for different values of t. The *area* under the curve $x(\tau)h(t-\tau)$, which also depends on t, is the value of $y(t)$ for that particular t.

To illustrate the convolution process, suppose that $h(t)$ is the impulse response of the R-C circuit shown in Figure A–1. To simplify the analysis, suppose $R = 1$ and $C = 1$. The transfer function for this system is

$$H(s) = \frac{\frac{1}{sC}}{R + \frac{1}{sC}} = \frac{1}{s+1} \qquad \text{(A–11)}$$

FIGURE A–1
R-C circuit. The input signal is $x(t)$; the output signal is $y(t)$.

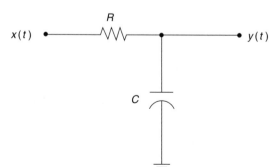

Taking the inverse Laplace transform yields the system impulse response

$$h(t) = \begin{cases} e^{-t}, t \geq 0 \\ 0, \text{otherwise} \end{cases} \tag{A-12}$$

Figure A–2(a) illustrates the function $h(t)$; Figure A–2(b) shows the function $h(\tau)$ (the name of the independent variable has changed). Figure A–2(c) shows the function $h(-\tau)$, which is a time-reversed version of Figure A–2(b). Figure A–2(d) shows the function $h(t - \tau)$, which is a shifted version of $h(-\tau)$. Note in particular that as t increases, $h(t - \tau)$ slides to the right on the τ axis. Figure A–3 shows how $h(t - \tau)$ appears for $t = -1, 0, 1$, and 3; note that $h(-\tau)$ can be thought of as $h(t - \tau)$ evaluated at $t = 0$.

The expression for $h(t - \tau)$ for the example at hand is

$$h(t - \tau) = \begin{cases} e^{-(t-\tau)}, \tau < t \\ 0, \text{otherwise} \end{cases} \tag{A-13}$$

Let us consider $y(t) = h(t) * x(t)$, where $h(t)$ is as given in Equation (A–12), and $x(t)$ is the pulse shown in Figure A–4(a). Changing the name of the independent variable to τ results in $x(\tau)$, as shown in Figure A–4(b). In general, we have

$$y(t) = \int_{-\infty}^{\infty} x(\tau)h(t - \tau)d\tau \tag{A-14}$$

but the actual limits of integration must be obtained (carefully!) for the problem at hand. These limits are defined by those intervals on the τ axis where the product $x(\tau)h(t - \tau)$ is nonzero, or in other words, where $x(\tau)$ and $h(t - \tau)$ *overlap*. Since $h(t - \tau)$ shifts position for different values of t, it is clear that the limits of integration (i.e., the overlap) will depend on t, and there may be several cases to consider. For the specific problem at hand, there are three distinct cases:

Case 1: $t < 0$

Case 2: $0 \leq t < 1$

Case 3: $t \geq 1$

For each case, the overlap is different, as shown in Figure A–5 and listed here for reference:

Case 1: *no overlap, thus* $y(t) = 0$

Case 2: *overlap* : $0 \leq \tau \leq t$

Case 3: *overlap* : $0 \leq \tau \leq 1$

Keeping in mind that for the problem at hand, $x(\tau) = 1$ in the interval of interest, we can set up the convolution integral for the three cases:

Case 1: $y(t) = 0, t < 0$

Case 2: $y(t) = \int_0^t (1)e^{-(t-\tau)}d\tau, 0 \leq t < 1$ $\qquad$ (A-15)

Case 3: $y(t) = \int_0^1 (1)e^{-(t-\tau)}d\tau, t \geq 1$

FIGURE A–2
(a) Impulse response $h(t)$ of the R-C circuit in Figure A–1. (b) Change the variable from t to τ. (c) Time-reversed version of $h(\tau)$. (d) Time reversal combined with time shifting of $h(\tau)$.

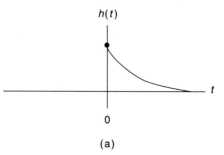

(a)

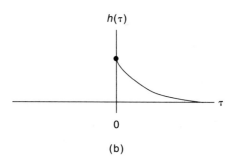

(b)

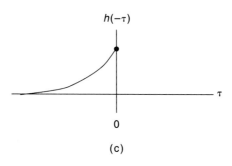

(c)

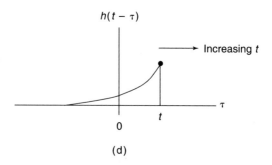

(d)

FIGURE A–3
The function $h(t - \tau)$ for different values of t.

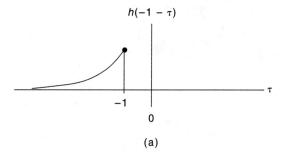

$h(-1 - \tau)$

(a)

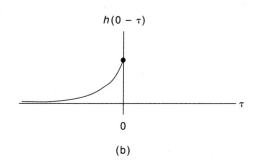

$h(0 - \tau)$

(b)

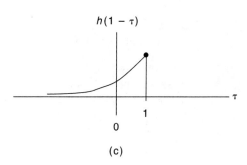

$h(1 - \tau)$

(c)

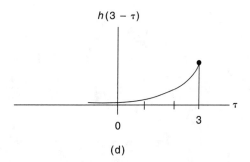

$h(3 - \tau)$

(d)

FIGURE A–4
(a) Input signal $x(t)$ for the
R-C circuit of Figure A–1.
(b) Change the variable from
t to τ.

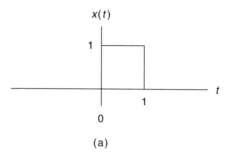

(a)

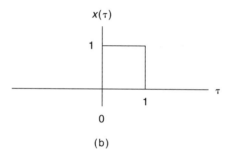

(b)

FIGURE A–5
Determining the limits of
integration for the convolution integral $h(t) * x(t)$. See
Figures A–2 and A–4 for $h(t)$
and $x(t)$, respectively.

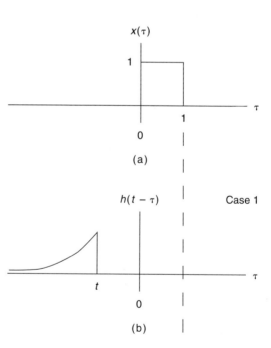

(a)

$h(t - \tau)$ Case 1

(b)

FIGURE A–5
(continued)

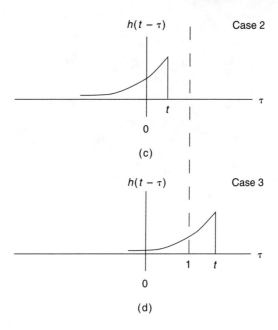

(c)

(d)

FIGURE A–6
Result of convolving $h(t)$
(Figure A–2[a]) with $x(t)$
(Figure A–4[a]).

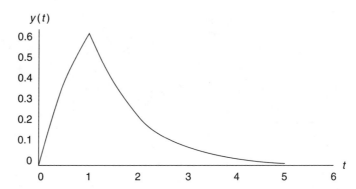

After evaluating the integrals, we obtain

$$y(t) = \begin{cases} 0, t < 0 \\ 1 - e^{-t}, 0 \le t < 1 \\ e^{1-t} - e^{-t}, t \ge 1 \end{cases} \tag{A–16}$$

$y(t)$ is shown in Figure A–6. The sharp edges of the original rectangular pulse, $x(t)$, are severely rounded off, and the resulting signal, $y(t)$, has a longer time duration than does $x(t)$. Both of these results are consistent with the fact that the R-C circuit in Figure A–1 is a simple lowpass filter.

B

PROGRAMMING FIR FILTER ALGORITHMS IN HIGHER-LEVEL LANGUAGES

Consider the following computer assignment: A student is given a file X.DAT containing sample values $x(n)$ from some signal. The first number in the file is $x(0)$, the next number in the file is $x(1)$, . . ., and the last number in the file is $x(L-1)$, where L is the total number of samples recorded. (In general, L could be quite large. For example, two seconds of speech, sampled at $f_s = 40$ kHz, would result in $L = 80{,}000$ samples.) The computer assignment is to write a program to process $x(n)$, $n = 0, 1, \ldots, L-1$ with an FIR filter algorithm:

$$y(n) = \sum_{k=0}^{N-1} h(k)x(n-k), n = 0, 1, \ldots, L-1 \qquad \text{(B–1)}$$

where $h(k)$, $k = 0, 1, \ldots, N-1$, are the coefficients of the N-point FIR filter, which perhaps was designed using one of the methods described in Chapter 9. The filter output $y(n)$, $n = 0, 1, \ldots, L-1$, is to be placed in a file named Y.DAT having the same format as X.DAT. (In general, the number of points in the input file (L) may not be specified in advance. Therefore, the program should continue to use Equation (B–1) to generate output values until the end of the file is detected.)

Equation (B–1) says that the current value of $y(n)$ is calculated as a linear combination of the current value of $x(n)$ and $N-1$ past values $x(n-1)$, $x(n-2)$, . . ., $x[n-(N-1)]$. Therefore, a computer program which implements Equation (B–1) must have a way of storing these past values and updating this information at each iteration (n). To simplify this discussion, let's consider the case $N = 4$. The algorithm of Equation (B–1) then becomes

$$y(n) = h(0)x(n) + h(1)x(n-1) + h(2)x(n-2) + h(3)x(n-3) \qquad \text{(B–2)}$$

$n = 0, 1, \ldots$ until the input file is exhausted

One way to handle the storing of past values problem is to create an array, such as $X(\ )$, having four elements in this case, that contains the current input and three past values. In other words, at the nth iteration,

$X(0)$ contains $x(n)$

$X(1)$ contains $x(n-1)$

$X(2)$ contains $x(n-2)$

$X(3)$ contains $x(n-3)$

Let's also assume that the coefficients $h(k)$ have been placed in an array named $H(\ )$, such that $H(K)$ contains coefficient $h(k)$. The algorithm of Equation (B–2) then becomes

$$Y = H(0)X(0) + H(1)X(1) + H(2)X(2) + H(3)X(3) \tag{B–3}$$

where Y is the current output value. The contents of array $X(\ )$ must be *shifted* at the end of each iteration. Thus at the end of each iteration, the contents of $X(3)$ are discarded, the contents of $X(2)$ are moved to $X(3)$, the contents of $X(1)$ are moved to $X(2)$, and the contents of $X(0)$ are moved to $X(1)$. At the beginning of the next iteration, the new input value is placed in $X(0)$.

The following pseudocode suggests one way to implement the FIR filter computer assignment based on the scheme outlined above (shown for an arbitrary filter length of N points):

```
Obtain FIR filter coefficients
Place coefficients in array H(K), K = 0,1, ..., N - 1
Initialize: set X(K) = 0, K = 0, 1, ..., N - 1
START
Get input value x from file X.DAT
X(0) = x
Y = 0               (initialize the sum to zero)
For K = 0, 1, ..., N - 1
       Y = Y + H(K)X(K)
Next K
Output Y to Y.DAT
Update the X array:
For K = N - 1, N - 2, ..., 2, 1
       X(K) = X(K - 1)
Next K
Loop back to START.
(Continue until the input file is exhausted.)
```

This method works well enough as long as the filter length (N) is small. However, if N is large, the time required for the X array update at each iteration can be very significant. What is clearly needed is a more efficient way of storing past values of the input. Therefore, let us consider the following alternative method. Instead of putting the current input value in location $X(0)$, that is, at the top of the array, consider filling up the array from top to bottom as the inputs are obtained. This way the location of the current input in the array shifts at each iteration instead of always being at the top of the array. After the bottom of the array is reached, we return to the top of the array and repeat the process, overwriting

TABLE B–1
Past Value Storage Array

Iteration Number	Location of Current Input Value
0	$X(0)$
1	$X(1)$
2	$X(2)$
3	$X(3)$
4	$X(0)$
5	$X(1)$
6	$X(2)$
7	$X(3)$
8	$X(0)$
9	$X(1)$
10	$X(2)$
11	$X(3)$
12	$X(0)$
13	$X(1)$
14	$X(2)$
.	.
.	.
.	.

the numbers previously stored. (This is sometimes called a *circular buffer.*) If the array has N locations, $X(0)$ to $X(N-1)$, the current value of $x(n)$ and past values back to $x[n-(N-1)]$ will always be available; past values even further back in time are simply overwritten. *This scheme does not require taking the time to do memory shifting during each iteration.* Of course, it *does* require keeping track of a pointer to the current location, which makes the program somewhat more complicated. However, If N is large (say 50 or more), the program will run significantly faster, so the extra programming effort is well worth the trouble.

Let's use $N = 4$ again as an example. The past value storage array has four locations: $X(0)$, $X(1)$, $X(2)$, and $X(3)$. Table B–1 shows how the inputs are stored, starting with the first iteration ($n = 0$).

To see how the algorithm must be altered to accommodate this new past value storage scheme, let's consider some specific iteration numbers. At iteration number 4, the algorithm must calculate

$$Y = H(0)X(0) + H(1)X(3) + H(2)X(2) + H(3)X(1) \tag{B–4}$$

since at iteration number 4, the current input is in $X(0)$, the previous input is in $X(3)$, the input before that is in $X(2)$, etc. Let's consider another example. At iteration number 10, the algorithm must calculate

$$Y = H(0)X(2) + H(1)X(1) + H(2)X(0) + H(3)X(3) \tag{B–5}$$

since at iteration number 10, the current input is in $X(2)$, the previous input is in $X(1)$, etc. In general, if the current input is located in $X(M)$ (where $M = 0, 1, 2,$ or 3 in this example), then the algorithm must calculate the current output (Y) as follows:

$$
\begin{aligned}
Y = {} & H(0)X(M) \\
& + H(1)X\big[M - 1(+4, \text{if } M - 1 < 0)\big] \\
& + H(2)X\big[M - 2(+4, \text{if } M - 2 < 0)\big] \\
& + H(3)X\big[M - 3(+4, \text{if } M - 3 < 0)\big]
\end{aligned}
\tag{B-6}
$$

The following pseudocode suggests a possible way to implement this alternative efficient past value storage scheme (written for an arbitrary filter length N):

```
Obtain N FIR filter coeffients
Place filter coefficients in array H(K), K = 0,1,..., N - 1
Initialize:  X(K) = 0, K = 0,1, ..., N - 1
START
For M = 0,1, ..., N - 1          (M is the "pointer")
        Read input value x from X.DAT
        X(M) = x
        Y = 0          (initialize the sum to zero)
        For K = 0,1, ..., N - 1
                J = M - K
                IF J < 0,  THEN  J = J + N
                Y = Y + H(K)X(J)
        Next K
        Output Y to Y.DAT
Next M
Loop back to START until the input file is exhausted.
```

Of course, if the end of the input file is detected while the program is inside the M loop, the program should be designed to terminate in an orderly fashion.

To understand this pseudocode, the reader is encouraged to go through it using an $N = 4$ case and see how it agrees with Equations (B–4), (B–5), and (B–6).

C

DERIVATION OF EQUATION (10–8)

To begin, we know that the frequency response function and transfer function of a continuous time system are related by

$$H_a(f) = \hat{H}(s)\Big|_{s=j2\pi f} \tag{C-1}$$

This relationship is reversible:

$$\hat{H}(s) = H_a(f)\Big|_{f=\frac{s}{j2\pi}} \tag{C-2}$$

Suppose transfer function $H(z)$ is obtained from $\hat{H}(s)$ using the bilinear transformation:

$$H(z) = \hat{H}(s)\Big|_{s=\frac{1-z^{-1}}{1+z^{-1}}} \tag{C-3}$$

By combining Equations (C–2) and (C–3), we see that $H(z)$ is related to $H_a(f)$ as follows:

$$H(z) = H_a(f)\Big|_{f=\frac{1}{j2\pi}\left(\frac{1-z^{-1}}{1+z^{-1}}\right)} \tag{C-4}$$

The frequency response function of the discrete time system is related to $H(z)$ as follows:

$$H(e^{j\omega}) = H(z)\Big|_{z=e^{j\omega}} \tag{C-5}$$

By combining Equations (C–4) and (C–5), we obtain the following:

$$H\left(e^{j\omega}\right) = H_a(f)\Bigg|_{f=\frac{1}{j2\pi}\left(\frac{1-e^{-j\omega}}{1+e^{-j\omega}}\right)} \tag{C–6}$$

The relationship between f and ω on the right side of Equation (C–6) can be simplified:

$$f = \frac{1}{j2\pi}\left(\frac{1-e^{-j\omega}}{1+e^{-j\omega}}\right) = \frac{1}{j2\pi}\left[\frac{e^{-j\frac{\omega}{2}}\left(e^{j\frac{\omega}{2}}-e^{-j\frac{\omega}{2}}\right)}{e^{-j\frac{\omega}{2}}\left(e^{j\frac{\omega}{2}}+e^{-j\frac{\omega}{2}}\right)}\right]$$

$$f = \frac{1\left(e^{j\frac{\omega}{2}}-e^{-j\frac{\omega}{2}}\right)}{j2\pi\left(e^{j\frac{\omega}{2}}+e^{-j\frac{\omega}{2}}\right)} = \frac{1\left[2j\sin\left(\frac{\omega}{2}\right)\right]}{j2\pi\left[2\cos\left(\frac{\omega}{2}\right)\right]} \tag{C–7}$$

$$f = \frac{1\left[\sin\left(\frac{\omega}{2}\right)\right]}{2\pi\left[\cos\left(\frac{\omega}{2}\right)\right]} = \frac{1}{2\pi}\tan\left(\frac{\omega}{2}\right)$$

Therefore, Equation (C–6) can be expressed as

$$H\left(e^{j\omega}\right) = H_a(f)\Bigg|_{f=\frac{1}{2\pi}\tan\left(\frac{\omega}{2}\right)} \tag{C–8}$$

As they say, QED, *quod erat demonstrandum,* that is, we have shown that "which was to be demonstrated."

D

PARSEVAL'S THEOREM

Parseval's theorem can be stated as follows: Let $x(n)$ be a real valued discrete time signal having DTFT $X(e^{j\omega})$. We will show that

$$\sum_{n=-\infty}^{\infty} x^2(n) = \frac{1}{2\pi} \int_{-\pi}^{\pi} \left| X\left(e^{j\omega}\right) \right|^2 d\omega \qquad \text{(D–1)}$$

assuming that the sum on the left side is finite. The DTFT is defined as usual:

$$X\left(e^{j\omega}\right) = \sum_{n=-\infty}^{\infty} x(n)e^{-j\omega n} \qquad \text{(D–2)}$$

Equation (D–1) is Parseval's theorem. The sum on the left in Equation (D–1) is defined as the *energy* of the discrete time signal. Therefore, $(1/2\pi)|X(e^{j\omega})|^2$ can be thought of as an *energy spectral density function.*

We begin the proof with a brief review of some facts about complex numbers. First of all, suppose z is a complex number and α is a real number. Using the superscript * to denote complex conjugation, we have

$$\left(\alpha z\right)^* = \alpha z^* \qquad \text{(D–3)}$$

and also

$$zz^* = |z|^2 \qquad \text{(D–4)}$$

If there is a set of complex numbers indexed by i, denoted as z_i, we note that

$$\sum_i z_i^* = \left(\sum_i z_i\right)^* \qquad \text{(D–5)}$$

We also note the following fact about the complex exponential $\exp(j\theta)$:

$$\left(e^{-j\theta}\right)^* = e^{j\theta} \tag{D-6}$$

These facts (Equations [D–3] through [D–6]) will be used below.

The next step is to consider the DTFT of the *time-reversed* signal $x(-n)$. Using the definition of the DTFT, we have

$$\textbf{DTFT}[x(-n)] = \sum_{n=-\infty}^{\infty} x(-n)e^{-j\omega n} \tag{D-7}$$

Since the summation is from $n = -\infty$ to $n = \infty$ and since the order of summation doesn't change the sum (commutative property of addition), Equation (D–7) can be expressed as

$$\textbf{DTFT}[x(-n)] = \sum_{n=-\infty}^{\infty} x(n)e^{j\omega n} \tag{D-8}$$

Using the properties given in Equations (D–3) through (D–6), we can write

$$\textbf{DTFT}[x(-n)] = \sum_{n=-\infty}^{\infty} x(n)\left(e^{-j\omega n}\right)^*$$

$$= \sum_{n=-\infty}^{\infty} \left(x(n)e^{-j\omega n}\right)^* \tag{D-9}$$

$$= \left(\sum_{n=-\infty}^{\infty} x(n)e^{-j\omega n}\right)^*$$

If the last line of Equation (D–9) is compared to Equation (D–2), it will be observed that

$$\textbf{DTFT}[x(-n)] = X^*\left(e^{j\omega}\right) \tag{D-10}$$

The next step is to recall that the *convolution* of two discrete time signals, say $g(n)$ and $x(n)$, is given by

$$g(n) * x(n) = \sum_{k=-\infty}^{\infty} g(k)x(n-k) \tag{D-11}$$

Recall also that convolution in the time domain becomes multiplication in the frequency domain; that is,

$$g(n) * x(n) \xleftarrow[\text{DTFT}]{} G\left(e^{j\omega}\right)X\left(e^{j\omega}\right) \tag{D-12}$$

With this in mind, consider the convolution $x(-n) * x(n)$. Using the definition of convolution (Equation [D–11]), we have

$$x(-n) * x(n) = \sum_{k=-\infty}^{\infty} x(-k)x(n-k) \tag{D-13}$$

Since the summation runs from $k = -\infty$ to $k = \infty$, and since the order of summation can be reversed, this can be written as

$$x(-n) * x(n) = \sum_{k=-\infty}^{\infty} x(k)x(n+k) \tag{D-14}$$

Using Equations (D–10), (D–12), and (D–14), we can write

$$x(-n) * x(n) = \sum_{k=-\infty}^{\infty} x(k)x(n+k) \xleftrightarrow[\text{DTFT}]{} X^*\left(e^{j\omega}\right)X\left(e^{j\omega}\right) = \left|X\left(e^{j\omega}\right)\right|^2 \qquad \textbf{(D–15)}$$

Therefore, by using the inverse DTFT integral (Chapter 6, Equation [6–34] on page 166), the following relationship is obtained:

$$\sum_{k=-\infty}^{\infty} x(k)x(n+k) = \frac{1}{2\pi}\int_{-\pi}^{\pi}\left|X\left(e^{j\omega}\right)\right|^2 e^{jn\omega}\,d\omega \qquad \textbf{(D–16)}$$

If we let $n = 0$ in Equation (D–16), Parseval's theorem emerges:

$$\sum_{k=-\infty}^{\infty} x(k)x(k) = \frac{1}{2\pi}\int_{-\pi}^{\pi}\left|X\left(e^{j\omega}\right)\right|^2 d\omega \qquad \textbf{(D–17)}$$

To make Equation (D–17) look like Equation (D–1), change the summation index from k to n.

The Fourier transform, Fourier series, and discrete Fourier transform (DFT) also have versions of Parseval's theorem associated with them. These are presented here without proof.

For a continuous time signal $x_a(t)$ having a Fourier transform $X_a(f)$, Parseval's theorem is

$$\int_{-\infty}^{\infty} x_a^2(t)\,dt = \int_{-\infty}^{\infty}\left|X_a(f)\right|^2 df \qquad \textbf{(D–18)}$$

assuming the integral on the left side converges.

For a periodic continuous time signal $x_a(t)$ expressed as a Fourier series:

$$x_a(t) = \sum_{n=-\infty}^{\infty} c_n e^{j2\pi n f_o t} \qquad \textbf{(D–19)}$$

the Parseval's theorem relationship is

$$\frac{1}{T_O}\int_0^{T_O} x_a^2(t)\,dt = \sum_{n=-\infty}^{\infty}\left|c_n\right|^2 \qquad \textbf{(D–20)}$$

where $T_o = 1/f_o$. The right side of Equation (D–20) is the average power in the periodic signal.

For an N-point discrete time signal $x(n)$, $n = 0, 1, \ldots, N-1$, having a DFT $X(k)$, the Parseval's theorem relationship can be shown to be

$$\sum_{n=0}^{N-1} x^2(n) = \frac{1}{N}\sum_{k=0}^{N-1}\left|X(k)\right|^2 \qquad \textbf{(D–21)}$$

BIBLIOGRAPHY

Aziz, P.M., H.V. Sorensen, and J. Van Der Spiegel. "An Overview of Sigma Delta Converters." *IEEE Signal Processing Magazine.* 13, January 1996, pp. 61–84.

Cooley, J.W., and J.W. Tukey. "An Algorithm for the Machine Computation of Complex Fourier Series." *Mathematics of Computation.* 19, April 1965, pp. 297–301.

Hamming, R.W. *Digital Filters.* 3rd ed. Englewood Cliffs, NJ: Prentice Hall, 1989.

Laiser, J.F. "Nonrecursive Digital Filter Design Using the I_0-Sinh Window Function," *Proceedings of 1974 IEEE International Symposium on Circuits and Systems,* April 1974, San Francisco, pp. 20–23.

Ludeman, L.C. *Fundamentals of Digital Signal Processing.* New York: Harper and Row, 1986.

Oppenheim, A.V., and R.W. Schafer. *Discrete Time Signal Processing.* Englewood Cliffs, NJ: Prentice Hall, 1989.

Papoulis, A. *Circuits and Systems.* New York: Holt, Rinehart and Winston, 1980.

Parks, T.W., and C.S. Burrus. *Digital Filter Design.* New York: John Wiley & Sons, 1987.

Parks, T.W., and J.H. McClellan. "Chebyshev Approximation for Nonrecursive Digital Filters with Linear Phase." *IEEE Transcripts Circuit Theory.* CT-19, March 1972, pp. 189–194.

Parks, T.W., and J.H. McClellan. "A Program for the Design of Linear Phase Finite Impulse Response Filters." *IEEE Transcripts Audio Electroacoustics.* AU-20, August 1972, pp. 195–99.

Proakis, J.G., and D.G. Manolakis. *Digital Signal Processing.* 2nd ed. New York: Macmillan, 1992.

Proakis, J.G., C.M. Rader, F. Ling, and C.L. Nikias. *Advanced Digital Signal Processing.* New York: Macmillan, 1992.

Rabiner, L.R., and R.W. Schafer. *Digital Processing of Speech Signals.* Englewood Cliffs, NJ: Prentice Hall, 1978.

Roberts, R.A., and C.T. Mullis. *Digital Signal Processing.* Reading, MA: Addison Wesley, 1987.

Van Valkenburg, M.E. *Network Analysis.* 3rd ed. Englewood Cliffs, NJ: Prentice Hall, 1974.

Widrow, B., and S.D. Stearns. *Adaptive Signal Processing.* Englewood Cliffs, NJ: Prentice Hall, 1985.

Ziemer, R.E., and W.H. Tranter. *Principles of Communications.* 2nd ed. Boston: Houghton Mifflin, 1985.

INDEX